Genetic Improvement of Seed Proteins

PROCEEDINGS OF A WORKSHOP

Washington, D.C.
March 18–20, 1974

BOARD ON AGRICULTURE AND
RENEWABLE RESOURCES
Commission on Natural Resources
National Research Council

NATIONAL ACADEMY OF SCIENCES
Washington, D.C. 1976

NOTICE: The project that is the subject of this report was approved by the Governing Board of the National Research Council, whose members are drawn from the Councils of the National Academy of Sciences, the National Academy of Engineering, and the Institute of Medicine. The members of the Committee responsible for the report were chosen for their special competences and with regard for appropriate balance.

This report has been reviewed by a group other than the authors according to procedures approved by a Report Review Committee consisting of members of the National Academy of Sciences, the National Academy of Engineering, and the Institute of Medicine.

This study was supported by the U.S. Atomic Energy Commission and the National Science Foundation.

Library of Congress Cataloging in Publication Data

Main entry under title:

Genetic improvement of seed proteins.

Includes bibliographies.

1. Selection (Plant breeding)—Congresses. 2. Plant proteins—Congresses. 3. Protein metabolism—Congresses. I. National Research Council. Board on Agriculture and Renewable Resources.

SB123.G38 633'.1'043 76-17097

ISBN 0-309-02421-8

Printed in the United States of America

80 79 78 77 76 10 9 8 7 6 5 4 3 2 1

Preface

The workshop reported herein was held under the sponsorship of the Board on Agriculture and Renewable Resources, Commission on Natural Resources, National Research Council, on March 18–20, 1974, in the Lecture Room at the National Academy of Sciences in Washington, D.C.

The purpose of the workshop was to present and consider current information on genetic improvement in quality and quantity of seed protein in relation to human nutrition. The opportunities for utilizing environmental and physiological variables, e.g., O_2 and CO_2, were considered.

Experts from a diversity of related scientific fields, including plant physiology, plant breeding, molecular biology, biochemical genetics, and mammalian nutrition, focused on the basic cellular and biochemical mechanisms that are involved in plant protein synthesis. Applicability of genetic engineering to plant breeding research was explored.

The organizers are grateful to the specialists who appeared on the program and to the participants in the discussions.

Contents

Introduction 1
V. A. Johnson

I SEED PROTEIN USE, DEVELOPMENT, AND MODIFICATION

The Protein–Calorie Trade-off 5
Aaron M. Altschul
Discussion 16

Production of Vacuolar Protein Deposits in Developing Seeds and Seed Protein Homology 18
Julius W. Dieckert and *Marilyne C. Dieckert*
Discussion 52

Case Histories of Existing Models 57
Edwin T. Mertz

Effects of Genes That Change the Amino Acid Composition of Barley Endosperm 71
L. Munck and *D. V. Wettstein*
Discussion 79

Plant Breeding, Molecular Genetics, and Biology 83
G. F. Sprague
Discussion 96

II BIOCHEMICAL AND BIOPHYSICAL LIMITATIONS

Biochemical and Biophysical Limitations: Introduction 101
Robert Rabson

Nitrate and Nitrate Reductase as Factors Limiting Protein Synthesis 103
R. H. Hageman, R. J. Lambert, Dale Loussaert, M. Dalling, and *L. A. Klepper*
Discussion 131

Metabolic Control of Biosynthesis of Nutritionally Essential Amino Acids 135
B. J. Miflin
Discussion 155

Stress Relationships in Protein Synthesis: Water and Temperature 159
J. S. Boyer
Discussion 167

Interrelationships between Carbohydrate and Nitrogen Metabolism 172
David T. Canvin
Discussion 191

Opportunities for Improved Seed Yield and Protein Production: N_2 Fixation, CO_2, Fixation, and O_2 Control of Reproductive Growth 196
R. W. F. Hardy, U. D. Havelka, and *B. Quebedeaux*
Discussion 228

Biochemistry of Protein Synthesis in Seeds 231
Donald Boulter
Discussion 246

III GENETIC REGULATORY MECHANISMS

Genetic Regulatory Mechanisms: Introduction 253
Sterling B. Hendricks
Some Aspects of the Regulation of Gene Expression 255
Geoffrey Zubay
Discussion 271

Posttranscriptional Regulation of Protein Synthesis 274
John M. Clark, Jr.
Discussion 286

Hormone-Regulated Synthesis of Tissue-Specific Proteins 288
Anthony R. Means, Savio Woo, Cassius Bordelon, John P. Comstock, and *Bert W. O'Malley*
Discussion 305

Characterization of Protein Metabolism in Cereal Grain 309
J. E. Varner, D. Flint, and *R. Mitra*
Discussion 327

Messenger RNA Synthesis and Utilization in Seed Development and Germination 329
Leon S. Dure III

Asexual Approaches, Including Transgenosis and Somatic Cell Hybridization, to the Modification of Plant Genotypes and Phenotypes 341
Colin H. Doy
Discussion 357

Cloning of Eukaryotic DNA as an Approach to the Analysis of Chromosome Structure and Function 359
Herbert W. Boyer, Robert C. Tait, Brian J. McCarthy, and *Howard M. Goodman*
Discussion 367

IV REVIEW AND SUMMARY

Interpretive Summary and Review 371
Hamish N. Munro
Discussion 381

Interpretive Summary and Review 383
Oliver E. Nelson
Discussion 391

List of Contributors 393

V. A. JOHNSON

Introduction

As biologists, geneticists, physiologists, plant breeders, we are aware of the important role of seed proteins in the world struggle against malnutrition, so I need not dwell on this fact, nor is it the intent of this workshop to dwell on the malnutrition problem as such. I do submit that against this backdrop of malnutrition our deliberations will be most timely and appropriate.

Scientists, several of whom are here as participants, are concerned by mounting evidence that genetic manipulation of seed protein by plant breeders may be seriously impeded by gaps in our knowledge of protein metabolism and, perhaps, by lack of effective communication between the biological disciplines involved.

These disciplines, particularly molecular biology and molecular genetics, need to focus on the problem. In this meeting, leading plant breeders, molecular biologists, geneticists, biochemists, and physiologists will examine plant protein metabolism in depth and, it is hoped, formulate some innovative approaches to acceleration of favorable modification of storage proteins in the food species of plants.

If the food species of plants are to be improved nutritionally, the job will ultimately fall to the plant breeder. The plant breeders have already responded to pressure for more and better plant protein with some notable successes.

This is a new kind of activity for the plant breeder. It is a difficult one, because guidelines are frequently absent or, at best, indistinct. It has been suggested that some breeders have a euphoric attitude about protein improvement. The implication is that they may be unaware of the serious biological constraints on plant protein manipulation.

The plant breeder attacks the problem by measuring protein, the end product of a long and complex chain of metabolic events. The breeder does this, for the most part, because he must do it. This is the method that is readily available to him. Is this what he should be doing? I suspect that in some cases, it may not be.

Some of the questions that must be answered are these:

What are the points in protein metabolism that offer the best opportunity for effecting useful changes in the amount or composition of seed proteins?

What are the most serious constraints imposed by biological or physiological phenomena in protein metabolism that breeders know little about? More important, what are the opportunities provided by these phenomena?

Can the molecular biologist or the molecular geneticist help the breeder? If so, when and how?

Where are the gaps in our knowledge of protein metabolism that will most seriously impede breeding progress?

What key cellular biochemical mechanisms are involved? At what point in the chain of metabolic events do they function?

Are there tests for protein precursors, or enzymes, or for key proteins themselves that would be useful selection tools in the hands of the plant breeders? Breeders may suggest that there are few such useful tests available to them at this time. The physiologists, on the other hand, may suggest that some potentially useful tests are now available but are not being used by the breeders. If this is so, why?

Are there cooperative investigations that should be instituted to provide such information?

The National Academy of Sciences has provided the forum. Communications across the disciplines represented here may be difficult because we speak somewhat different languages, but such communications must occur.

I

SEED PROTEIN USE, DEVELOPMENT, AND MODIFICATION

AARON M. ALTSCHUL

The Protein–Calorie Trade-off

The purpose of this presentation is to examine the relationships between protein supply and total food supply with a view to providing an orientation for the ensuing discussions on the biological aspects of protein synthesis in cereals.

The basic question that must be faced is whether or not there is a protein problem, for if there is no protein problem, then there is no point in worrying about more or less protein in cereals. Perhaps a more pertinent question is whether one can talk about either a protein problem or food problem as two separate problems. As the analysis proceeds, it will turn out that the protein problem and the food problem are two aspects of the same problem: Neither can be discussed without reference to the other.

Historically, the protein problem had its origins in medical considerations. Kwashiorkor was discovered and identified as a disease of extreme protein deprivation. It occurred in children of poor people in poor countries whose diet contained predominantly cereals with no animal or legume protein or, worse yet, whose diets contained large proportions of roots such as cassava. It was quite clear that the observed clinical cases represented the tip of the iceberg, and that, for each child seen in a hospital, there were tens or hundreds or thousands of children whose growth was stunted by inadequate protein in their food.

The realization of the crucial role of protein in the lives of vulnerable groups such as children and pregnant and lactating women led such

United Nations organizations as UNICEF to ship milk all over the world and to attempt to develop alternative cheaper sources of protein products suitable for these groups. It led the U.S. Agency for International Development (AID) and the U.S. Department of Agriculture to develop and ship huge amounts of a mixture called CSM, which was predominantly corn and soybean flour. It led AID to initiate large-scale demonstrations of fortification of cereals with amino acids in Tunisia, Thailand, and Guatemala; these projects are in progress. It led AID, in another effort, to try to stimulate the food industry to introduce protein into common foods in a way that would provide additional protein from hitherto untapped resources. And it stimulated people (e.g., Mertz and Nelson, this volume) to look for varieties of cereals with higher protein value.

But, obviously, it is not possible to divorce protein considerations from the rest of the food issues. What was originally called kwashiorkor is more properly termed protein–calorie malnutrition. It was asked whether shipments of protein foods alone would solve the medical problems. The medical analysis was unequal to the task of quantifying the problem. It was large, it was severe, but hard numbers were not available.

The logical approach toward quantification is to determine, it seems, what the human needs for protein are; compare these to available supplies; and ascertain the relationship between protein utilization and total calorie intake. This is an ongoing effort; periodically, expert committees of UN agencies and like committees in several nations have met to codify the state of knowledge in this area.

It is clear that protein requirements cannot be determined or presented in isolation and that any figures for protein requirements must presume an adequate supply of calories and other nutrients. Moreover, since proteins are sources of amino acids, protein requirements cannot be described without specifying the type of protein: digestibility, percentage of essential amino acids as part of total amino acids, and pattern of essential amino acids as related to requirements. It is in infancy that the human being is most sensitive to protein requirements, needing more protein per unit of weight than at any other age, a greater proportion of essential amino acids as part of the total amino acids, and the widest spread among the essential amino acids.

In the last several decades there have been a number of revisions of the amino acid requirements. The trend has been downward; the requirements are now lower than had earlier been thought (FAO/WHO Report, 1973).

What can one do with this kind of information? It is possible to say that the average person who receives these amounts of protein along with adequate amounts of other nutrients will get enough protein to avoid showing evidence of protein deficiency. They can identify people whose protein intake is grossly deficient. They can be used in designing diets for people in well-defined clinical situations such as parenteral feeding. And they can, in a general way, predict how much protein is required, given adequate availability of other nutrients, to feed human beings under rigidly controlled conditions, approaching the feeding of animals in pens.

What is not provided by these kinds of data? They do not deal with those issues in which human requirements differ from those of animals. The question of optimum nutrition is not as easily defined as it relates to humans because it is part of a quality of life that includes many other aspects of life including sensory needs, performance, happiness, emotional stability, and length of useful life. More important, these data do not predict what people will do when faced with choices brought about either by restricted food supply or by increasing purchasing power.

When this kind of consideration is applied to public policy, it leads to ambiguities. The same information can be used to support the allegation that there is adequate protein or to substantiate the charge that for certain groups there is a protein deficiency. The same information can allow some to conclude that there is a protein problem and others to conclude that there is a general food problem, that *more of the same kind of food* is needed rather than particular emphasis on protein foods. Thus, the nutrition community is divided on this issue, and this diminishes interest in the protein supply question as a specific issue (Altschul, 1974a,b,c).

Further examination of the concept that postulates *more of the same kind of food* shows that of the choices made when people become poorer or richer, this is less likely than others (see Figure 1). When people become poorer or when the ratio of resources to population in a nation decreases, *the same amount of food of a different kind* or *less food of a different kind* are more likely alternatives. And when a person's income increases or the ratio of resources to population in a nation increases, the most likely alternative is *more food of a different kind*. These come about because protein foods are more expensive than calorie foods. As resources decrease, the expensive foods suffer first; and as resources increase, more expensive (and tastier) foods enter into the mix of the diet.

Hence, medical considerations alone or nutritional considerations alone are insufficient to provide a basis for public policy. Cost and

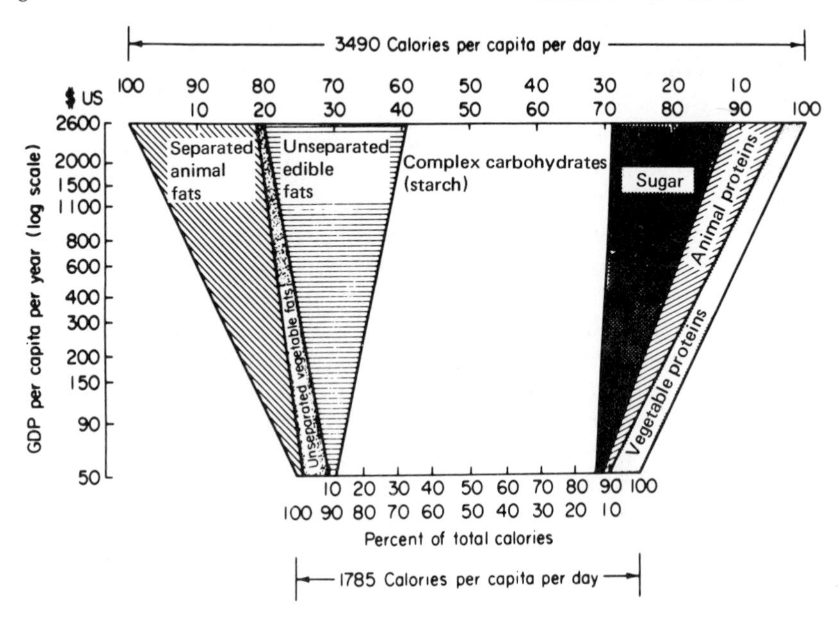

FIGURE 1 Effect of income on energy intake and distribution of major sources of food energy. The total calories and proportion of calories derived from each food type are shown according to country income for 1962. The ranges in total calorie intake represent the lowest and highest values recorded. Figures are based on national food balance sheets. (Reprinted with permission from Altschul, 1974a.)

esthetic considerations are equally important; actually, they usually override medical or nutritional considerations.

We can illustrate these points by considering two cases:

1. Where there is a threat to the current level of food supply in any given country
2. Where income is increasing in a specific country

In the first case, the threat could arise from increased population, or changing conditions of energy constraints, or a combination of the two. We must remember that the two major constraints on total food production are availability of land and energy. The Green Revolution made it possible to apply more fossil energy to optimize photosynthetic energy. With the sharp increase in energy costs, some areas may not be able to continue the generous energy inputs that are required to maintain the gains of the Green Revolution. The other case, the one where income is increasing, constitutes a pressure for foods of higher esthetic quality, particularly meat.

Case No. 1 When there is a threat to the current level of food supply, maintenance of total calorie supply is the first interest. One could almost say that everything is sacrificed to maintain calorie intake to avoid starvation. And since calories cost less than protein, poverty favors less and poorer protein. For example, increased demand for calories favors cereals over legumes since the yield of calories per acre is greater for cereals than for legumes. An increased demand for calories favors both of them over animal protein since this is the most expensive of all kinds of protein. Moreover, even greater poverty favors starchy roots over cereals, and these contain very little protein.

Let us take the case of India, for example. The Green Revolution was quite successful there. For the decade of the 1960's, calorie availability per capita remained constant despite the increase in population, and so did the protein content per capita (Figures 2 and 3). There was, however, a major rearrangement: The percentage of

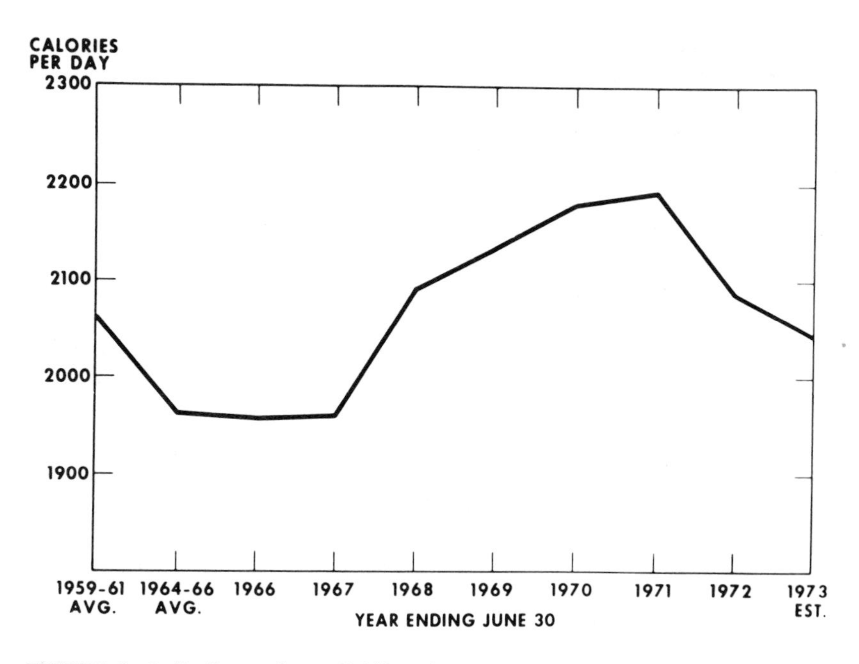

FIGURE 2 India: Per capita availability of energy 1960–1973. (Reprinted with permission from Schertz, 1973.)

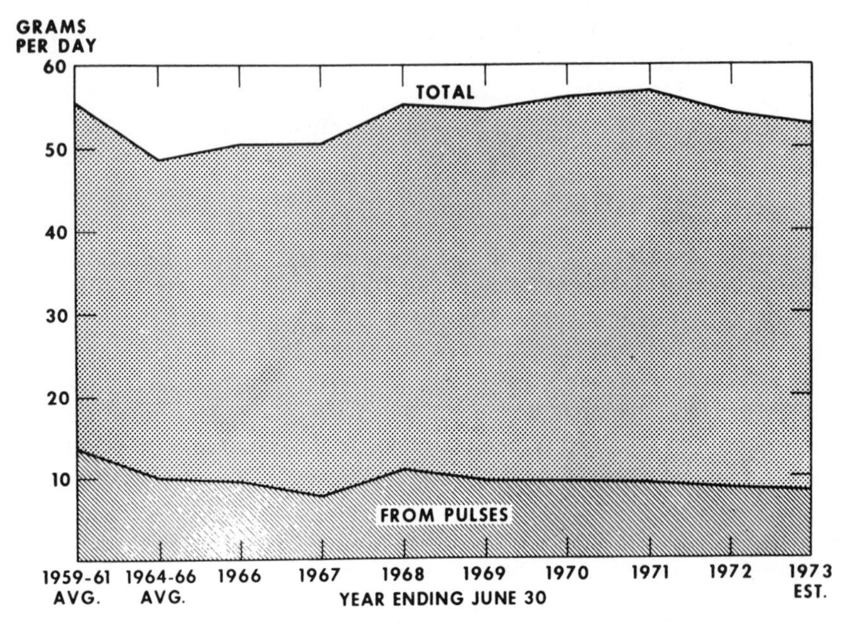

FIGURE 3 India: Per capita availability of protein, total and from pulses. (Reprinted with permission from Schertz, 1973.)

protein from pulses was reduced from about 24 percent to 15 percent. If there is no concomitant increase in other sources of excellent protein such as milk, this represents an alarming trend.

To better understand the nature of the problem we can look at the figure derived from Bressani and Elias (1974) showing the complementary effects of corn and legume protein (Figure 4). Neither corn alone nor legume alone is a completely effective source of protein, corn because it is deficient in lysine and tryptophan and legume because it is deficient in methionine. But they complement each other. Therefore, the combination of the two is far better than either one alone and is entirely adequate as a protein source. It turns out that in the best combination 50 percent of the protein is supplied by corn and 50 percent by the legume. Poor people generally do not have enough resources to make that kind of a combination; the mixture is generally on the side of less than adequate legume protein, and that means some lysine deficiency. In the case of India the percentage of legume protein was 24 percent at the beginning of the 1960's and this was, no doubt, augmented by a certain amount of milk protein. By the end of the 1960's, the legume protein contribution was 14 percent.

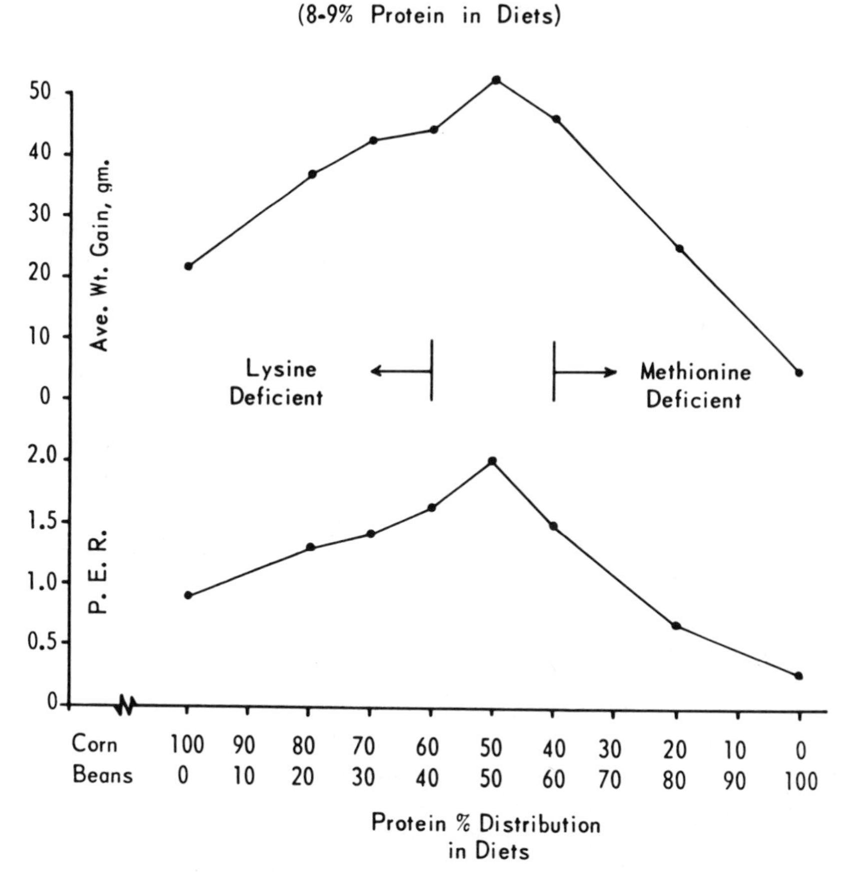

FIGURE 4 Protein nutrients value of mixtures of corn and beans. (Reprinted with permission from Bressani and Elias, 1974.)

The Food and Agriculture Organization of the United Nations (FAO) in its projections of protein supply for Southeast Asia (FAO, 1974) has indicated a decrease in percentage of calories from protein from over 10 percent to less than 10 percent. Since this decrease means a decrease in the amount of higher protein concentrate in favor of more cereal protein, there has to be a concomitant reduction in protein quality as well as quantity (Table 1).

These considerations did not take into account the explosive increase in the cost of energy, which constitutes nothing but a disaster to the underdeveloped countries, which require energy for fertilizer, for irrigation, and for cultivation and harvesting.

TABLE 1 Trend of Protein–Calorie Ratio Relative to Projected Supplies (Percent)

Region	1962	1975	1985
South America	10.6	10.5	10.6
Central America	10.2	9.9	9.7
Africa (south of the Sahara)	10.9	10.8	10.9
North Africa	11.3	11.3	11.2
Near East	12.1	12.0	12.0
Asia	10.3	10.1	9.9

SOURCE: FAO/WHO (1969).

One could make a good case that with the present technologies and habits one can expect less protein and less protein quality in underdeveloped countries. The situation in some countries will get bad enough to be insufficient in calories as well. But protein reduction will precede calorie reduction. This will unquestionably show as an increased incidence of protein–calorie malnutrition as well as increased incidences of starvation.

Case No. 2 On the other side of the coin, there are countries in Western Europe and elsewhere where income is increasing and where there is an increase in the consumption of animal protein. But consumption of animal protein requires a high level of grain utilization, which converts to high land and energy requirements. In the United States approximately a ton of grain is available per capita but only about 150 pounds is eaten directly; the remainder is converted into animals. It is possible to estimate the grain equivalent calories required to support a particular nutritional way of life. This would be the sum of the direct calories eaten plus the calories required to feed the animals that are eaten. Such a calculation is shown in Figure 5. The American meat consumption pattern requires about 11,000 grain equivalent kilocalories per day, while in many underdeveloped countries less than 3,000 grain equivalent kilocalories are consumed per day. There is no evidence that the trend in the developed countries will not continue. This will require more grain for the animals and more protein concentrates such as soy meal, and since neither will be in sufficient quantity, more of the amino acids methionine and lysine will be required to gain optimum protein efficiency. It is predicted that the annual lysine consumption, which is now about 16,000 tons worldwide, will increase

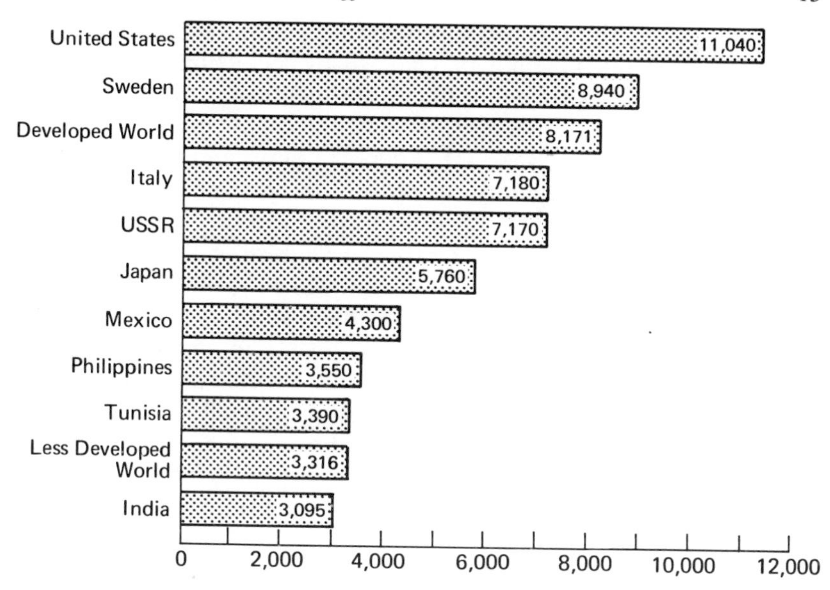

FIGURE 5 National food intake grain equivalent kilocalories per capita per day. (Reprinted with permission from Altschul, 1974a.)

to about 70,000 tons by 1980, of which 50,000 tons will be required in the animal-feeding industry.

The developed countries will, therefore, become major competitors for the food and protein supply that can be eaten directly by humans but that will be instead diverted to animal production. In these developed countries it is not a question of meeting nutritional requirements, since in all cases average consumption already exceeds requirements. The driving force is for more esthetics in food, not more nutrition, but this driving force is most potent, second only to the requirements of preventing famine.

The two world trends just described will combine to strain total food supply. For the lowest economic groups in some countries or in the underdeveloped countries as a group, the first result will be to impoverish protein supply and this will be exacerbated by the energy crisis. The chances are that the rich countries cannot rectify this situation by large-scale transfers of food without agreeing to lower their own quality of life. This means to agree to eat less meat and to make more grain available for direct human consumption.

We come to the point of asking what technology can do to change this situation and these prospects. The fundamental issues of poverty of peoples or nations cannot be changed by technology alone: They require fundamental economic, social, and political readjustment. But technology can provide greater flexibility and more options. For some nations, at least, it can make the difference between food success or failure. For others, it can accelerate the benefits of social readjustments.

New sources of energy can change the nature of the energy constraints on food production. Since energy will be more expensive, the rich will still retain their advantage over the poor.

Another type of new technology is a shift in the efficiency of utilization of present food sources. It should become possible through technology to diminish the requirements for animals as a source of protein nutrition and esthetics and also to diminish the requirements for legumes as a source of protein nutrition in favor of a greater contribution of protein from the cereals. Some examples of this kind of technology are the following:

- New animal protein analogues on meat and milk models from oilseeds and other vegetable sources will reduce the pressure for meat and milk and reduce the cost of those models. The best example of progress in this area is the availability in the marketplace of textured analogues from soybean and other vegetable protein concentrates which are being marketed as hamburger extenders or even as entirely nonanimal meat analogues. These are making a serious contribution to the supply of the meat model in the United States. They are allowed in the U.S. school lunch in amounts up to 30 percent of meat equivalent. Substantial quantities are now being offered as complete analogues. There have been various estimates of the probable rate of progress of this development. By 1980, 10 to 20 percent of meat equivalent in the United States may be these textured products. My own guess, on the basis of available technology, is that at least half of the meat equivalent in processed meat products will be these textured products by 1980. Another estimate has indicated that the market in 1972 was 80 million dollars for these kinds of products; in 1976 it is estimated to be 300 million dollars; by 1980 it is estimated to be a market of one and a half billion dollars. These products are also proliferating in other countries of the developed world and can constitute a counterforce to the increase in demand for meat products.
- Another example of new technologies is the addition of isolated nutrients to food. Vitamins and minerals are no longer a food issue. We hear proposals that children be given vitamin A injections that can

last for 6 months. Vitamin and mineral supplements are sold over the counter in the United States. The addition of calcium to foods is not at all new. The precipitation of soy curd in the Orient was by calcium sulfate and this added calcium to the diet. In Latin America, lime-soaking of corn is standard practice, and this is another way of increasing the calcium content of foods.

Of more recent origin is the addition of pure amino acids to foods. Infant formulas for those not able to tolerate milk are now made from soy protein to which methionine is added. The addition of methionine converts the soy protein to a food equivalent in performance to milk, meat, and egg protein. Amino acids added to cereals transform the protein impact of breads or other cereal products. The modified cereals are transformed thereby into a cereal-legume model. It has been estimated that in Tunisia, for example, the fortification of a million tons of wheat would add 20 million tons of good protein to the diet. In an estimate of energy costs of production of lysine versus the energy cost of the production of milk on less than optimum pasturage, Slesser (1973) calculated that the same amount of energy input would yield, when provided as a lysine supplement to wheat, 5.6 pounds of additional good protein as compared to 1 pound of milk protein from pasture. The addition of these synthetic nutrients does not eliminate energy constraints, but it allows concentration on yield of calories unencumbered by more than the usual present protein considerations.

• The third aspect of technology is breeding; I cite two cases. First is the breakthrough in distribution of protein in certain varieties, such as high-lysine corn, achieved by changing the protein mix within the seed so as to improve the protein quality and the protein impact. This approach will be discussed further in this conference. I would only point out here that the improvement achieved in this type of corn is not entirely free of cost: It might require more land for equivalent calories; it might require more energy input in terms of higher level of fertilization; it might require adjustment in processing to provide products suitable for human consumption; and it might require some social organization to ensure that the proper seeds are distributed and planted. The ultimate success or failure of such a development will depend on an economic comparison with the alternatives of supplementing with synthetic micronutrients. Or perhaps there will be a union of the two techniques for optimum energy and land utilization.

The other approach is to explore the possibility of reducing both the land and energy constraints by genetic intervention in the fundamental photosynthetic process, in the conversion of inorganic nitrogen to protein and in the further concentration of protein, as in legumes. If this is done, a better balance of protein and calories can be maintained in face of the anticipated greater demands for calories. This model differs from the immediately previous breeding model in that the energy would be free—that is, the energy would come from increased efficiency of photosynthetic utilization of available solar energy without requiring additional fossil energy inputs.

A technological fix such as envisioned by this final approach, if achieved, could stand along the Neolithic agricultural revolution as one of the great events of history.

REFERENCES

Altschul, A. M. 1974a. *In* New Protein Foods; Vol. 1A, Technology (A. M. Altschul, ed.). Academic Press, New York, p. 1.

Altschul, A. M. 1974b. Nature 248:643.

Altschul, A. M. 1974c. Lancet ii:532.

Bressani, R., and L. G. Elias. 1974. *In* New Protein Foods, Vol. 1A, Technology (A. M. Altschul, ed.). Academic Press, New York, p. 282.

FAO. 1974. Food and Agriculture Situation. Rome.

FAO/WHO. 1969. Provisional World Food Plan. Geneva.

FAO/WHO. 1973. Energy and Protein Requirements. WHO Technical Report Series No. 522, WHO, Geneva.

Schertz, L. P. 1973. Nutrition Realities in Low Income Countries. Economic Research Service, U.S. Dept. of Agriculture, Washington, D.C.

Slesser, M. 1973. J. Sci. Food Agr. 24:1193.

DISCUSSION

DR. KING: Dr. Altschul, you mentioned grain and soybean protein. With the advent of the liquid cyclone process, could you comment on the feasibility of using cottonseed protein now as a supplement to the animal protein?

DR. ALTSCHUL: The process of texturizing vegetable protein, which is the process that allows you to get the esthetic equivalent of meat, is equally applicable to other oilseed proteins. It might also be applicable in time to proteins from microorganisms and other sources. This is not strictly a soy issue.

DR. MACKE: I am interested to hear your comments on the potential use of protein extracted from vegetative tissue, such as leaf, directly instead of through the seed.

DR. ALTSCHUL: The proteins of leaves are generally better than the proteins of seeds. However, I do not think that nutrition is an issue anymore, since one can add amino acids. Nutrition is not the issue.

I think the major issue is this: Can you make a product that people will eat, and that people will like, at a price that is competitive? At the moment, with the present technologies, the cheapest source of vegetable protein is the soybean. The next cheapest sources are likely to be other oilseed proteins like peanuts and cottonseed, and perhaps others. The next one may be proteins from microorganisms.

DR. INGLETT: We work with both soybeans and cereals. Oilseed proteins have an economic advantage at the present time. There are about 30 years of science and technology already accumulated in the area. Within the past 5 years, technology on the use of cereal proteins has increased, especially oat and wheat proteins. I am not quite sure where they would fall in the priority area. But certainly, they will be and are being considered.

DR. ALTSCHUL: You are right. Among the products on the U.S. market, there are some that now include wheat gluten. It is probable that corn proteins will be utilized more than they are now.

JULIUS W. DIECKERT *and*
MARILYNE C. DIECKERT

Production of Vacuolar Protein Deposits in Developing Seeds and Seed Protein Homology

Seed proteins were among the first proteins investigated. Brohult and Sandegren (1954) credit Beccari with initiating work on seed proteins in 1745, when gluten was extracted from wheat flour. Ritthausen and Osborne are two important pioneers in the field of seed protein chemistry. Osborne (1924) summarized his and Ritthausen's outlook on seed proteins: Ritthausen seemed to believe that a comparatively small number of seed proteins existed, while Osborne felt that seeds of no two species contained proteins that could not readily be shown to be chemically different, except seeds closely related botanically. Osborne used differences in solubility as a primary criterion for identifying these proteins. For instance, in his classification scheme water-soluble proteins are albumins; proteins insoluble in water but soluble in dilute salt solutions are globulins; ethanol-soluble proteins are prolamins; and proteins insoluble in neutral salt or in aqueous alcohol but soluble in dilute alkali or acid are glutelins.

Other early workers looked at the seed proteins in a different way. Pfeffer (1872) pointed out that dormant seeds contain numerous intracellular protein granules. Hartig (1855) had earlier isolated these granules by a nonaqueous technique and showed that they contain protein. He later named the granules "aleurone grains." Pfeffer (1872) noted that the aleurone grains contain metal and phosphorus compounds. More recently, Dieckert *et al.* (1962) isolated aleurone grains from peanut seeds by a nonaqueous technique and demonstrated that they contain essentially all of the phytic acid of the seed, a large fraction of the cellular potassium, magnesium, manganese, and copper, in addition to protein.

A logical next step was to classify seed proteins on the basis of intracellular location. To this end Altschul *et al.* (1964, 1966) proposed that all proteins found in the aleurone grains be called "aleurins." The following seed proteins are examples of aleurins: arachin, from peanut seed (Altschul *et al.*, 1964; Daussant *et al.*, 1969), edestin from hempseed (St. Angelo *et al.*, 1968), vicilin and legumin from pea seeds (Varner and Schidlovsky, 1963), vicilin and legumin from *Vicia faba* (Graham and Gunning, 1970), zein in corn endosperm (Duvick, 1961), and 7S globulin and glycinin in soybeans (Koshiyama, 1972).

Protein granules in plant cells may originate from a variety of subcellular organelles. For instance, protein deposits develop in dilations of the endoplasmic reticulum in the root cells of radish (Bonnet and Newcomb, 1965) and in the plastids of bean root tips (Newcomb, 1967). Guilliermond (1941) considered that aleurone grains arise from vacuoles. The view is strengthened by the observation that mature aleurone grains of cotton (Engleman, 1966; Yatsu, 1965), shepherd's purse, and peanut (Dieckert and Dieckert, 1972) are enclosed in a characteristic single-unit membrane similar to the tonoplast.

Protein granules also develop in many types of specialized animal cells. A well-studied case is the formation of zymogen granules in the acinar cells of the pancreas. A brief annotated review of the process is given by Dieckert (1971). Correlated biochemical and cytological work showed that the cells of the exocrine pancreas adjoining an acinus contain numerous protein granules, each bounded by a cytomembrane of the unit type. These are called zymogen granules and derive their name from the fact that zymogens of certain digestive enzymes are located in them. The proteins of the zymogen granule are synthesized by polyribosomes attached to the rough endoplasmic reticulum and concentrated in elements of the Golgi apparatus. The protein droplets thus generated fuse to form zymogen granules. Under certain stimuli, the limiting membrane of the zymogen granules fuses with the plasma membrane, leaving the contents of the zymogen granule on the outside of the cell. Evidence cited in the present paper and elsewhere (Dieckert, 1969, 1971; Dieckert and Dieckert, 1972) suggests that a similar sequence of events leads to aleurone grain formation in developing seeds. Under this model the aleurone vacuoles are expected to contain a comparatively few proteins.

Biochemical genetics offers another powerful way of examining the aleurins. In 1964 a colinear relationship between the base sequence of a gene and the amino acid sequence of a polypeptide was first established experimentally for tryptophan synthetase from *E. coli* (Yanofsky *et al.*, 1964) and for the head protein of the T_4 phage (Sarabhai *et al.*, 1964). In so doing, the one gene, one enzyme principle of Beadle (1959) was verified at the molecular level. Earlier, Crick (1958) had proposed that the folding of a protein is determined by the amino acid sequence of the polypeptide. In 1961 Anfinsen showed that ribonuclease activity was determined by the amino acid sequence of the enzyme (Haber and Anfinsen, 1961; see Anfinsen, 1973, for a recent review). These considerations led Neurath *et al.* (1967), Smith (1970), and Nolan and Margoliash (1968) and others to utilize comparative studies of the amino acid sequences of proteins thought to be derived from a common ancestral gene (homologous proteins) to assess structure–function relationships of enzymes (proteins) and evolutionary relationships between organisms synthesizing homologous proteins. Almost nothing is known yet about the amino acid sequence of any major aleurin. Therefore, it is not possible at present to use the criterion of sequence similarities to identify homologous series of aleurins. The study of homologous aleurins is important for the purposes of this conference, however, because it should provide information on the variability in amino acid sequence that is acceptable to the plant and hence to be expected in the gene pool. Consequently, plausible conjectures about homologies between selected, quantitatively important aleurins will be set forth in this paper on the basis of the best evidence available today.

ORIGIN OF THE ALEURONE GRAINS

MATERIALS AND METHODS

Embryos at different stages of development were obtained from shepherd's purse (*Capsella bursa-pastoris* Medic.), peanut (*Arachis hypogaea* L.), cotton (*Gossypium hirsutum* L., variety Delta Pineland, No. 14), and coconut endosperm (*Cocos nucifera* L.). Shepherd's purse embryos were taken from wild plants growing on the Texas A&M University campus; peanut and cotton embryos were obtained from plants cultivated in a greenhouse; and coconuts were taken from a grove made available to the authors by the Virgin Islands National Park on St. John Island. The details of the procedures for the preparation of tissues from shepherd's purse, peanut, and cotton for light and electron microscopy are given elsewhere (Dieckert and Dieckert, 1972). The coconut endosperm was fixed first in a mixed aldehyde fixative containing 3 percent glutaraldehyde, 1.5 percent paraformaldehyde,

2.9 percent acrolein, 0.05 M cacodylate buffer, pH 7.0, then postfixed in 1 percent osmium tetroxide, 0.05 M cacodylate buffer, pH 7.0, at 0–4°C (Mollenhauer and Totten, 1971; Sarabhai *et al.*, 1963). The tissue was embedded in epoxy resin according to the method of Spurr (1969). Sections were made with glass knives on an LKB Ultratome III. Sections were stained with 0.5 percent uranyl acetate (Watson, 1958) and phosphotungstic acid (Locke and Krishnan, 1971). Electron micrographs were made with a Hitachi HU11A electron microscope.

The intracellular distribution of protein in cotton embryos was determined as described by Dieckert and Dieckert (1972). A brief description of the procedure follows: Embryonic tissue was fixed in glutaraldehyde and embedded in epoxy resin according to the methods of Winborn (1965) and Spurlock *et al.* (1963). Serial sections were separated and the plastic loosened with hydrogen peroxide. Alternate sections were then treated with pronase to digest protein (Anderson and Andre, 1968). All sections were then stained with magnesium uranyl acetate (Frasca and Parks, 1965) and were mounted on slot grids covered with a carbon-stabilized formvar film.

Rabbit antisera against the total antigens of peanut seed aleurone grains and α-conarachin were prepared for use in the analysis of extracts of developing peanut embryos as described by Dieckert and Dieckert (1972). Peanut embryos were classified into four groups according to the length of the embryo: Group 1, 2–5 mm; Group 2, 6–8 mm; Group 3, 10–13 mm; and Group 4, 14–19 mm. Extracts were made and analyzed by micro double immunodiffusion, as described by Dieckert and Dieckert (1972).

ALEURONE GRAINS OF MATURE SEEDS

The structure of aleurone grains from mature seeds seems to be the same for all species that become desiccated at maturity. A light micrograph of a section from the cotyledons of mature shepherd's purse is shown in Figure 1. The aleurone grain stains heavily with the mercuric bromphenol blue reagent (Mazie *et al.*, 1953), demonstrating the presence of protein. The nucleus and nucleolus stain less densely. The aleurone grains contain characteristic globoids. An electron micrograph of a similar cell fixed in osmium tetroxide is shown in Figure 2. The most striking feature of electron micrographs of cells of a mature embryo is the drastic reduction in cytoplasm and increase in ergastic materials. This implies that the protein population has shifted from one in which cytoplasmic proteins are dominant and the vacuolar proteins are minor to a population in which the reverse is true.

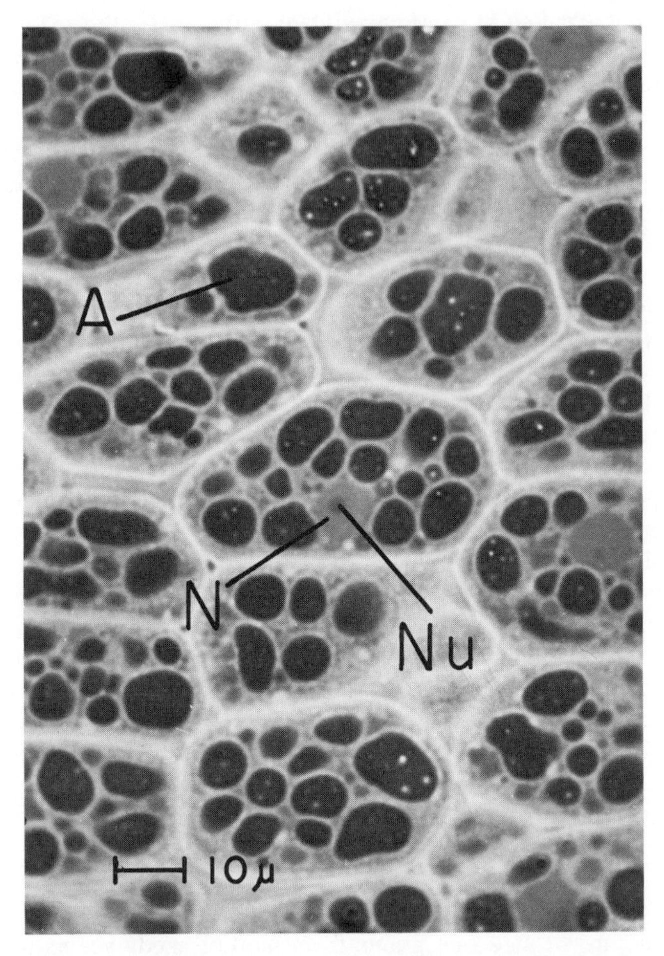

FIGURE 1 Shepherd's purse embryo. Cells of mature embryo fixed in glutaraldehyde only showing nucleus (*N*), nucleolus (*Nu*). Aleurone grains (*A*) stained deeply with mercuric bromphenol blue. Light micrograph.

Aleurone grains from all species examined so far are bounded by a unit membrane similar to the tonoplast. Figure 5*a* shows a cross-sectional view of an aleurone grain from a peanut seed. The unit membrane is clearly shown in the inset in Figure 5*a*. The same structure is found in shepherd's purse (Dieckert and Dieckert, 1972), cotton (Engleman, 1966; Yatsu, 1965), and broad bean (Briarty *et al.*, 1969).

The vacuolar space represents an extracellular region enclosed in the cell. Therefore, the sequestration of aleurins in the vacuole can be thought of as a secretion phenomenon. The production and secretion of zymogens by the acinar cells of the exocrine pancreas is an animal system that has received extensive study and appears to be analogous to the system in plant embryos for producing and secreting aleurins into the aleurone vacuoles. Figure 3 is a diagram of a pancreatic acinar cell. The basal region of the acinar cell is rich in rough endoplasmic

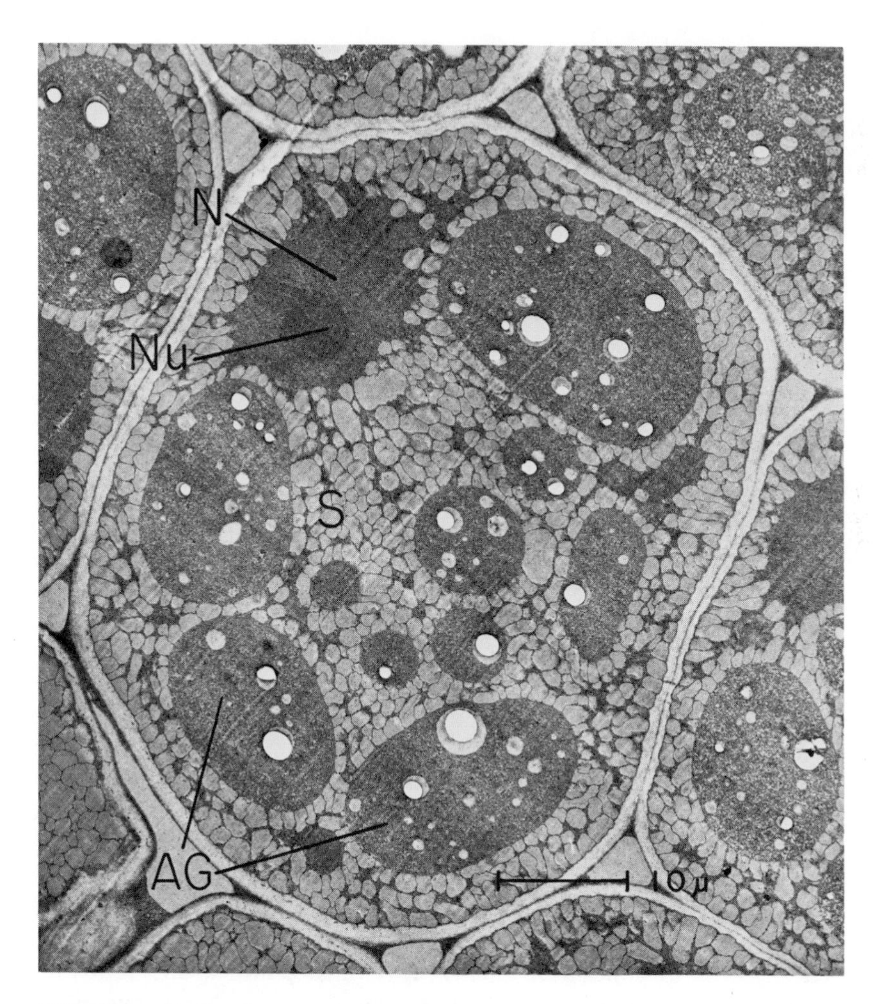

FIGURE 2 Shepherd's purse embryo. Cell of mature embryo fixed in osmium tetroxide showing mature aleurone grains (*AG*), numerous spherosomes (*S*), and a nucleus (*N*) containing a nucleolus (*Nu*). Electron micrograph.

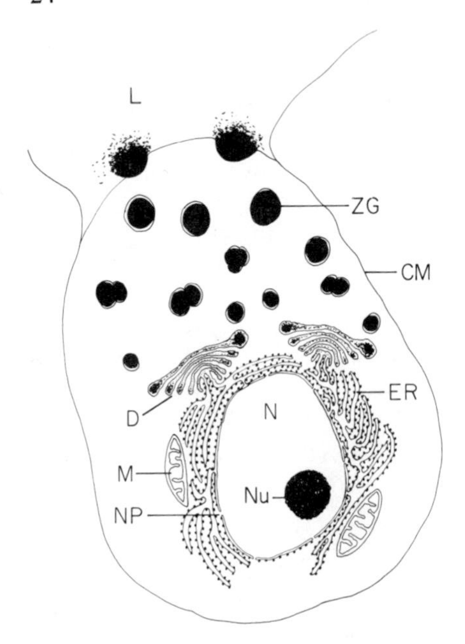

FIGURE 3 Diagram of an acinar cell of the exocrine pancreas secreting protein. Zymogen granules (*ZG*) are shown migrating toward and depositing their protein into the lumen (*L*) of the acinus. The endoplasmic reticulum (*ER*) is indicated in its relationship to the dictyosomes (*D*), in which the proteins are packaged into vesicles prior to their release into the cytoplasm as zymogen granules. A nuclear pore (*NP*) is shown in the nuclear membrane surrounding the nucleus (*N*), which contains a nucleolus (*Nu*). The cell membrane (*CM*) and mitochondria (*M*) are indicated.

reticulum, and, at certain stages in the secretion cycle, the distal regions of the cell are filled with zymogen granules containing pancreatic lipase, amylase, proenzymes of trypsin, chymotrypsin A, chymotrypsin B, carboxypeptidase A, and carboxypeptidase B. The zymogens of the digestive enzymes are synthesized by polyribosomes bound to the cytoplasmic side of the endoplasmic reticulum. The protein is transferred to the Golgi bodies, where it is concentrated into droplets and perhaps modified biochemically. The protein droplets coalesce to form the zymogen granules. In the secretion phase the zymogen granules make contact with the cell membrane facing the acinus. The membrane of the zymogen granule fuses with the cell membrane. The protein of the zymogen granule is transferred to the acinus without the cytoplasm being exposed to the extracellular environment.

If a similar scheme is operative in developing seeds that are producing aleurins, then one should find cytomorphological similarities between cells of the seeds and the acinar cells of the exocrine pancreas. Figure 4*b* shows the ultrastructure of a cell from the cotyledon of a cotton embryo that is actively producing and sequestering aleurins. The cytoplasm contains a highly developed endoplasmic reticulum coated with ribosomes. The dictyosomes, the plant homologue of the Golgi bodies of animal cells, are often filled with a substance that stains like the aleurins in the aleurone vacuole. In the peanut embryo a similar

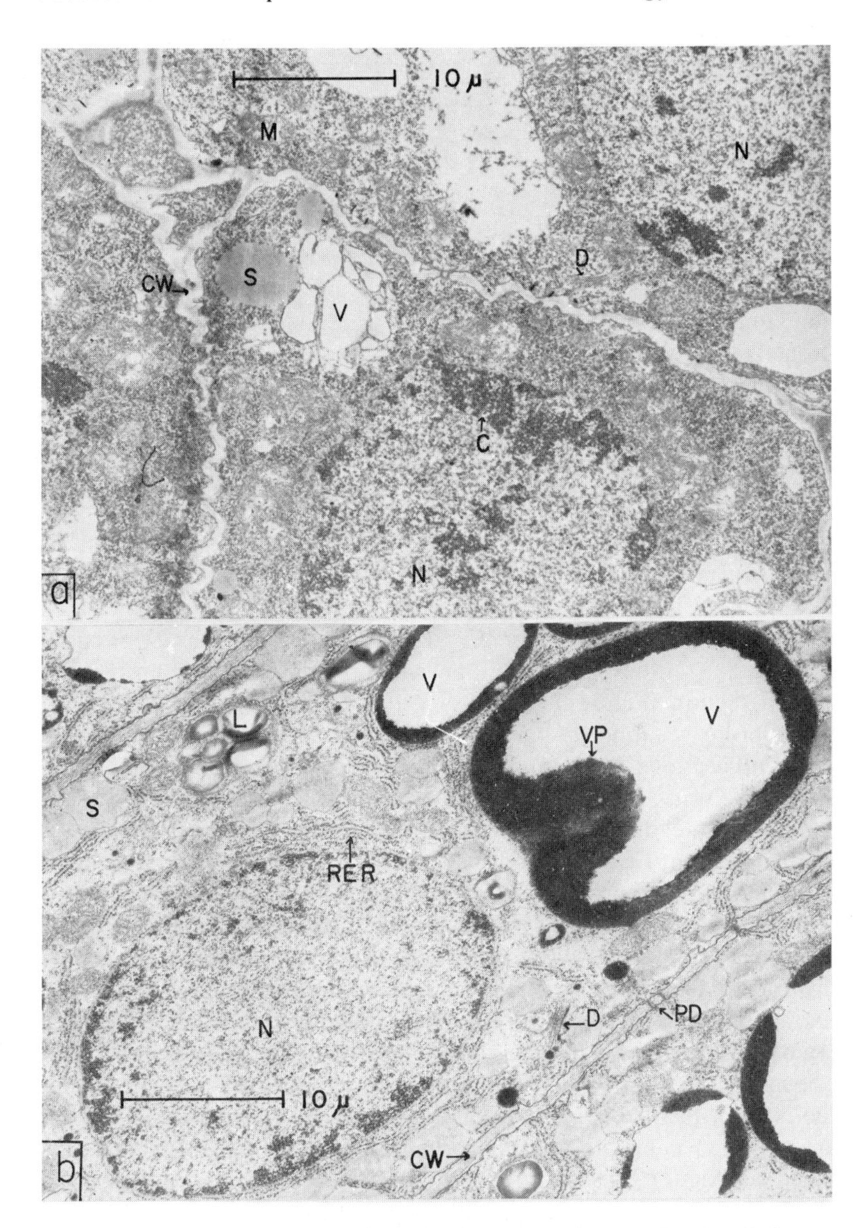

FIGURE 4 Time of appearance of aleurins in cotton embryo. *a*. Barely rolled stage. The vacuoles (*V*) do not contain aleurins, and the dictyosomes (*D*) do not have electron-dense vesicles. *b*. Loose scroll stage. The vacuoles and dictyosomal vesicles contain protein (*VP*). The rough endoplasmic reticulum (*RER*) is highly developed.

ultrastructure exists (Figure 5). The rough endoplasmic reticulum is highly developed (Figure 5*c*); polyribosomes are attached to the rough endoplasmic reticulum (Figure 5*b*); aleurone vacuoles contain densely stained aleurin deposits (Figure 5*a*). In cotton embryos protein droplets are not observed in the lumina of the rough endoplasmic reticulum, but such protein droplets are observed in the saccules of rough endoplasmic reticulum of peanuts (Figure 6*a*). Protein droplets are also found in the cisternae of the dictyosomes in peanuts (Figure 6*b*,*c*). In Figure 6*b* a series of dictyosomal protein droplets can be seen in the neighborhood of the aleurone vacuole. Droplet G_1 is still in the cytoplasm while droplets G_2 and G_3 are clearly just inside the vacuolar membrane. It is tempting to think that the direction of flow of the aleurins deposited (Figure 6*b*) is from the dictyosome to the vacuole. A similar relationship is shown in Figure 6*c*. The same ultrastructural pattern was observed for shepherd's purse embryos (Dieckert, 1969; Dieckert and Dieckert, 1972).

The electron stains employed are not specific for proteins, so cytochemical experiments were made to verify that proteins are indeed located in the aleurone vacuoles, saccules of the dictyosome, and small cytoplasmic protein droplets. Figure 7 shows the results of such an experiment. Three serial sections were treated independently. The middle section (Figure 7*b*) was digested with pronase, a mixture of serine proteases, and the neighboring sections (Figure 7*a*,*c*) were not. Then, all sections were stained for protein. The results show that the stained deposits in the aleurone vacuole, the saccules of the dictyosome (*D*), the dictyosomal vesicles, and a small free vesicle (G_1) contain protein. Figure 8 shows a second experiment of the same kind. Chromatin (*C*), ribosomes (*RER*) and most of the dictyosomal figures seem relatively unaffected by pronase, but the stained material in the aleurone vacuole and dictyosomal vesicles (G_1 through G_5) and the drop attached to the aleurone vacuole (G_6) contain protein. What is not known at present is precisely which proteins are located in these regions. Experiments are being attempted to elucidate this problem. The imagery with the electron microscope supports the hypothesis that the cells of plant embryos produce and deposit aleurins in a fashion analogous to zymogen granule formation in the cells of the exocrine pancreas. Figure 9 summarizes the hypothesis in schematic form. If the dictyosomes are in the process line of aleurin synthesis and sequestration, they may modify the aleurins passing through them. For example, the dictyosomes may modify the fundamental polypeptides produced by the rough endoplasmic reticulum by adding groups or deleting sections of the polypeptide chains (the insulin analogy).

All of the plants discussed so far are dicotyledons. The situation in

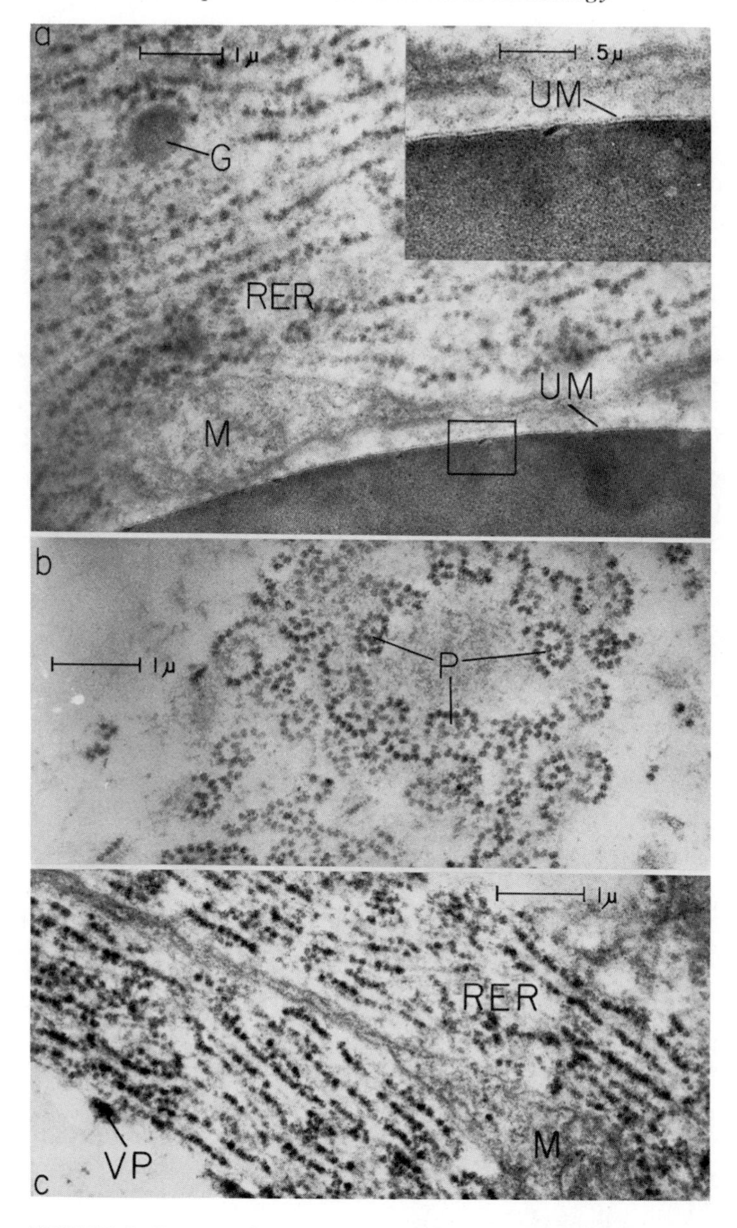

FIGURE 5 Peanut embryo. *a*. Mature aleurone grain (*G*) with closely appressed unit membrane (*UM*). Inset shows the unit membrane. *b*. Section coplanar with the membrane of the rough endoplasmic reticulum (*RER*) showing the polyribosomes (*P*). *c*. Cross section of the rough endoplasmic reticulum (*RER*) showing stacks of cisternae and protein within a vacuole (*VP*).

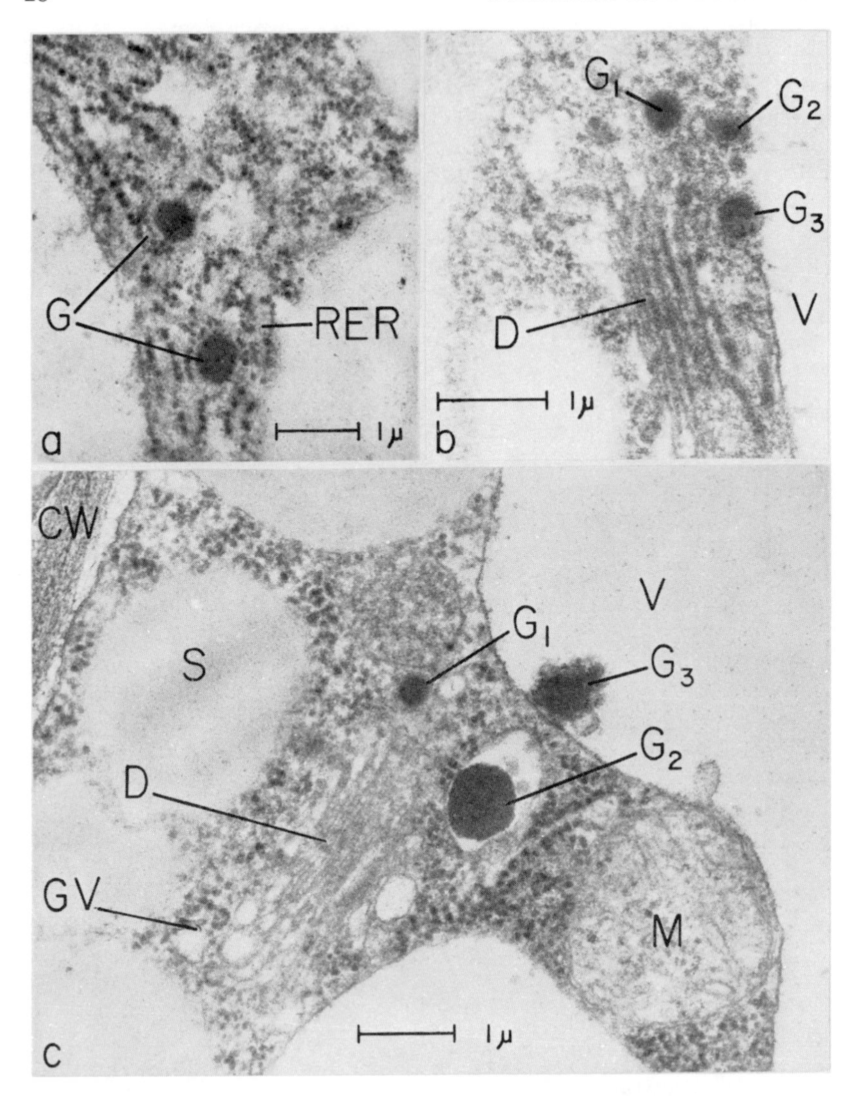

FIGURE 6 Maturing peanut embryo. *a*. Dense granules (*G*) in saccules of the rough endoplasmic reticulum (*RER*). *b*. Dictyosome (*D*) with dense granule (G_1). Also note that dense granules (G_2 and G_3) are inside vacuole (*V*). Stained with alcoholic uranyl acetate (Stempak and Ward, 1964). *c*. Cross section of dictyosome (*D*). Dense granule (G_1) in dictyosomal vesicle. Same material in larger vesicle (G_2) and one (G_3) in the vacuole (*V*).

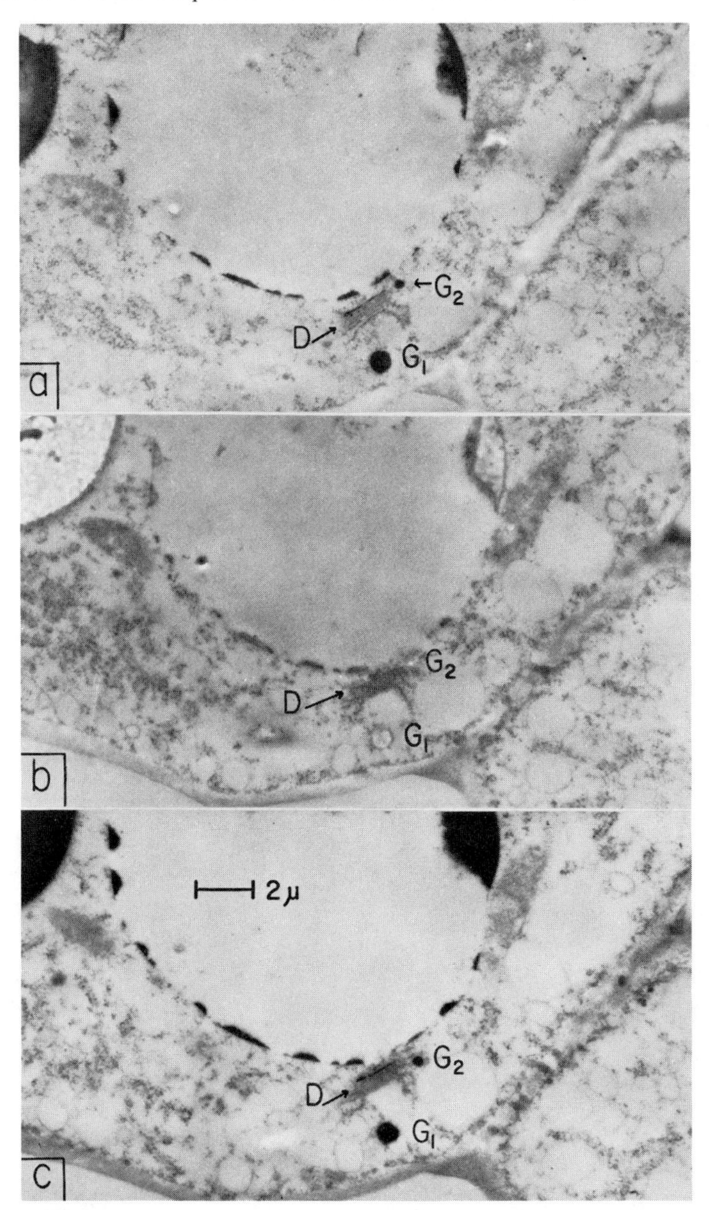

FIGURE 7 Effect of pronase on cotton embryo cells at the loose scroll stage. Serial sections. *a*. Untreated section of a cell containing an electron-dense vacuolar substance, a dictyosome (*D*), and cytoplasmic dense granules (G_1, G_2, G_3). *b*. Pronase-treated section. Staining is abolished for dense granules (G_1, G_2, G_3), the dictyosome (*D*), and the vacuolar protein. *c*. Same as *a*.

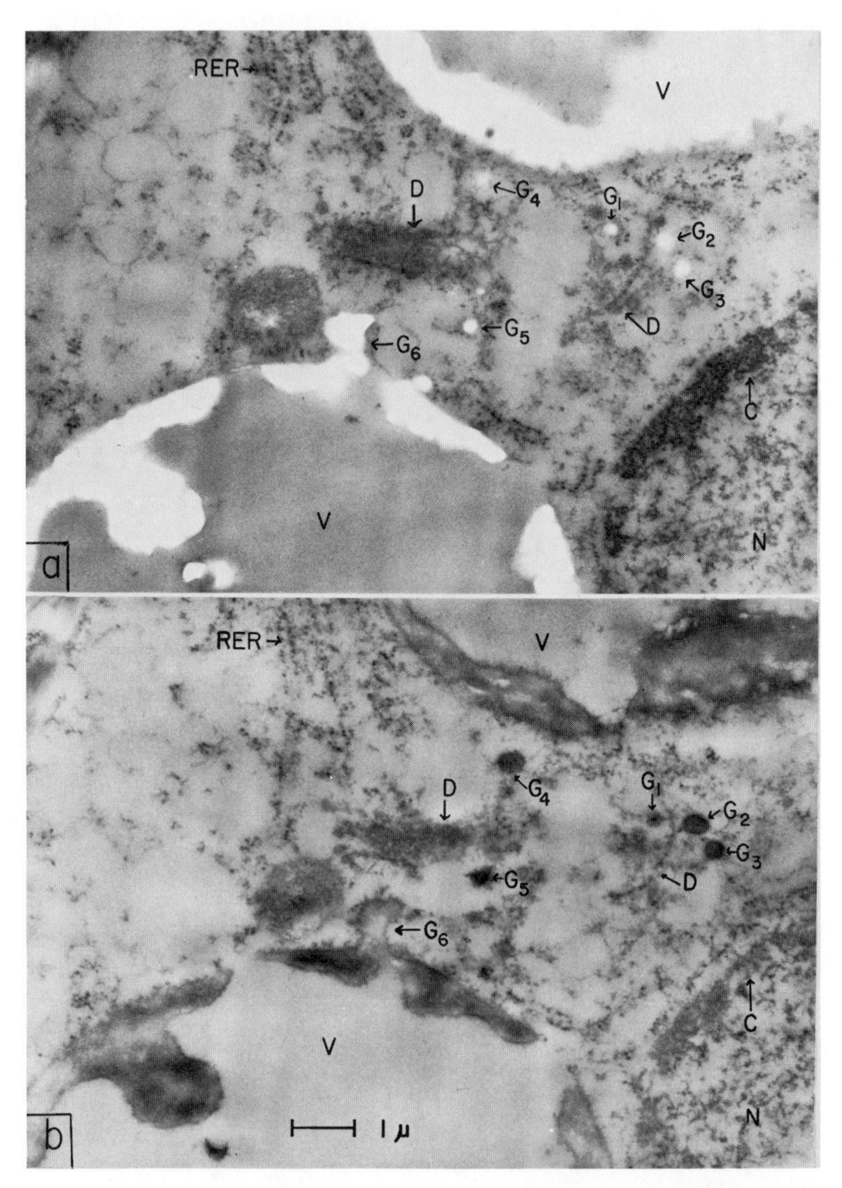

FIGURE 8 Effect of pronase on cotton embryo cells at the loose scroll stage. Serial sections. *a*. Pronase-treated section showing the absence of stain in the dense granules (G_1, G_2, G_3, G_4, G_5, G_6) containing protein. *b*. Untreated section showing the aleurone vacuole (V) containing stainable material, dictyosomes (D), rough endoplasmic reticulum (RER), and electron-dense granules (G_1, G_2, G_3, G_4, G_5, G_6).

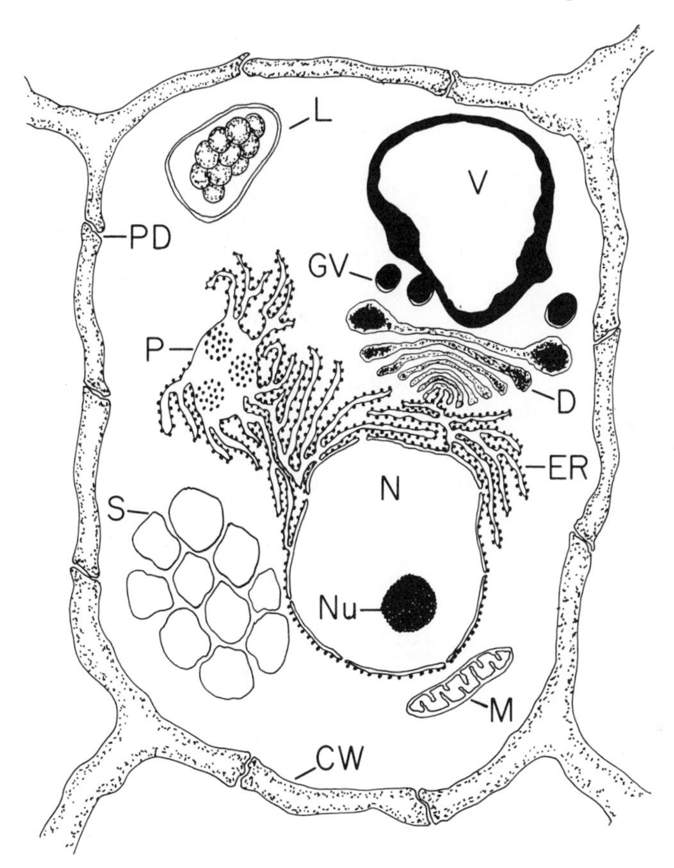

FIGURE 9 Diagram of a plant cell. This plant cell is pictured as actively synthesizing, packaging, and depositing protein in a vacuole (*V*) that is destined to become an aleurone grain. The close relationship between the dictyosome (*D*) and endoplasmic reticulum (*ER*) is illustrated, as the Golgi vesicles (*GV*) are formed and released into the cytoplasm. A coplanar view of the endoplasmic reticulum reveals the arrangement of the polyribosomes (*P*). A nucleus (*N*) with nucleolus (*Nu*), a mitochondrion, and a leucoplast are shown in a cell enclosed by a cell wall (*CW*) complete with numerous plasmadesmata (*PD*).

the monocotyledons with respect to the production and deposition of the aleurins is not so clear. Khoo and Wolf (1970), working with maize endosperm, found that protein droplets developed in vesicles produced by the endoplasmic reticulum and were formed as small localized cisternal dilations and as enlargements at the ends of the endoplasmic reticulum. They also reported that the dictyosomes produced protein granule vesicles. It is not clear from their work whether the dictyosomes are obligatory components in the production and deposition process.

Dieckert and Dieckert (unpublished data) are studying the process in coconut endosperm. The coconut is one of the largest seeds in the world and takes one year or a little more to reach maturity. Figure 10 shows a diagram of a longitudinal section through a coconut fruit at about 11 months of development. The solid endosperm is cellular and is attached to the seed coat lining the central cavity of the endocarp. At 11 months the endosperm is usually about 11 to 12 mm thick. The electron micrographs shown here (Figures 11 through 14) were taken from a wedge of endosperm located midway along the long axis of the seed. The wedge was subdivided into the outer, middle, and inner regions.

Figure 11*a* shows the cells of the seed coat one cell layer removed from the seed coat–solid endosperm interface. The sectioning plane is normal to the seed coat–endosperm interface. Figures 11*b* and 12*a* are similarly oriented sections through the seed coat–endosperm interface. The structural differences in the two areas are apparent. Of greatest interest here is that the cells of the first layer of endosperm are richly endowed with mitochondria, and the cytoplasm contains numerous densely osmiophilic fat droplets. There are no aleurone vacuoles. Six cells in from the endosperm–seed coat interface, remarkable changes in ultrastructure occur (Figures 12*b*, 13*a*). Aleurone vacuoles are present and numerous, and less osmiophilic spherosomes appear.

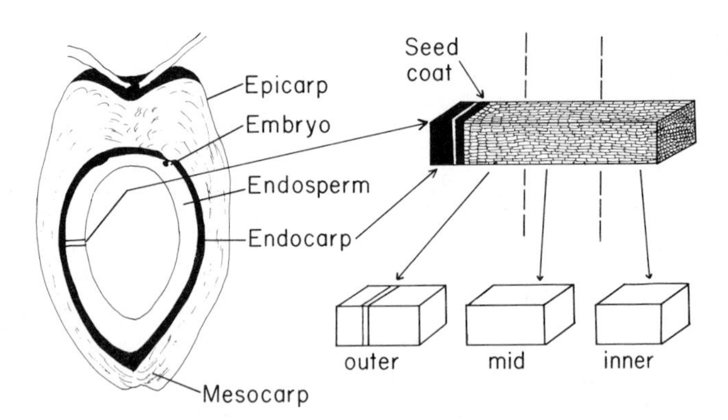

FIGURE 10 Coconut dissection scheme. Tissue wedge from middle region of solid endosperm is subdivided into three parts prior to embedment for ultrastructural study.

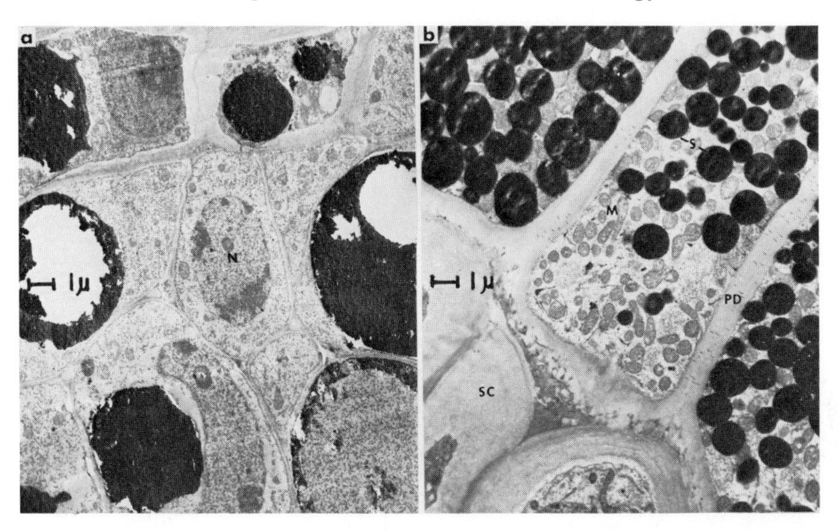

FIGURE 11 Coconut seed-coat–endosperm interface (inflorescence 11). *a*. Cross section of cells of the seed coat, one cell layer removed from the seed-coat–solid-endosperm interface, with a well-defined nucleus in evidence (*N*). *b*. Cross section of cells at the seed-coat–endosperm interface, showing a layer of differentiated cells of the seed coat (*SC*) at upper left and columnar cells of the solid endosperm containing osmiophilic spherosomes (*S*) and numerous mitochondria (*M*) at the lower right. Plasmadesmata (*PD*) are shown in the cell walls between cells of the endosperm.

Still further changes in the ultrastructure appear in the middle section of the wedge of endosperm (Figure 13*b*). The space of the spherosomes originally occupied by triglycerides seems to be empty. There is a reason for this: Coconut oil triglycerides contain mostly saturated fatty acids that do not react with osmium tetroxide. The lipids are extracted in the embedding process. The aleurins are present in large watery vacuoles and often exist in crystalline form. Figure 14*a* shows another section through the same region. The aleurone vacuoles contain crystalline aleurins surrounded by noncrystalline amorphous protein matrix.

The cells of the inner region (Figure 14*b*) exhibit the same general ultrastructure, but the impression is that there is less protein and fat here. The watery vacuoles are larger and the spherosomes are smaller. The electron micrographs shown here give no clues as to the process of aleurin synthesis and deposition, except that most of the seed proteins are located in vacuoles of the usual type. Apparently, aleurin synthesis and deposition was completed in this coconut by the 11-month stage. The conspicuous mitochondrial activity of the endosperm cells at the

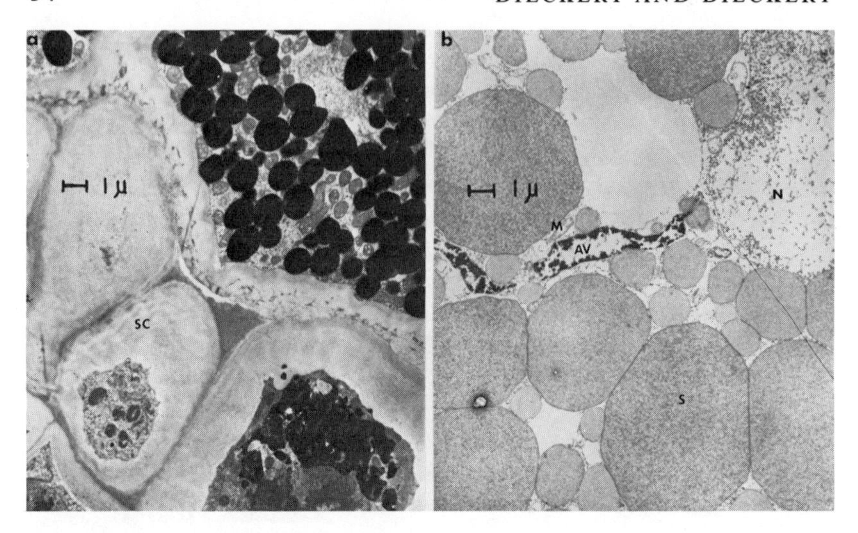

FIGURE 12 Coconut seed-coat–endosperm interface (inflorescence 11). *a*. Section of cells at the interface between the seed coat layer (*SC*) and the solid endosperm (lower right). *b*. Section of a portion of a cell of the solid endosperm, six cells in from the seed coat, showing an aleurone vacuole (*AV*) containing protein, a nucleus (*N*), mitochondrion (*M*), and numerous spherosomes (*S*).

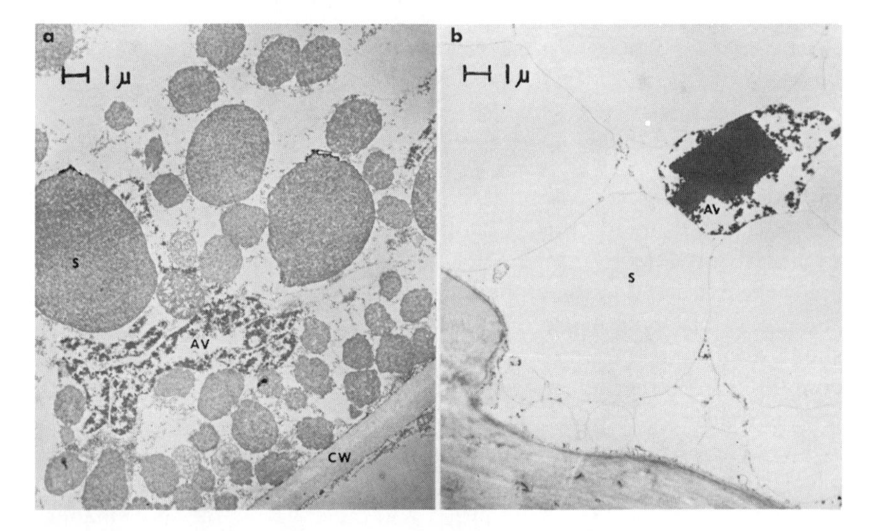

FIGURE 13 Coconut endosperm (inflorescence 11). *a*. Section of a portion of a cell of the solid endosperm, six cells in from the seed coat, showing an aleurone vacuole (*AV*) containing protein and numerous spherosomes (*S*), which appear less osmiophilic than those in the outer region. *b*. Section of a portion of a cell of the solid endosperm in the midregion between the seed coat and the central cavity of the coconut. A protein crystal is seen in the aleurone vacuole, and many spherosomes (*S*) are present that appear to have lost their affinity for osmium tetroxide.

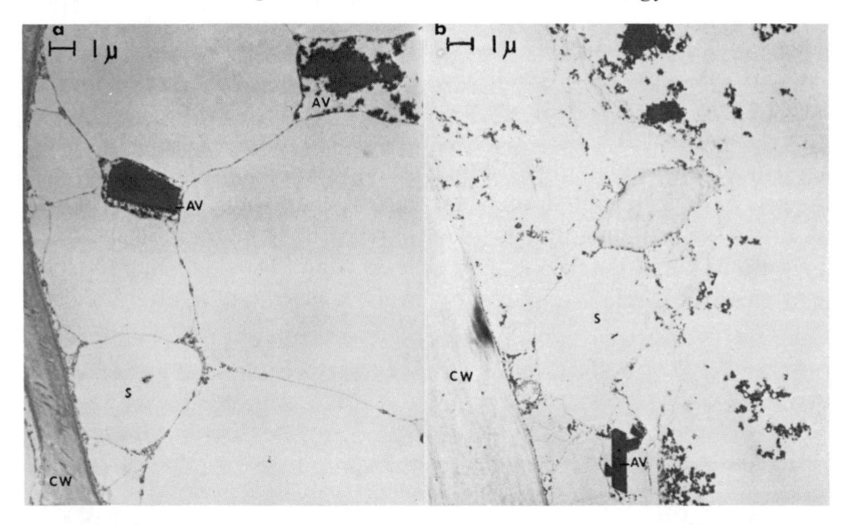

FIGURE 14 Coconut endosperm (inflorescence 11). *a*. Section of a portion of a cell of the solid endosperm in the midregion between the seed coat and the central cavity of the coconut, showing aleurone vacuoles (*AV*) containing protein in a crystalline and amorphous state. The cell wall (*CW*) borders a cell packed with spherosomes (*S*) that have lost their affinity for osmium tetroxide. *b*. Section of a portion of a cell of the solid endosperm in the inner region of solid endosperm but not adjoining the central cavity of the coconut. A small protein crystal is seen in an aleurone vacuole (*AV*), and the spherosomes (*S*) have a different appearance from those in middle and outer regions.

seed coat–endosperm interface is probably related to continued triglyceride synthesis occurring at this stage.

INITIATION OF ALEURONE GRAIN FORMATION

Aleurin synthesis and deposition starts only after a certain stage of development is reached by the seed. Cotton embryos at the barely rolled stage of development (Figure 4*a*) do not show a highly developed rough endoplasmic reticulum; the vacuoles lack aleurin deposits, as do the cisternae and saccules of the dictyosomes. At the loose-scroll stage, however, the aleurin vacuoles are present and contain conspicuous deposits of protein; the dictyosomal vesicles and saccules contain protein deposits; and the rough endoplasmic reticulum is highly developed (Figure 4*b*). A similar pattern is found in developing shepherd's purse embryos (Dieckert, 1969; Dieckert and Dieckert, 1972). At the walking-stick stage the shepherd's purse embryo does not contain visible vacuolar protein, but by the upturned-U stage abundant aleurin deposits are visible. The other changes described for cotton are also present. In the peanut, antigens of the arachin type were first detectable by micro double immunodiffusion analysis by the 10–13 mm stage (Figure 15). The relative concentration of these antigens increased progressively with the average size of the embryo. Ultrastructural analysis of the same series of peanut embryos showed general agreement with the immunochemical analysis, except that the start of the process was detected one stage earlier. Khoo and Wolf (1970) first observed the protein granules (aleurin deposits) in maize endosperm 12 days after pollination. In coconut (Dieckert and Dieckert, unpublished data) the endosperm does not show massive growth until after the seed

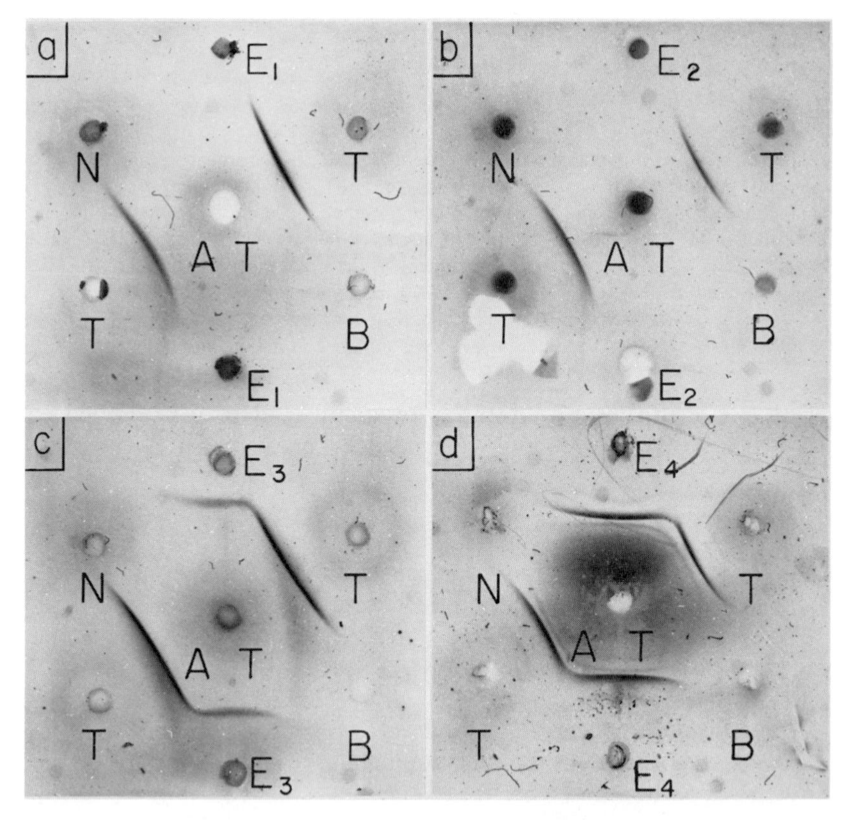

FIGURE 15 Micro double diffusion analysis of developing embryos of the peanut. *a*. Analysis of an extract (E_1) of 2–5-mm embryos with A/T. The extract (*T*) of mature aleurone grains served as reference; a nonimmune serum (*N*) and the buffer (*B*) served as controls. *b*. Analysis of an extract (E_2) of 6–8-mm embryos with A/T. *c*. Analysis of an extract (E_3) of 10–13-mm embryos with A/T. *d*. Analysis of an extract (E_4) of 14–19-mm embryos with A/T.

is 5–6 months old; aleurins do not seem to appear until sometime after the gelatinous endosperm stage.

PROPOSED HOMOLOGIES BETWEEN ALEURINS

As mentioned earlier, one of the most impressive features of an electron micrograph of a cell of a mature dry seed is the reduced cytoplasm and the expanded volume of ergastic protein. Subcellular organelles such as plastids, mitochondria, rough endoplasmic reticulum, and dictyosomes are rarer in such a cell than in a similar cell actively depositing aleurins. The implication is that proteins found in the cell organelles and cytoplasmic sap of the growing embryo are absent or much reduced in quantity in the mature cell. In view of the model of aleurone grain formation outlined in the previous section, the population of proteins has changed during seed maturation from one dominated by enzymatic proteins of diverse kinds to one dominated by nonenzymatic reserve proteins. This is not to say that enzyme proteins and other specialized proteins are absent. A variety of enzymes are present in seeds. Some of those of particular interest here are the proteolytic enzymes and their inhibitors (Ryan, 1973) and the interesting cell-agglutinating, often sugar-specific, proteins, the lectins (Sharon and Lis, 1972). Usually, these and the other enzyme proteins are present in low concentrations. Some of the proteolytic enzymes, their inhibitors, and the lectins may be located in the aleurone vacuoles. If so, they are aleurins.

Proteins and enzymes are commonly compared in terms of their similarities in function, structure, and evolutionary antecedents. The similarities may be considered as "analogies" or "homologies." Neurath *et al.* (1967) uses the term "analogy" to denote similarities in function without regard to structure. The assumed functional analogy among the principal seed proteins is that they all reserve proteins dedicated to the nutrition of the germinating seed. Nolan and Margoliash (1968) use the term "analogous" somewhat differently: The structures of analogous proteins are similar but the evolutionary antecedents of the proteins are different. Polypeptides are said to be homologous if the genes coding for them evolved from a common ancestral gene (Nolan and Margoliash, 1968). The question naturally arises as to how many ancestral genes are involved in the evolution of the principal aleurins. Since the aleurins are vacuolar proteins, this number is probably small.

In plant cells the interior of the vacuole can be considered to represent an extracellular space. The sequestration of aleurins in vacuoles of the developing seed is, therefore, analogous to the secretion of zymogens, peptide hormones, and other secretory polypeptides by gland cells in animals. It is instructive in this connection to examine the known homologies among the proteins produced by the exocrine pancreas. Some of the enzymatic activities assigned to the zymogen granules are amylase, lipase, trypsinogen chymotrypsinogens A and B, and procarboxypeptidases A and B. The best available criterion for homology between proteins is the demonstration of extensive areas of identity in amino acid sequence of the polypeptides over and above the minimum required for common activity (Nelson and Margoliash, 1968). Under this criterion there is great homology in the sequences of trypsin and chymotrypsins A and B. [See Neurath *et al.* (1967) and Smith (1970) for review.] The above homologies are all serine proteinases. The carboxypeptidases are metallopeptidases. And, although there is less extensive data on the subject than for the serine proteases, the carboxypeptidases are thought to be homologous (Neurath *et al.*, 1967). The above analysis indicates that four ancestral genes can account for the digestive enzymes of pancreatic juice.

Almost no data on the amino acid sequence exist for any major aleurin. There are several other weaker criteria for recognizing homologous polypeptides. Similarities in peptide maps suggest similarities in amino acid sequence for extensive regions of the parent polypeptides. Thus, similarity in peptide maps is used to help identify homologous proteins. There are peptide maps published for a few legume globulins. Since homologous proteins generally have similar amino acid compositions, similar amino acid compositions serve as weak indications of homology. Amino acid compositions are available for a wide variety of seed proteins. In some cases the data are subject to question since the proteins are difficult to purify. However, the present search for homology among the principal aleurins will depend heavily on these data. Another useful indication of homology is serological cross reactivity between suspected homologues. A negative result here leaves the question of how divergent two homologues can be and still show cross reactions. In addition to the above criteria, other similarities between the aleurins will be considered, including molecular weight of the parent and the molecular weight distribution of the subunits as determined by sodium dodecyl sulfate (SDS) gel electrophoresis.

SEED GLOBULINS

The aleurins considered in this paper represent globulins, glutelins, and prolamins. The globulins are grouped into two series, the legumins and vicilins. The legumin group includes edestin, glycinin, legumin from *Vicia faba*, arachin, and cocosin. Cocosin is here defined as the globulin fraction of coconut endosperm that is soluble in dilute salt solution but precipitates in 65 percent saturation of ammonium sulfate. The vicilin group includes α-conarachin, vicilin from *Phaseolus aureus*, vicilin from *P. sativum*, vicilin from *V. faba* and α-concocosin from coconut endosperm. The last globulin is the fraction of coconut globulins that does not precipitate at 65 percent saturation with ammonium sulfate but precipitates at 85 percent saturation. Acalin A, from cotton seed, is included in both groups. Glutenin of wheat is the sole representative of the glutelin class and eight gliadins from wheat represent the prolamins. In the tables that follow, the source of the data for each protein is given by the reference cited.

First, let us consider possible homologies between the vicilin and legumin types of seed globulins from closely related species. The best evidence adduced so far is that of Jackson *et al.* (1969) for the vicilins and legumins isolated from *P. sativum, V. faba,* and *Cicer arietinum*. Fingerprint patterns of the trypsin digests of the vicilin and legumin from *P. sativum* were similar, indicating that considerable portions of the amino acid sequences of these proteins are the same or very similar. Seventy percent of the legumin spot pattern was common to the vicilin pattern and 80 percent of the vicilin spot pattern was present in the legumin pattern, thus indicating homology. Similar data indicated that the "vicilin" fractions of *V. faba* and *C. arietinum* are homologous to vicilin from *P. sativum* and the "legumins" from the former are also homologous to legumin from *P. sativum*. Homologous vicilins and homologous legumins from the three species resembled each other more closely than did the vicilins and legumins from any single species.

Homologous proteins from closely related species are expected to have similar amino acid compositions. For comparison, the amino acid compositions of the vicilins and several other vicilinlike proteins are given in Table 1. As expected, the amino acid composition of the vicilins from *V. faba* and *P. sativum* are very similar. Data from another source on the amino acid composition of the vicilin from *P. aureus* shows that this vicilin is also probably homologous with *P. sativum* vicilin. The amino acid composition of α-conarachin is shown in the first column of Table 1. The similarities in amino acid composition between α-conarachin and the vicilins are striking. However, there are differences, as might be expected. The peanut is a legume, but it is not as closely related to *Pisum* as *Vicia* and *Phaseolus*. α-Concocosin is a globulin from coconut endosperm. The coconut is a more distant relative of *P. sativum* than is *Arachis*, so one might expect homologous

TABLE 1 Comparison of the Amino Acid Compositions[a] of the Vicilin, α-Conarachin, and α-Concocosin Series of Seed Globulins

		Vicilin				
Amino Acid	α-Conarachin: *A. hypogaea*[b]	*P. aureus*[c]	*P. sativum*[d]	*V. faba*[d]	α-Concocosin *C. nucifera*[e]	Acalin A: *G. hirsutum*[f]
Ile	4.7	4.5	4.8	5.1	2.9	2.3
Leu	7.1	9.4	8.9	9.2	6.3	5.1
Lys	5.5	6.0	6.3	7.1	4.4	4.8
Met	1.2	0.3	0.20	0.31	1.17	0.62
Phe	4.5	6.1	4.7	5.2	3.7	5.9
Thr	3.1	3.0	3.5	3.3	2.7	3.3
Trp	0.77	—	—	—	0.56	0.50
Val	6.5	6.6	4.9	4.9	5.6	4.8
Ala	6.0	5.6	4.5	4.9	5.6	6.1
Arg	7.9	5.5	6.4	5.6	12.7	8.7
Asp	13.0	13.4	11.2	11.6	8.1	10.2
Cys	0.97	—	0.39	0.31	3.3	0.87
Glu	17.7	19.9	15.8	15.3	21.3	19.2
Gly	7.0	5.4	5.7	5.0	8.7	7.8
His	2.5	2.0	2.1	2.0	2.3	3.6
Pro	4.0	2.9	—	—	4.3	5.7
Ser	6.7	7.3	6.9	6.6	4.9	7.9
Tyr	0.96	1.9	2.6	2.6	1.59	2.6
NH_3	10.9	—	—	—	11.07	18.6
Recovery	83%	98–99%	96%	104%	99.6%	

[a] Composition in mole %.
[b] Altschul (1964).
[c] Ericson and Chrispeels (1973).
[d] Jackson *et al.* (1969).
[e] Khaund (1971).
[f] Rossi-Fanelli (1968).

TABLE 2 Properties of Some Proteins of the Vicilin Type

Protein	Source	Molecular Weight		
		Parent		SDS Subunits
α-Conarachin	*A. hypogaea*	142,000	7.8S[a]	70,000[b]
Vicilin	*P. aureus*	—	8.0S[c]	50,000[c]
α-Concocosin	*C. nucifera*	110,000	7.6S[d]	56,000[d]

[a] Dechary *et al.* (1961).
[b] Yu and Dieckert, unpublished data.
[c] Ericson and Chrispeels (1973).
[d] Khaund (1971).

proteins from these two sources to show differences in amino composition. The amino acid composition of α-concocosin is different from that of the vicilins, but there are important similarities as well. For example, both proteins contain high content of arginine + lysine and aspartic + glutamic acids. It is interesting that acalin A from cotton seed also shows similarities in amino acid composition with α-concocosin and the vicilins.

Table 2 shows a comparison of some other properties of proteins of the vicilin type. The data here are more limited. Even so, the molecular weights (or sedimentation coefficients) for the parent proteins are similar in value, and the parent molecule consists of two apparently identical subunits in each case as determined by SDS gel electrophoresis by the method of Weber and Osborn (1969). The molecular weights of the SDS subunits differ, however. The data at present are admittedly meager but, considered *in toto*, support the view that the α-conarachin and α-concocosin series are homologues. Acalin A may also be homologous, but there is more doubt here.

The arguments for homology between the vicilin-type seed globulins are also applicable to the edestin-to-cocosin (legumin-type) series of seed proteins. The amino acid composition and membership of the series is given in Table 3. The similarities for the set are surprisingly good. It is interesting to note here that Altschul (1964) pointed out the similarities in amino acid composition of edestin and α-conarachin. Acalin A also seems to fit the same amino acid compositional pattern. Some additional properties of these proteins are detailed in Table 4. The proteins of the series representing the legumes have similar molecular weights, while the representatives from the more distant relatives seem to be of lower molecular weight. All of the proteins

TABLE 3 Comparison of Amino Acid Composition of Legumin-Type Proteins[a]

Amino Acid	Edestin: *C. sativa*[b]	Glycinin: *G. hispida*[c]	Legumin: *V. faba*[d]	Arachin: *A. hypogaea*[e]	Cocosin: *C. nucifera*[f]	Acalin A: *G. hirsutum*[g]
Ile	3.2	4.2	4.3	3.4	3.5	2.3
Leu	5.1	7.0	7.3	6.0	6.6	5.0
Lys	2.3	5.0	4.7	2.5	3.3	5.3
Met	0.88	1.16	0.38	0.53	2.1	0.69
Phe	3.2	5.0	4.6	5.7	4.2	7.3
Thr	2.7	3.5	3.4	2.5	3.1	2.9
Trp	1.55	—	1.13	—	0.9	0.91
Val	3.9	4.2	4.4	4.3	5.2	4.2
Ala	3.3	3.2	3.2	4.0	3.7	4.0
Arg	12.1	8.0	9.6	11.6	15.7	11.4
Asp	10.3	12.0	10.6	11.2	8.2	10.0
Cys	0.75	—	1.26	1.31	2.3	1.25
Glu	17.1	22.1	15.7	17.3	23.0	21.7
Gly	3.2	3.8	3.4	3.8	4.2	4.4
His	2.1	2.3	2.9	2.5	1.7	4.2
Pro	3.1	5.8	6.4	4.6	3.3	4.9
Ser	4.3	5.4	4.2	4.6	4.2	6.2
Tyr	3.0	4.1	4.9	4.2	3.0	3.6
NH_3	2.1	—	—	2.2	1.7	—
TOTAL	88.3	—	92.36	90.88	99.9	—

[a] Composition in g residue per 100 g protein.
[b] Altschul (1964).
[c] Catsimpoolas *et al*. (1971).
[d] Bailey & Boulter (1970).
[e] Dawson (1971).
[f] Khaund (1971).
[g] Rossi-Fanelli (1968).

contain multiple subunits of several molecular weight classes. All of the representatives have similar subunit molecular weight patterns except acalin A. Another interesting common property is cryoprecipitability. Glycinin, arachin, edestin, cocosin, and acalin A are all cryoprecipitable from dilute salt solutions. The legumins are probably cryoprecipitable as well. All the proteins of this series are aleurins. Again, the evidence for homology in this series is not perfect, but it is suggestive. These proteins seem to belong to one homologous series, which I call the legumin series.

The question of homology between the vicilins and legumins was considered earlier. In Table 5 a comparison is made between the amino acid compositions of the vicilin–legumin pairs for each of several seed globulins. There are obvious differences and similarities in the amino acid compositions of the pairs. If homologies already proposed are correct, then it seems likely that the pairs from each species are also homologous. In effect, then, it seems that one ancestral gene is the progenitor of all of the current genes for the vicilin and legumin types of aleurins. This could happen by gene duplication of the ancestral gene with subsequent divergent evolution. [See Smith (1970) for review.] The question of the number of genes present today in, for instance, *V. faba* for homologous vicilin and legumin is open.

In the case of glycinin, the possibility of heterogeneity in the two principal molecular weight classes of subunits was suggested by the work of Catsimpoolas *et al.* (1971). Isoelectric focusing experiments revealed that the 22,300 molecular weight class is a set of three polypeptides (the B series) differing in charge. Similarly, the 37,200 molecular weight class is a mixture of three polypeptides (A series) with different charges. These subunits were isolated and a partial amino acid composition determined (Table 6). The authors noted difficulties with the amino acid analyses but considered these subunits to be authentic and different in composition. Apparently, the A series and B series form two homologous sets. There must be a family of glycinins in the soybean extracts.

GLUTELINS AND PROLAMINS

The grasses provide an interesting new set of problems for evolutionary homology. Wheat will serve as the example. The globulin fraction of wheat accounts for only about 6 percent of the total protein (Danielsson, 1949). Two globulins designated α and γ were found. Both the α and γ globulins were found in the aleurone layers, but only γ was found in the embryo proper. The molecular weights are: α = 29,000

TABLE 4 Properties of Some Proteins of the Legumin Type

Protein (source)	Molecular Weight					
	Parent		SDS Subunits			
Legumin: *V. faba*	320,000[a]		23,000*		42,000*	56,000[a]
Legumin: *P. aureus*	n.d.	11.3S[b]	16,500[b]		44,000[b]	56,000[b]
Glycinin: *G. hispida*	363,000[c]	11S[d]	22,300*		37,200*	44,200[e]
Arachin: *A. hypogaea*	330,000[f]		24,000*		43,000*	Others[g]
Edestin: *C. sativa*	212,000[h]		18,000*		36,000*	Others[i]
Cocosin: *C. nucifera*	208,000[j]		26,000*		34,000*	56,000[k]
Acalin A: *G. hirsutum*	180,000[l]		46,000	48,000*	54,000*	Others[i]

*Major subunit.
n.d. = Not determined.
[a] Bailey and Boulter (1970).
[b] Ericson and Chrispeels (1973).
[c] Anderson and Andre (1968).
[d] Wolf and Briggs (1959).
[e] Catsimpoolas *et al*. (1971).
[f] Johnson and Shooter (1950).
[g] Singh and Dieckert (1973).
[h] Svedberg and Stamm (1929).
[i] Wallace and Dieckert, unpublished data.
[j] Sjogren and Spychalski (1930).
[k] Khaund (1971).
[l] Rossi-Fanelli *et al*. (1964).

(2.5S) and $\gamma = 210{,}000$ (8.7S, $D = 3.6 \times 10^{-7}$ cm g^{-1} s^{-1}). Glutenin and gliadin account for 51 percent and 43 percent, respectively, of the wheat protein. The amino acid composition of gliadin and glutenin is given in Table 7. Both of these classes are complex mixtures of subunits and, therefore, the amino acid compositions are averages. Even so, the similarities in amino acid composition are significant. A number of the wheat gliadins have been isolated, and their amino acid compositions and molecular weights determined (Ewart, 1973). The results, which are given in Table 8, show significant similarities in the amino acid composition. Ewart says that the similarities in amino acid composition imply close relationship between the molecular structures of the gliadins. He points out that these facts, when added to earlier work showing that common amino acid sequences exist in glutenin and gliadin, are strong support for the working hypothesis that glutenin polypeptides and gliadins have evolved from a common precursor; in other words, they are homologous.

Boulter *et al.* (1972) developed a phylogeny of higher plants based on the amino acid sequences of cytochrome C isolated from selected species. The indications are that the monocotyledonous species diverged from the main dicotyledonous stock at about the same time as the Compositae. Apparently, the divergence occurred after the Malvaceae diverged. Wheat and coconut are both monocots; so, presumably they are more closely related to each other than to the legumes. It is of interest, therefore, to compare the aleurins of wheat to the aleurins of coconut endosperm. As already pointed out in the preceding paragraph, wheat has a very low content of globulins (6 percent). Coconut endosperm, on the other hand, has mostly globulins and little, if any, glutelins or prolamins. In Table 7 the amino acid compositions of gliadin and glutenin are compared with those of α-concocosin and cocosin. Except for six residues, there are striking similarities in the amino acid composition of glutenin and α-concocosin. The differences in mole percent for the six residues are: −3 percent Lys, −11 percent Arg, −5 percent Asp, −2 percent Cys, +12 percent Glu, and +9 percent Pro. Two important questions arise: What happened to the genes for the missing globulins? What are the gene antecedents for the newly dominant aleurins, the glutelins and prolamins? It is tempting to propose that the glutelins and the globulins of the vicilin or legumin types have a common ancestral gene. One possibility is that a subsequence of an ancestral gene in the monocotyledonous branch was deleted, resulting in the loss of the corresponding globulins and the gain of the new class of proteins, the glutelins.

TABLE 5 Comparison of Amino Acid Compositions[a] of the Principal Seed Proteins of Three Closely Related Dicots and One Monocot

Amino Acid	α-Conarachin: *A. hypogaea*[b]	Arachin: *A. hypogaea*[c]	Vicilin: *V. faba*[d]	Legumin: *V. faba*[e]	α-Concocosin: *C. nucifera*[f]	Cocosin: *C. nucifera*[f]
Ile	3.7	3.4	5.4	4.3	2.8	3.5
Leu	5.6	6.0	9.6	7.3	6.0	6.6
Lys	4.9	2.5	8.5	4.7	4.8	3.3
Met	1.1	0.53	0.38	0.38	1.3	2.1
Phe	4.6	5.7	7.1	4.6	4.6	4.2
Thr	2.2	2.5	3.1	3.4	2.3	3.1
Trp	1.0	—	—	1.13	0.9	0.9
Val	4.5	4.3	4.5	4.4	4.7	5.2
Ala	3.0	4.0	3.2	3.2	3.4	3.7
Arg	8.6	11.6	8.1	9.6	16.9	15.7
Asp	10.5	11.2	12.4	10.6	7.9	8.2
Cys	0.7	1.31	0.30	1.26	2.9	2.3
Glu	16.0	17.3	18.3	15.7	23.3	23.0
Gly	2.8	3.8	2.6	3.4	4.2	4.2
His	2.4	2.5	2.5	2.9	2.7	1.7
Pro	2.7	4.6	—	6.4	3.5	3.3
Ser	4.1	4.6	5.3	4.2	3.6	4.2
Tyr	1.1	4.2	3.9	4.9	2.2	3.0
NH_3	1.3	2.2	—	—	1.6	1.7
Recovery	83%	90.88%	104%	92.36%	99.6%	99.9%

[a] The compositions are given in g residue per 100 g protein.
[b] Altschul (1964).
[c] Dawson (1971).
[d] Jackson *et al.* (1969).
[e] Bailey and Boulter (1970).
[f] Khaund (1971).

TABLE 6 Apparent Variations in Amino Acid Composition of Glycinin Subunits[a]

Amino Acid	A_1	A_2	A_3	B_1	B_2	B_3	Glycinin[b]
Ile	4.0	4.6	3.5	4.7	4.8	4.0	4.2
Leu	5.8	6.5	4.8	9.7	9.1	8.5	7.0
Lys	6.9	7.4	6.7	4.7	6.7	5.0	5.0
Met	0.3	0.75	0.60	1.41	0.87	0.72	1.16
Phe	4.5	4.6	3.7	6.6	6.0	5.1	5.0
Thr	2.7	3.5	3.0	4.0	4.1	3.6	3.5
Trp	—	—	—	—	—	—	—
Val	3.6	3.5	3.4	5.5	5.3	5.9	4.2
Ala	2.5	3.1	2.0	5.4	5.0	4.8	3.2
Arg	10.0	7.2	8.1	7.6	6.8	8.3	8.0
Asp	11.9	12.0	14.5	15.2	14.1	13.2	12.0
Cys	—	—	—	—	—	—	—
Glu	27.3	25.9	29.0	16.4	18.4	17.8	22.1
Gly	4.4	4.8	4.0	3.5	3.8	4.3	3.8
His	3.8	2.5	2.9	2.0	0.0	4.3	2.3
Pro	5.9	6.1	6.6	5.0	5.0	4.5	5.8
Ser	4.2	5.1	4.9	5.2	6.0	5.9	5.4
Tyr	3.2	3.0	2.7	3.2	3.9	4.5	4.1
NH_3	—	—	—	—	—	—	—

[a] Composition in g of residue per 100 g of amino acid recovered. Adapted from Catsimpoolas *et al.* (1971).
[b] g residue per 100 g protein.

ALEURINS AS LIBERAL PROTEINS

In the preceding discussion the principal aleurins are thought of as a branching series of homologous proteins. It is proposed that the genes for the principal aleurins evolved from one or at most a very few ancestral genes. If the homologies are correctly drawn, then why is there so much variability in these proteins? The answer may be that the constraints on the structure of the aleurins are much less strict than for a specialized and functionally essential protein such as cytochrome C. Consequently, the genes for the principal aleurins can accept more point mutations, codon deletions, chain shortenings, chain elongations, and so on than can genes coding for a protein such as cytochrome C without being lethal to the organism. Under this model the principal aleurins are considered to be liberal proteins, while proteins of the cytochrome C type are conservative proteins. There is some experi-

TABLE 7 Comparison of Amino Acid Composition[a] of Prolamins and Glutelins of a Grass with the Principal Globulins of a Palm Seed

Amino Acid	Purified Gliadin[b]	Glutenin[c]	α-Concococosin[d]	Cocosin[d]
Ile	3.9	3.3	2.9	3.6
Leu	7.4	6.7	6.3	6.8
Lys	0.6	1.4	4.4	3.0
Met	1.1	1.4	1.2	1.9
Phe	4.4	3.2	3.7	3.3
Thr	1.8	3.1	2.7	3.6
Trp	—	—	0.56	0.56
Val	4.2	4.8	5.6	6.1
Ala	2.9	4.0	5.6	6.1
Arg	1.9	2.2	12.7	11.7
Asp	2.5	2.7	8.1	8.3
Cys	2.1	1.4	3.3	2.6
Glu	39.4	32.8	21.3	20.8
Gly	2.5	9.2	8.7	8.6
His	1.9	1.5	2.3	1.4
Pro	16.3	13.5	4.3	4.0
Ser	5.1	5.9	4.9	5.6
Tyr	2.0	2.7	1.6	2.1

[a] Composition in mole %.
[b] Bietz and Wall (1973).
[c] Wu and Dimler (1963).
[d] Khaund (1971).

mental basis for this interpretation in the case of animal proteins. [See Smith (1970) for a recent review.]

What hypothetical constraints might be required for a successful aleurin? From a functional point of view, the aleurins appear to be nonenzymatic proteins that serve as a source of nutrition for the germinating seed. Therefore, the structural constraints required to maintain enzymatic activity are absent. Since the aleurins are vacuolar proteins synthesized by the polyribosomes bound to the rough endoplasmic reticulum and packaged and/or modified by the dictyosomes, there may be some restrictions imposed on their structure by the details of the synthesis and sequestration processes. High molecular weight of the associated form of the aleurins is almost certainly a structural imperative. Wetlaufer (1973) points out that the association of subunits to form multisubunit assemblies reduces the osmotic

TABLE 8 Wheat Gliadins: Residues of Amino Acid per Polypeptide Chain[a]

Amino Acid	Polypeptide Chain							
	α_{1B}	α_{1C}	α_2	HSβ	β_3	γ_1	γ_2	γ_3
Ile	12	15	14	14	15	15	19	17
Leu	23	25	25	25	24	24	27	24
Lys	2	1	1	1	2	—	3	2
Met	2	2	3	2	2	3	6	4
Phe	10	12	11	15	12	12	19	22
Thr	4	5	5	6	6	6	7	8
Trp	1	1	1	2	2	1	3	3
Val	13	14	14	15	15	14	17	16
Ala	7	8	8	10	11	11	11	13
Arg	5	5	5	7	5	5	5	5
Asp	8	9	9	8	10	10	7	7
Cys	6	7	6	6	8	7	8	7
Glu	105	120	121	118	119	132	144	151
Gly	7	8	8	10	9	8	11	10
His	6	7	6	9	5	5	5	5
Pro	40	46	45	52	55	51	68	72
Ser	15	16	17	19	19	18	19	17
Tyr	9	10	10	8	11	9	3	1
TOTAL	275	311	309	327	330	331	382	384
Calc. mol wt	32,000	36,200	35,900	37,800	38,000	38,200	43,800	44,100

[a] Adapted from Ewart (1973).

pressure that a cell must bear and prevents the cell from rupturing. The aleurins are present in high concentrations in the vacuoles of the cells; thus, high molecular weight seems to be an important requirement for survival of the organism. Finally, the amino acid composition of the aleurins should provide suitable nutrition for the germinating seed. On the surface these seem to be modest requirements for an acceptable aleurin and could explain why they are liberal proteins.

If the hypothesis is correct that the aleurins are liberal proteins, then the gene pool for a given species may contain many variants for each aleurin. Some of the variant genes may code for proteins with significantly better amino acid composition than the average. If such variants can be identified, it may be possible to breed them into commercial lines suitable for food production. Nelson (1969) felt that the chances are limited for affecting the overall amino acid composition of a natural protein mixture by achieving a mutation in a structural gene for a given protein. He was thinking of tissues composed of large numbers of conservative proteins. In seeds there appear to be a relatively few quantitatively important proteins, and they are liberal proteins. Nelson's pessimism may not be justified for seeds. There are other possibilities for the genetic manipulation of genes determining what aleurins are synthesized. Some of these will be discussed in later papers.

REFERENCES

Altschul, A. M. 1964. *In* Symposium on Food Proteins and Their Reactions (H. W. Schultz and H. F. Anglemier, eds.). The Avi Publishing Co., Inc., Westport, Conn.

Altschul, A. M., N. J. Neucere, A. A. Woodham, and J. M. Dechary. 1964. Nature 203:501.

Altschul, A. M., L. Y. Yatsu, R. L. Ory, and E. M. Engleman. 1966. Ann. Rev. Plant Physiol. 17:113.

Anderson, W. A., and J. Andre. 1968. J. Microscopie (Paris) 7:343.

Anfinsen, C. B. 1973. Science 181:223.

Bailey, C. J., and D. Boulter. 1970. Eur. J. Biochem. 17:460.

Beadle, G. 1959. Science 129:1715.

Bietz, J. A., and J. S. Wall. 1973. Cereal Chem. 50(5):537.

Bonnet, H. T., Jr., and E. H. Newcomb. 1965. J. Cell Biol. 27:423.

Boulter, D., J. A. M. Ramshaw, E. W. Thompson, M. Richardson, and R. H. Brown. 1972. Proc. R. Soc. London Ser. B. 181:441.

Briarty, L. G., D. A. Coult, and D. Boulter. 1969. J. Exp. Bot. 20:358.

Brohult, S., and E. Sandegren. 1954. *In* The Proteins, Vol. II, Part A (H. Neurath and K. Bailey, eds.). Academic Press, Inc., New York.

Catsimpoolas, N., J. A. Kenney, E. W. Meyer, and B. F. Szuhaj. 1971. J. Sci. Food Agr. 22:448.

Crick, F. H. C. 1958. Symp. Soc. Exp. Biol. 12:138.

Danielsson, C. E. 1949. Biochem. J. 44:387.

Daussant, J., N. J. Neucere, and L. Y. Yatsu. 1969. Plant Physiol. 44:471.

Dawson, R. 1971. Anal. Biochem. 41(2):305.

Dechary, J. M., K. F. Talluto, W. J. Evans, W. B. Carney, and A. M. Altschul. 1961. Nature 190:1125.

Dieckert, J. W. 1969. Abstr. XI Intern. Botan. Congr., Seattle, Wash.

Dieckert, J. W. 1971. *In* Chemistry of the Cell Interface (H. D. Brown, ed.). Academic Press, New York.

Dieckert, J. W., and M. C. Dieckert. 1972. *In* Symposium: Seed Proteins (G. E. Inglett, ed.). The Avi Publishing Co., Inc., Westport, Conn.

Dieckert, J. W., J. E. Snowden, Jr., A. T. Moore, D. C. Heinzelman, and A. M. Altschul. 1962. J. Food Sci. 27:321

Duvick, D. N. 1961. Cereal Chem. 38:374.

Engleman, E. M. 1966. Am. J. Bot. 53:231.

Ericson, M. C., and M. J. Chrispeels. 1973. Plant Physiol. 52:98.

Ewart, J. A. D. 1973. J. Sci. Food Agr. 24:685.

Frasca, J. M., and V. R. Parks. 1965. J. Cell Biol. 25:157.

Graham, T. A., and B. E. S. Gunning. 1970. Nature 228:81.
Guilliermond, A. 1941. The Cytoplasm of the Plant Cell. Chronica Botanica Co., Waltham, Mass.
Haber, E., and C. B. Anfinsen. 1961. J. Biol. Chem. 236(2):422.
Hartig, T. 1855. Bot. Z. 13:881.
Jackson, P., D. Boulter, and D. A. Thurman. 1969. New Phytol. 68:25.
Johnson, P., and E. M. Shooter. 1950. Biochim. Biophys. Acta 5:361.
Khaund, R. N. 1971. Ph.D. Dissertation. Texas A&M University, College Station, Texas.
Khoo, U., and M. Wolf. 1970. Am. J. Bot. 57:1042.
Koshiyama, I. 1972. Agr. Biol. Chem. 36(1):63.
Locke, M., and N. Krishnan. 1971. J. Cell Biol. 50:550.
Mazia, D., P. A. Brewer, and M. Alfert. 1953. Biol. Bull. 104:57.
Mollenhauer, H. H., and C. Totten. 1971. J. Cell Biol. 48(2):387.
Nelson, O. E. 1969. Adv. Agron. 21:171.
Neurath, H., K. A. Walsh, and W. P. Winter. 1967. Science 158:1638.
Newcomb, E. H. 1967. J. Cell Biol. 33:143.
Nolan, C., and E. Margoliash. 1968. Ann. Rev. Biochem. 37:727.
Osborne, T. B. 1924. The Vegetable Proteins. 2d ed. Longmans, Green and Co., London.
Pfeffer, W. 1872. Jahrb. Wiss. Bot. 8:429.
Rossi-Fanelli, A. 1968. Project UR-15-(40)-33, Final Research Report, Istituto di Chimica Biologica, Universita di Roma, Rome, Italy.
Rossi-Fanelli, A., E. Antonini, M. Brunori, M. R. Bruzzesi, A. Caputo, and F. Satriani. 1964. Biochem. Biophys. Res. Commun. 15:110.
Ryan, C. A. 1973. Ann. Rev. Plant Physiol. 24:173.
Sabatini, D. D., K. Bensch, and R. J. Barrnett. 1963. J. Cell Biol. 17:19.
Sarabhai, A. S., A. O. W. Stretton, S. Brenner, and A. Bolle. 1964. Nature 201:13.
Sharon, N., and H. Lis. 1972. Science 177:949.
Singh, J., and J. W. Dieckert. 1973. Prep. Biochem. 3(1):73.
Sjogren, B., and R. Spychalski. 1930. J. Am. Chem. Soc. 52:4400.
Smith, E. L. 1970. *In* The Enzyme, Vol. I. Structure and Control (P. D. Boyer, ed.). Academic Press, Inc., New York.
Spurlock, B. O., V. C. Kattine, and J. A. Freeman. 1963. J. Cell Biol. 17:203.
Spurr, A. R. 1969. J. Ultrastruct. Res. 26:31.
St. Angelo, A. J., L. Y. Yatsu, and A. M. Altschul. 1968. Arch. Biochem. Biophys. 124:199.
Stempak, J. G., and R. T. Ward. 1964. J. Cell Biol. 22:697.
Svedberg, T., and A. J. Stamm. 1929. J. Am. Chem. Soc. 51(2):2170.
Varner, J. E., and G. Schidlovsky. 1963. Intracellular distribution of proteins in pea cotyledons. Plant Physiol. 38:139.
Watson, M. L. 1958. J. Biophys. Biochem. Cytol. 4:475.
Weber, K., and M. Osborn. 1969. J. Biol. Chem. 244(16):4406.
Wetlaufer, D. B. 1973. J. Food Sci. 38:740.
Winborn, W. B. 1965. Stain Technol. 40:227.
Wolf, W. J., and D. R. Briggs. 1959. Arch. Biochem. Biophys. 85:186.
Wu, Y. V., and R. J. Dimler. 1963. Arch. Biochem. Biophys. 103:310.
Yanofsky, C., B. C. Carlton, J. R. Guest, D. R. Helinski, and U. Henning. 1964. Proc. Nat. Acad. Sci. U.S.A. 51:266.
Yatsu, L. Y. 1965. J. Cell Biol. 25:193.

DISCUSSION*

DR. MUNRO: Do these storage proteins have carbohydrates attached during the process as in many secreted mammalian proteins?

DR. DIECKERT: There is a report that some of the proteins of *Vicia faba* do, but I am not convinced that that is correct. I think for the most part they do not.

DR. MUNRO: Have you examined by electron microscope autoradiography the pathway you describe?

DR. DIECKERT: No. This is in the works, but we have not done it yet.

DR. WILSON: I have a question about whether the system that you have described for oilseeds is different from the system of endosperm proteins in cereals. First of all, there is a little confusion with your nomenclature. You were talking about aleurone bodies and aleurins, which I see by Webster's dictionary is correct. We have the confusion in cereals that the term aleurone has been applied to a specific tissue, yet most of the protein is in the endosperm, not the aleurone tissues. I think that the protein bodies in the endosperm are of a different nature. You mentioned the Khoo and Wolf paper (U. Khoo and M. J. Wolf. 1970. Am. J. Bot. 57:1042–1050), which suggested that there are differences and that we should not transfer the picture that you are finding with oilseeds over to cereal endosperms.

I do not think anyone has really carried through the work of Morton in Australia that was done some 10 years ago. It really needs to be repeated to determine whether the protein synthesis takes place within the protein bodies of cereal grains.

DR. DIECKERT: I have opinions on this point, but I must admit that the data are sketchy. I think the processes are essentially the same, that is, that proteins of these endospermous tissues are produced in the rough ER, but they may or may not make it through the Golgi bodies and into the specialized aleurone vacuoles.

I would like to still use the term "aleurones" to cover those proteins that remain in the dilations of the rough ER. All of the proteins of this general type I just like to see classified in this way, but the dictyosomes may very well modify these proteins. I do not know whether there is a definitive answer to your question at this point. It is something that ought to be considered.

DR. INGLETT: We studied seeds of various kinds, and we found protein bodies in practically all of them except in mature wheat. However, in immature wheat, there is a phase in which protein bodies are observed. I would like to know how the nutrition of these protein bodies would relate to their possible use by the seed, because there seems to be such a variation between the composition of these protein bodies—all the way from zein bodies in corn, which has very poor nutritive value, to those in soybeans, which have very good amino acid composition.

* Dr. Julius W. Dieckert presented the paper and participated in the discussions.

DR. DIECKERT: Cyanoficin granules are simply copolymers of glutamic acid and arginine. Yet those copolymers with their high molecular weight, 20,000 to 100,000, serve as efficient nitrogen sources for the developing blue-green algae. So, my point is that I do not think that the germinating seed is too fastidious in its requirement for these nitrogen compounds. That probably is why the nutritional requirements for the seedling are not a severe constraint on the proteins.

As for the absence of formal spherical aleurone grains in wheat, I do not think they have to adopt that form. I do not even think they have to wind up in a classical vacuole at all except that they probably are in membrane-bound sacs, but they might not have the same composition as the vacuole or membrane. This process may stop, for example, in the dilations of the ER. As far as I can tell, and I have read about this wheat case a little bit, the consensus seems to be that the rough ER is involved in elaboration of these proteins and that they do find their way on the inside.

Now, it is what happens to them after that that seems to be the subject of disagreement, and I just do not have any further information. My speculation would be that the Golgi is not getting as serious a crack at these proteins as it does in the case of some of the oilseeds. That is no answer, but that is either my lack of knowledge or the general lack of knowledge in the field. Someone else may wish to make a comment on that.

DR. WILSON: One thing that came out from the work that Dr. Sodek did in my laboratory (D. M. Wilson and L. Sodek. 1970. Arch. Biochem. Biophys. 140:29) was that plants seem to have within the individual cells considerable ability to put the nitrogen on the carbon skeletons that they need in those particular cells. Even if we look back to Morton's wheat work, if you look at the levels of proteins as they are synthesized and the levels of amino acids, it looks as if leucine, one of the major amino acids, is present in low supply or is supplied in low amounts to a seed, and the seed must make it. At least we will assume the seed makes the completed leucine molecule, and yet, in turn, at germination it exports what would be excess leucine for the developing seedling, and this developing seedling takes that nitrogen and uses it for something else. The opposite would be true in the case of lysine. The developing seed is furnished with excess lysine and uses the nitrogen elsewhere, if it comes in. The germinating seed has a lysine deficiency, and the developing seedling now makes lysine.

Maybe the constraint is in the best way of transporting nitrogen around a plant as the seed develops, and then what is the best way of getting nitrogen from the seed out to the growing points. This is one of the areas that we really do not know much about. How is nitrogen transported to the seeds, and what compounds are transported out of the seed to the growing parts?

DR. DIECKERT: I would like to make one more comment in answer to Dr. Inglett's question. I think one of the principal reasons seed proteins are used as a storage form of nitrogen is that they do have high molecular weights, and the cells can avoid osmotic stress problems.

Wetlaufer (1973. J. Food Sci. 38:740) recently pointed out that the probable reason for multistranded enzymes is to reduce the osmotic pressure on the inside of the cells and prevent disasters from occurring in such cells.

DR. BOULTER: I think that there are considerable constraints on the amino acid composition of storage proteins, one of them being that their amino acid composition must mesh with the enzyme capability of the germinating seed.

The amino acid composition of the storage protein has to be such that when the protein is broken down into its constituent amino acids, the enzymic machinery of the germinating seed can use the amino acids to build all the new nitrogen building blocks, i.e., for new nucleic acids, proteins, and so on. For example, in peas and beans the storage proteins have a high arginine content, and the enzymes in the germinating seed can use arginine to generate the necessary building blocks for the developing seedling.

In the case of cereals—in barley, for instance, where it is well worked out—you have a similar situation. The storage proteins are high in proline, and proline is the amino acid that is mainly used by the enzymes of the germinating seed to generate the nitrogen building blocks it needs.

Also, there are a series of constraints with regard to the drying out of the seed and its subsequent utilization of protein. Storage proteins have large molecular weights and are made up of subunits. For example, legumin of soybean is a dimer with 12 subunits (N. Catsimpoolas. 1969. FEBS Let. 4:259). There are various constraints when two polypeptide chains interact, since this requires specific amino acid interaction sites and it is not a random process. Another series of constraints is due to the fact that storage proteins are made on the endoplasmic reticulum. Therefore, they have to enter the endoplasmic reticulum so as to be moved to where they are going to end up in the protein bodies. It is wrong, in my opinion, to look at storage proteins simply as a store of nitrogen. They are much more specific than that. If you actually fingerprint these proteins or sequence them, and we are only just at the beginning, you find that they are relatively conserved. So, while I will agree that something like cytochrome C may have even more structural constraints, storage proteins nevertheless have plenty of constraints on them.

The other point that I would like to make with regard to considering homology of these proteins is that, for example, in the legumes Danielsson (C. E. Danielsson. 1949. Biochem. J. 44:387–400) used ultracentrifugational separations to separate legumin and vicilin in different legumes. Legumin is a 12S protein and vicilin is a 7S protein, but we know now, for example, that there are several 7S proteins in various legumes, and so you have to compare the right 7S protein. There is a lot more characterization work that still has to be done, in order to sort out the situation. Nevertheless, a lot of that work is being done, and much of it is unpublished. It is possible to trace an 11S protein, characterized not only by its sedimentation value but also by its subunits, amino acid composition, end terminal amino acids, and peptide maps. This 11S protein is present in many legumes. There are 7S proteins that also occur widely among the legumes, but that is a more complicated story.

As far as the phylogenetic tree of the cytochromes that Dieckert put up is concerned, we have now looked at more cytochromes [D. Boulter, 1973. The use of amino acid sequence data in the classification of higher plants, pp. 211–216. *In* Nobel Symposium 25, Chemistry in Botanical Classification (eds. G. Bendz and J. Santesson). Nobel Foundation, Stockholm, Sweden; Academic Press, New York.] Hemp, for example, comes way up the tree. I do not want to get into a discussion about molecular evolution of plants, but the point about the plants versus the animals is in my view that the angiosperms are a relic group, and present-day families are the tips of the branches. Most of the ancestral families along the branches are no longer in existence. It is not like the situation with animals. Most of the present-day groups have had a long separate evolutionary history.

DR. CANVIN: In the developing oilseeds, the proplastids are very prominent bodies in these cells and, as far as I am concerned, play an extremely important part in lipid biosynthesis. Do they, in fact, play any role in protein biosynthesis in these seeds?

DR. DIECKERT: I have not seen any data on this point. I keep looking for the possibility that they might contribute some of the proteins to the system, but I have no data for it.

DR. MARCUS: I have been intrigued with the thought that seemed to follow from one line of questioning. That is that perhaps the germinating seed does a much better job in terms of making the right kind of nutritional proteins than the developing seed. At least one basis for this would be that obviously it, itself, grows, and so it takes what appears to be a convenient type of storage protein and converts it into something that is better for actual growth.

Is it possible that we could actually use that trick instead of worrying about trying to insert new genes—let the seed germinate, convert the storage proteins into better proteins and then harvest the protein? You do not have to let the seedling germinate. A lot of the metabolism of the protein will probably go on in the cotyledon by itself. There is new enzyme synthesis in the endosperm of cotyledons that might work so that you could perhaps even inhibit germination in some way so that you would not have the technical problem of isolating stuff from a growing seedling. In other words, you could turn it on but not let it go too far and thus keep it within the realm of technical isolation.

DR. DIECKERT: There are people who are thinking along these lines. There is one student at Texas A&M who is thinking about doing something like this. My personal opinion is that, first, they are going to introduce some rather curious flavors that are going to be difficult to get rid of. Second, it seems to me that the amino acid composition changes to something you cannot use. The amino acid composition is worse when you get through with this process than when you started as far as nutrition is concerned. Now, that may be in error.

DR. DALBY: We have been doing something on this for several months, and it is indeed possible in corn, for example, to demonstrate an increase in lysine and tryptophan in whole germinating seed. The increase would appear to be not in the endosperm, but in the young plant.

DR. MUNRO: Could you give us some quantitation of the proportion of these proteins that you have described? You have given examples of how much they resemble one another in the different species. The question therefore is whether the proportions of storage proteins could be manipulated by regulating the genetic expression of the amounts made. What are the possibilities within this field? Can you reduce a low-lysine, low-methionine type of protein and increase another quantitatively?

DR. DIECKERT: In the case of peanut seeds, for example, approximately two-thirds of the protein of the whole seed is arachin, which is one more or less closely related family. The arachins all have about the same molecular weight. Alpha kinin arachin—that is another of these proteins—comprises about one-fourth of the total seed proteins.

EDWIN T. MERTZ

Case Histories of Existing Models

HIGH-LYSINE SYNDROME OF MAIZE

The first published report of a high-lysine syndrome in cereal grains appeared in the *Proceedings of the 18th Annual Hybrid Corn Research Industry Conference* (Mertz, 1964). At that conference I reported that "in a recent survey of nitrogen distribution and lysine levels in the endosperms of inbred corn samples supplied by Dr. O. E. Nelson, . . . we found one endosperm that had a relatively high level of lysine. . . ." A review of the events that led to this discovery was published in 1968 (Mertz, 1968). In the *Proceedings*, I reported the high-lysine corn only as selection 540, although it was known to be homozygous for the opaque-2 gene. Wishing to be cautious, I stated that "endosperm 540 may be the result of a fortuitous combination of genes, season and soil fertility that cannot be duplicated, or if duplicated, will be impossible to incorporate into hybrid seed. We do not know yet whether it can be duplicated."

That the high-lysine syndrome was indeed caused by the opaque-2 gene was proved a few months later (Mertz *et al.*, 1964). To make the critical test, a backcross progeny on a segregating ear was divided into opaque-2 and normal kernels. Quoting from the paper in *Science*, "The opaque-2 endosperms had a different amino acid pattern and 69 per cent more lysine than the normal seeds" (Mertz *et al.*, 1964). In 1966 (Mertz *et al.*, 1966) we reported that "the effect of the opaque-2 gene is . . . primarily (or perhaps exclusively) confined to the endosperm."

MAJOR SYMPTOMS OF THE HIGH-LYSINE SYNDROME

Table 1 lists the major symptoms that have been found to be associated with this peculiar syndrome in the opaque-2 maize endosperm. The endosperm is completely floury unless special efforts are made to modify the phenotype by selective breeding (Bauman, 1975; Vasal, 1975). Wolf and coworkers (1967, 1969) found that the protein bodies are greatly reduced in size. Zein, the major protein of normal endosperm, is reduced by one-half, and the glutelin proteins are doubled. At the same time, the saline-soluble fraction, consisting mainly of albumins, globulins, and free amino acids, is doubled.

The enzyme ribonuclease (Dalby and Davies, 1967; Dalby and Cagampang, 1970; Dalby *et al.*, 1972; Wilson and Alexander, 1967) and the trypsin inhibitor of maize (Mertz, 1972) are increased manyfold. Free amino acids increase twofold to tenfold, without any major change in the pattern (Mertz, 1972; Mertz *et al.*, 1974a), and peptide-

TABLE 1 Major Symptoms of the High-Lysine Syndrome in Opaque-2 Maize[a]

	High-Lysine	Normal	Reference
Physical texture	Floury	Vitreous	
Protein bodies	0.1 μ	2 μ	Wolf *et al.*, 1967
Zein (% of protein)	27	60	Misra *et al.*, 1972
Glutelin (% of protein)	29	14	Misra *et al.*, 1972
Saline-soluble fraction (% of protein)	13.6	5.8	Misra *et al.*, 1972
Ribonuclease (units/endosperm)	4,000	1,000	Mertz, 1972
Trypsin inhibitor (units/endosperm)	210	47	Mertz, 1972
Free amino acids (% of protein)	5.7	0.5	Mertz, 1972
Leucine (% of protein)	12	16	Misra *et al.*, 1972
Tryptophan (% of protein)	0.7–1	0.35–0.5	Misra *et al.*, 1972; Villegas and Mertz, 1971
Lysine (% of protein)	3–4	1.5–2.0	Misra *et al.*, 1972; Villegas and Mertz, 1971

[a] Endosperm of isogenic lines.

bound leucine decreases substantially as percent of protein (Misra *et al.*, 1972).

Both peptide-bound tryptophan and lysine are doubled as percent of protein, their ratio to each other (about 1:4) remaining essentially constant (Misra *et al.*, 1972; Villegas and Mertz, 1971). The changes in levels of peptide-bound amino acids can be attributed mainly (if not completely) to the changes in the proportions of zein, glutelin, albumins, and globulins (Misra *et al.*, 1975). There is no evidence of increased production of any new proteins.

OTHER MUTANT GENES THAT CAUSE THE HIGH-LYSINE SYNDROME

Table 2 lists eight other mutant genes that produce zein-deficiency symptoms and elevate the lysine level in maize endosperm. The floury-2 gene was identified at about the same time as the opaque-2 gene (Nelson *et al.*, 1965), but the opaque-7 and the six starch-modifying genes were identified only recently (Misra *et al.*, 1972, 1975). The increased lysine expressed as percent above the normal varies from 12 percent to 106 percent, as compared with 118 percent for opaque-2. Zein reduction also varies widely, as do the increases in glutelin and saline-soluble nitrogen.

ACUTE ZEIN DEFICIENCY SYMPTOMS IN CERTAIN DOUBLE MUTANTS

Even more striking are the zein-deficiency symptoms and increased lysine levels produced by combining the starch-modifying mutants listed in Table 2 with the opaque-2 gene (Table 3). The two floury mutants, floury-2 and opaque-7, when combined with opaque-2, give double mutants with lysine levels lower than that of opaque-2 alone. In contrast, the double mutants of opaque-2 with any one of the six starch-modifying mutants yield lysine contents that are higher than one finds in any single mutant. Zein deficiency symptoms are extremely acute, with almost, if not complete, blocking of zein formation, and a compensatory increased production of glutelin and saline-soluble nitrogen.

TABLE 2 Mutant Genes Causing High-Lysine Syndrome in Maize[a]

Mutant Gene	Chromosome	Lysine (g/100 g protein)	Increased Lysine (% above normal)	Zein (%)	Glutelin (%)	Saline Soluble (%)
Normal	—	1.6	—	59	13.8	5.8
Opaque-2	7	3.5	118	26.9	29.2	13.6
Floury-2	4	2.7	69	49.1	22.0	9.2
Opaque-7[b]	10	3.5	65	20.3	29.5	16.6
Sugary-1	4	1.8	12	27.1	22.8	11.9
Shrunken-1	9	1.9	19	43.7	16.3	8.2
Shrunken-2	3	2.7	69	29.4	23.6	12.3
Shrunken-4	5	3.0	87	30.8	23.6	25.7
Brittle-1	5	2.3	44	36.0	27.4	8.8
Brittle-2	4	3.3	106	26.1	27.9	12.1

[a] Near-isogenic lines in Oh43 inbred (Misra *et al.*, 1975).
[b] Compared with isogenic inbred W22 (40.6% zein, 21% glutelin, 6.9% saline-soluble, and 2.3 g lysine/100 g protein).

TABLE 3 High-Lysine Syndrome in Double Mutants of Maize[a]

Double Mutant	Lysine	Increased Lysine in Endosperm (% above normal)	Zein (%)	Glutelin (%)	Saline Soluble (%)
Normal	1.6	—	59	13.8	5.8
o_2fl_2	2.7	69	25	24.8	17.0
o_2o_7	3.5	52[b]	8.7	33.3	17.6
o_2su_1	3.9	144	3.0	45.3	22.7
o_2sh_1	4.8	200	1.8	32.2	39.9
o_2sh_2	4.2	162	1.2	35.4	25.3
o_2sh_4	4.0	150	6.5	26.8	43.3
o_2bt_1	4.8	200	2.7	50.2	23.3
o_2bt_2	5.3	230	2.9	48.0	22.3

[a] Misra *et al.* (1972, 1975).
[b] Compared with W22 normal endosperm (2.3 g lysine/100 g protein).

ABNORMAL PROTEIN METABOLISM IN DEVELOPING ENDOSPERMS

Figure 1 shows that zein (Mertz *et al.*, 1974b) is being produced in the normal counterpart at the 14th day postpollination, but in the brittle-2 and opaque-2 single mutants the first signs of zein formation are just appearing at this time. Since the normal has a substantial head start, the single mutants reach only about half the normal level of zein by the 49th day. Even more striking, however, is the complete absence of zein synthesis throughout endosperm development in the double mutant of brittle-2 and opaque-2. Acrylamide gel patterns of the zein fraction of the double mutant show no normal zein (Figure 2) (mol wt 22,000 and 25,000).

Figure 3 traces the production of glutelin (Mertz *et al.*, 1974b) during development. Glutelin production is rapid in the normal and single mutant during the first 21 days postpollination. It then reaches a plateau and drops somewhat in the normal and opaque-2. In the brittle-2 it plateaus and then rises moderately. In the double mutant, glutelin formation proceeds slowly at the start, following an S-shaped curve.

Figure 4 traces the production of albumin and globulin proteins and free amino acids (Mertz *et al.*, 1974b). Recent studies show that about one-third of the nitrogen in the normal counterpart and in the single mutants and one-half of the nitrogen in the double mutant can be accounted for as free amino acids. The remainder of the saline-soluble nitrogen (Misra *et al.*, 1975) is mainly albumin and globulin. Figure 4

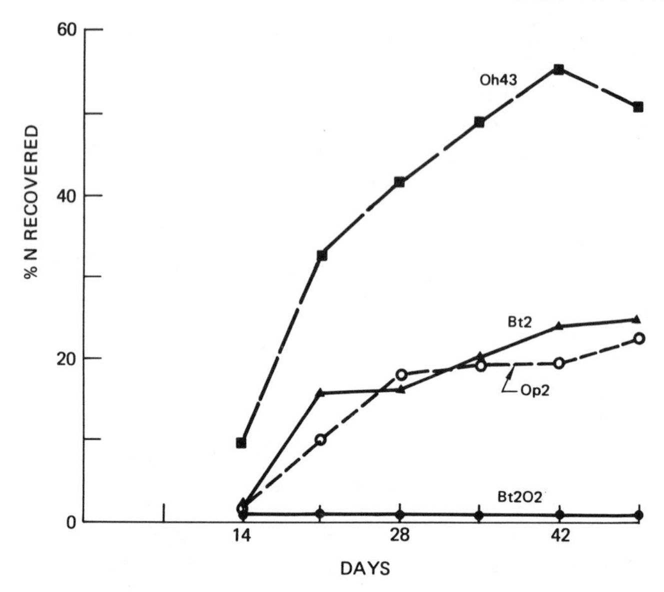

FIGURE 1 Zein synthesis in developing endosperms of opaque-2 and brittle-2 mutants.

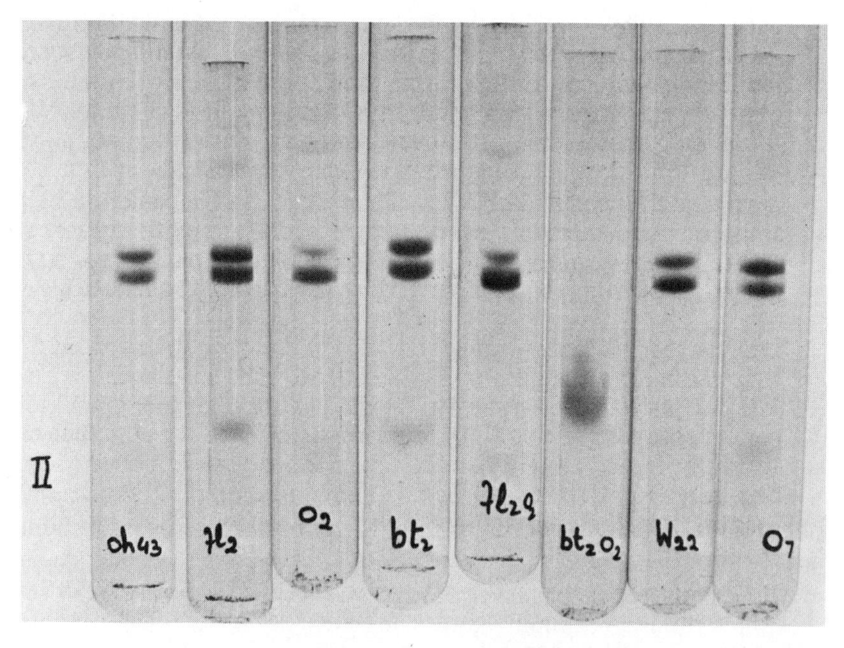

FIGURE 2 SDS-polyacrylamide gel patterns of zein fractions.

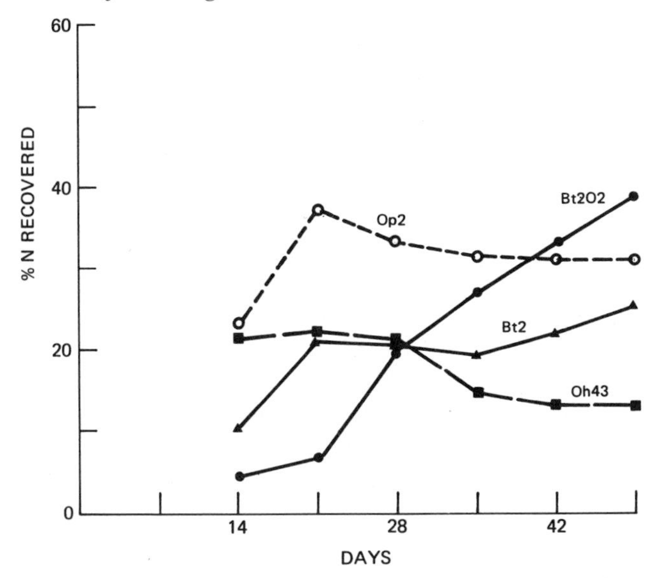

FIGURE 3 Glutelin synthesis in developing endosperms of opaque-2 and brittle-2 mutants.

indicates that these nitrogenous components are at a maximum level 14 days postpollination, with lowest levels in the normal and highest in the double mutant. There is a continuous fall-off in all endosperms between the fourteenth and forty-ninth days.

CHRONIC HYPERAMINOACIDITY IN HIGH-LYSINE MAIZE MUTANTS

In Figures 5 and 6 the free amino acids are expressed as *μM* leucine equivalents per 100 mg of protein. Figure 5 shows the free amino acid levels of the endosperm, and Figure 6 shows the levels in the whole kernel. In Figure 6, at 49 days postpollination, the free amino acid level is twice as high in opaque-2 and three times as high in the double mutant as in the normal Oh43.

We have used ninhydrin to detect this hyperaminoacidity. The simple color test quickly identifies hard endosperm opaque-2 as well as opaque-2 in a floury-1 background (Mertz *et al.*, 1974a). In some backgrounds the free amino acid level is unusually high. Thus, in the spontaneous mutant W64A opaque-2 the level in the endosperm is 10 times that in the isogenic normal counterpart (Mertz, 1972). In the

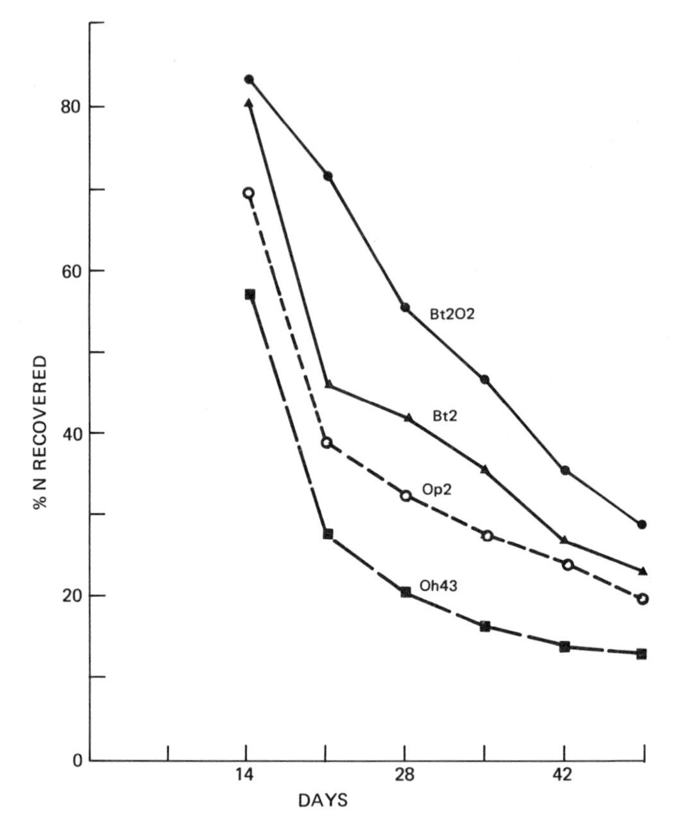

FIGURE 4 Saline-soluble nitrogen in developing endosperms of opaque-2 and brittle-2 mutants.

Oh43 line, we have found (Mertz *et al.*, 1974b) that free amino acids are elevated not only in the opaque-2 and opaque-2/brittle-2 endosperm, but in the embryo and pericarp as well (see Table 4).

HIGH-LYSINE SYNDROME IN OTHER CEREALS

BARLEY

Two outstanding examples of the high-lysine syndrome in barley are Hiproly (Munck, 1972; Tallberg, 1973), a spontaneous mutant from Ethiopia found by screening the world barley collection, and mutant 1508, produced by treating barley seeds with the chemical mutagen

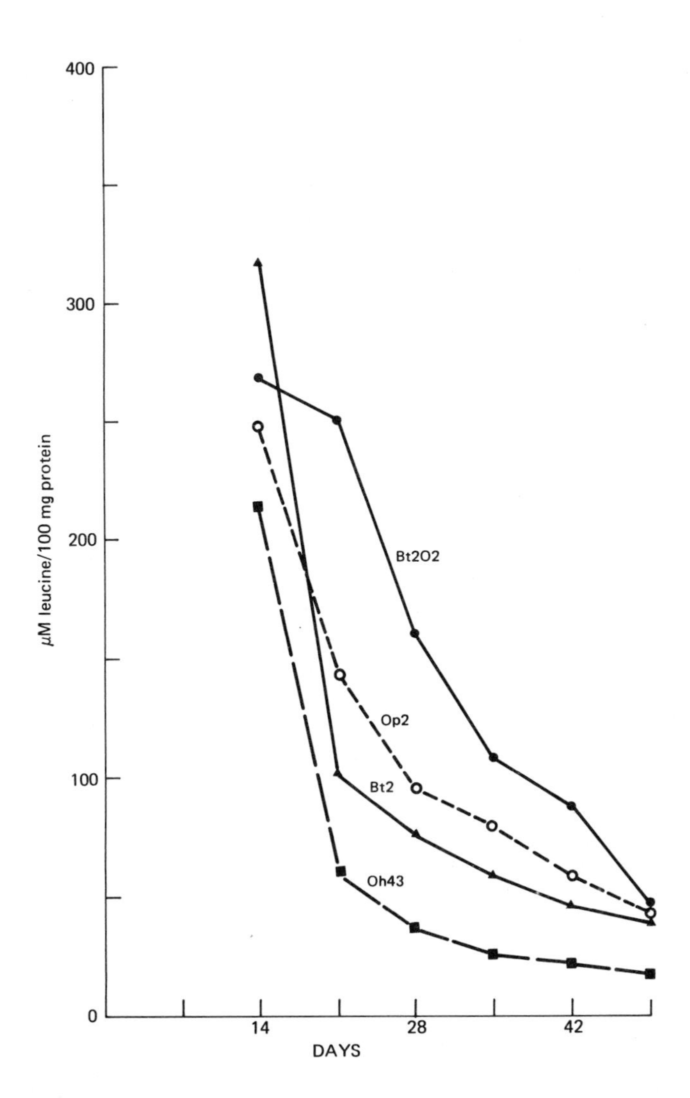

FIGURE 5 Free amino acids in developing endosperms of opaque-2 and brittle-2 mutants.

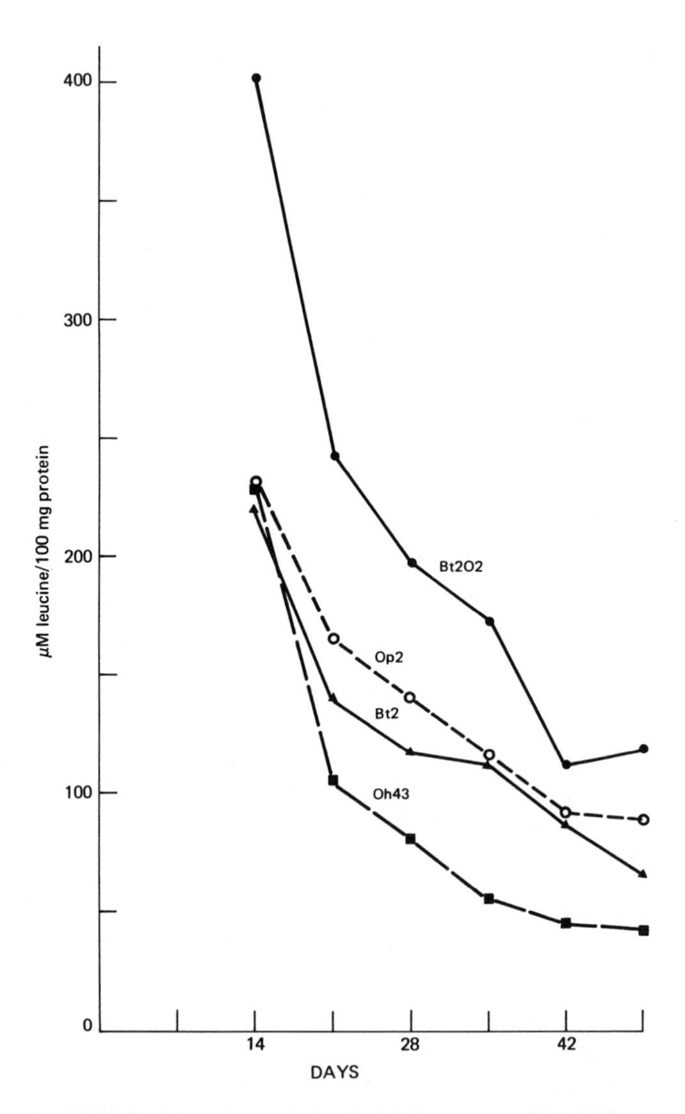

FIGURE 6 Free amino acids in developing whole kernels of opaque-2 and brittle-2 mutants.

TABLE 4 Free Amino Acids in Opaque-2 Maize Kernel Fractions[a]

Maize	Pericarp	Embryo	Endosperm
Normal	68	81	26
Opaque-2	90	104	41
Opaque-2/brittle-2	111	110	50

[a] Oh43 inbred line (Mertz *et al.*, 1974b). Values expressed as micromoles of leucine per 100 mg of protein.

ethyleneimine. In Hiproly, changes in levels of salt-soluble proteins account for the increased lysine level. Protein bodies and prolamins do not decrease. In mutant 1508, the higher levels can be accounted for mainly on the basis of an increase in albumin and globulins and a decrease in protein bodies and prolamines. The level of glutelins remains fairly constant. Table 5 compares the values obtained by Ingversen in Denmark (Ingversen *et al.*, 1973) with those obtained on these same lines of barley grown at Purdue in 1973 (Mertz *et al.*, 1974b). These data suggest that the regulation of protein synthesis in Hiproly barley differs from that in mutant 1508 and opaque-2 maize.

SORGHUM

By screening the world collection of sorghum, Singh and Axtell at Purdue (Singh and Axtell, 1973) found two spontaneous mutants from Ethiopia that were high in lysine. The data in Table 6 show that protein metabolism in the high-lysine mutants resembles that found in opaque-2 maize.

HYPERAMINOACIDITY IN OTHER CEREALS

The level of free amino acids in the high-lysine mutants of barley and sorghum resembles that found in opaque-2 maize. Mutant 1508 (barley) and IS 11758 (sorghum) contain 3–4 times more free amino acids than the normal counterparts (Mertz *et al.*, 1974a). These differences are responsible for much deeper ninhydrin colors when split high-lysine seeds are heated in an aqueous ninhydrin solution (see Figure 7).

TABLE 5 Protein Distribution in High-Lysine Barley[a]

Reference	Albumin/Globulin		Prolamine		Glutelin		Lysine (g/100 g)	
	Mutant	Normal	Mutant	Normal	Mutant	Normal	Mutant	Normal
Ingversen *et al*. (1973)	46	27	9	29	39	39	7.0	4.7
Mertz *et al*. (1974b)	48	27	3	33	32	28	5.7	3.3

[a] Percent of total nitrogen in fractions. Purdue values based on Landry-Moureaux fractions (Misra *et al*., 1972): albumin/globulin = Fraction I; prolamine = Fractions II + III; and glutelin = Fractions IV + V. Bomi barley (normal) and mutant 1508 kindly supplied by H. W. Ohm, Purdue University. The former contained 18.5 percent protein, the latter 19.8 percent. Ingversen's samples contained 10.2 percent and 10.1 percent, respectively.

TABLE 6 Protein Distribution in High-Lysine Sorghum[a]

Sorghum	Fraction I	II	III	IV	V
Normal	15.3	26.4	26.5	4.3	22.5
High-lysine	22.4	13.7	20.2	4.3	33.5

[a] High lysine values average of assays from five segregating heads using high lysine (hl hl hl) kernels from F_2 heads derived from crosses between normal plants and high-lysine sorghum line IS 11758 (Mertz *et al.*, 1974b). Lysine content of normal sorghum: 1.8 g/100 g protein; lysine content of high-lysine sorghum: 3.3 g/100 g protein.

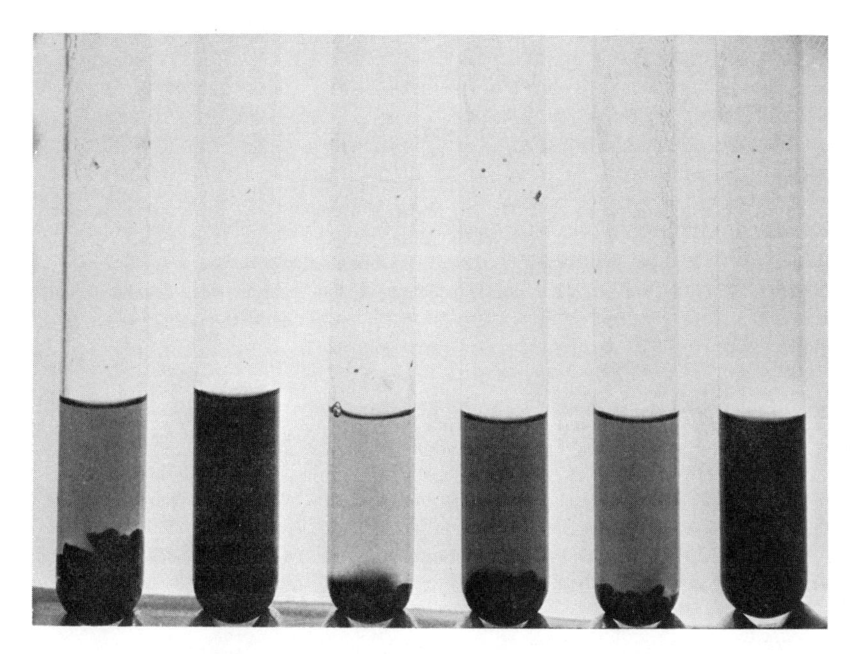

FIGURE 7 Ninhydrin color test. *Left to right*: normal and high-lysine maize, normal and high-lysine sorghum, and normal and high-lysine barley.

CONCLUSIONS

I have reviewed case histories of an inherited metabolic disease in cereal grains that could be called high-lysine syndrome, prolamine-deficiency disease, or chronic hyperaminoacidity. To date, this disease has been identified in three cereals: maize, barley, and sorghum. In maize it is caused by any one of nine specific mutations on six different chromosomes. The disease has not been studied as extensively in barley and sorghum, where it was discovered more recently. Plant breeders and geneticists are attempting at present to produce this disease in all of the common cereal species. Nutritionists are encouraging and supporting this effort because normal cereals, when consumed in major amounts by preschool-age children, often cause a "low-lysine" syndrome, kwashiorkor. Preschool-age children have been cured of kwashiorkor by feeding them opaque-2 maize, a cereal grain with the high-lysine genetic defect (Byrnes, 1969; Reddy and Gupta, 1974).

REFERENCES

Bauman, L. F. 1975. *In* High Quality Protein Maize. Dowden, Hutchison & Ross, Stroudsburg, Pa.

Byrnes, F. C. 1969. Rockefeller Found. Quart. I:4.

Dalby, A., and I. I. Davies. 1967. Science 155:1573.

Dalby, A., and G. B. Cagampang. 1970. Plant Physiol. 46:142.

Dalby, A., G. V. Cagampang, I. I. Davies, and J. J. Murphy. 1972. *In* Symposium: Seed Proteins. Avi Publishing Co., Westport, Conn., p. 39.

Ingversen, J., B. Køie, and H. Doll. 1973. Experimentia 29:1151.

Mertz, E. T. 1964. Proc. 18th Annual Hybrid Corn Research-Industry Conf. Dec. 12 and 13, 1963. Publ. 18, American Seed Trade Assoc., Washington, D.C., pp. 7–12.

Mertz, E. T. 1968. Agr. Sci. Rev. 6:2.

Mertz, E. T. 1972. *In* Symposium: Seed Proteins. Avi Publishing Co., Westport, Conn., p. 136.

Mertz, E. T., L. S. Bates, and O. E. Nelson. 1964. Science 45:279.

Mertz, E. T., P. S. Misra, and R. Jambunathan. 1974a. Cereal Chem. 51:304.

Mertz, E. T., P. S. Misra, D. V. Glover, and J. D. Axtell. 1974b. Unpublished data.

Mertz, E. T., O. E. Nelson, L. S. Bates, and O. A. Veron. 1966. Adv. Chem. 57:228.

Misra, P. S., E. T. Mertz, and D. V. Glover. 1975. *In* High Quality Protein Maize. Dowden, Hutchison and Ross, Stroudsburg, Pa.

Misra, P. S., R. Jambunathan, E. T. Mertz, D. V. Glover, H. M. Barbosa, and K. S. McWhirter. 1972. Science 176:1425.

Munck, L. 1972. Hereditas 72:1.

Nelson, O. E., E. T. Mertz, and L. S. Bates. 1965. Science 150:1465.

Reddy, V., and C. P. Gupta. 1974. Am. J. Clin. Nutr. 27:122.

Singh, R., and J. D. Axtell. 1973. Crop. Sci. 13:535.

Tallberg, A. 1973. Hereditas 75:195.

Vasal, S. K. 1975. *In* High Quality Protein Maize. Dowden, Hutchison & Ross, Stroudsburg, Pa.

Villegas, E., and E. T. Mertz. 1971. Chemical Screening Methods for Maize Protein Quality at CIMMYT. Research Bull. No. 20, CIMMYT, Mexico City.

Wilson, C. K., and D. E. Alexander. 1967. Science 157:556.

Wolf, M. J., U. Khoo, and H. L. Seckinger. 1967. Science 157:556.

Wolf, M. J., U. Khoo, and H. L. Seckinger. 1969. Cereal Chem. 46:253.

L. MUNCK *and* D. V. WETTSTEIN

Effects of Genes That Change the Amino Acid Composition of Barley Endosperm

In 1968 the first high-lysine barley, Hiproly, of Ethiopian origin, was isolated from 2,000 samples of the world barley collection by Hagberg, Karlsson, and Munck at the Swedish Seed Association, Svalöf (Munck *et al.*, 1970). It was selected for a high content of basic amino acids (lysine) and for a high dye-binding capacity (DBC) relative to its protein content as determined by Kjeldal analyses. This method has also been used by Doll, Ingversen, and Køie at the Agricultural Research Department of the Danish Atomic Energy Commission, Risø, Roskilde, to isolate 12 artificially induced mutants among the approximately 15,000 lines that they examined (Doll, personal communication). Additionally, a high-protein line displaying an increased lysine content at high levels of fertilizer application was found with an amide screening analysis by Toft-Viuf (1972) at the Department of Plant Husbandry of the Royal Veterinary and Agriculture University, Copenhagen.

Two lines—Hiproly from Svalöf (Hagberg *et al.*, 1970; Munck, 1972) and 1508 (mutant of the Bomi variety) from Risø (Ingversen *et al.*, 1973; Doll, 1973) have been studied in detail and compared to their mother varieties. The kernels of both lines display an increased content of the majority of essential amino acids such as lysine, methionine, and threonine (Table 1). Mutant 1508 contains about 45 percent more lysine than its mother variety Bomi, and Hiproly has about 30 percent more lysine than normal barley. With regard to the food value of these lines (Eggum, 1973), true digestibility of nitrogen (TD) is unchanged in

TABLE 1 Nutritional Value in Rat Nitrogen Balance Trials of High-Lysine Barley Mutants[a]

	Normal Barley	Hiproly Recombinant	1508	Normal Maize	Opaque-2
Lysine g/16 g N	3.63	4.33	5.26	2.74	4.48
Methionine g/16 g N	1.60	1.88	1.96	1.88	1.78
Cysteine g/16 g N	2.02	1.47	2.09	1.83	2.19
Threonine g/16 g N	3.10	3.47	4.16	3.31	3.46
True digestibility of N (%)	83	84	87	96	94
Biological value (%)	76	82	90	60	74
NPU (%)	63	69	78	57	70

[a]Analyses by Eggum (1973).

Hiproly and increased in 1508 (Table 1). In contrast to maize, barley contains a major fraction of proteins rich in essential amino acids, which are unavailable to the rat and the pig (Eggum, 1970b). For rats, true digestibility of lysine in barley is the lowest of all amino acids, with a figure of 70 percent of the total lysine. The true digestibility of lysine in Hiproly is 10 percent higher than in a normal reference line (Munck *et al.*, 1970). The biological value (BV) of 1508 is very favorable and approaches that of skim milk, whereas Hiproly has a BV intermediate between normal barley and 1508. In net protein utilization (NPU), which is the product of TD and BV, the high-lysine lines are about equal, and both are superior to standard barley varieties. The NPU value of Hiproly barley reaches the same level as that shown by opaque-2 maize (Table 1), while 1508 has a superior value.

The two high-lysine traits show a simple recessive Mendelian inheritance (Munck *et al.*, 1970; Karlsson, 1972; Doll, 1973). The two genes causing a high-lysine content are nonallelic (Tallberg, 1973). The expression of the Hiproly gene on lysine content is effectively modified by placing the gene into different genotypes (Munck, 1972).

Favorable changes in the amino acid composition of cereal proteins can be brought about either by increasing the lysine-rich nitrogen-containing fractions such as free amino acids, peptides, albumins, and globulins or by reducing the lysine-poor fractions such as the prolamins (hordeins). A combination of these two types of changes is an additional possibility.

The 1508 mutant is characterized by an increase in the low-molecular-weight nitrogen fraction and can thus be effectively screened for with the ninhydrin method as proposed by Mertz *et al.* (1974). It is not possible to select for the Hiproly gene with this rapid method because of the normal level of amino nitrogen in the water extract of that mutant.

Both 1508 and Hiproly barley are characterized by a 50 to 100 percent increase of the crude water- and salt-soluble protein fractions (Ingversen *et al.*, 1973; Munck, 1972). Polyacrylamide gel electrophoresis of the water-extractable proteins at pH 8.6 reveals a pronounced relative increase of four protein components in Hiproly (Munck, 1972). Ingversen and Køie (1973) have isolated these components by gel chromatography and electrophoresis and verified their high concentration in Hiproly. The lysine content of the four proteins was as high as 8.6 percent. Polyacrylamide gel electrophoresis at pH 8.6 of the water extract of the 1508 mutant gives a relative balanced pattern of the protein fractions similar to that of the mother variety Bomi, which is also true for the F_1 hybrid between Hiproly and 1508 (Tallberg, 1973).

The two mutations thus have different effects on the water-soluble proteins. The four prominent water-soluble proteins of Hiproly are produced at a maximal rate between 3 and 5 weeks after fertilization, i.e., at a time when the synthesis of hordein is at its peak (Munck, 1972). The above mentioned 10 percent increase of lysine availability in rat tests with Hiproly as compared with a normal barley is understandable in light of the finding that Hiproly contains a higher amount of water-soluble proteins with a high lysine content.

While hordein synthesis is only slightly affected in Hiproly (Munck, 1972), the hordein content of the ripe kernels of mutant 1508 is reduced to one-third of that found in Bomi (Ingversen *et al.*, 1973). The lysine-poor hordein proteins are more easily digested than the other barley proteins (Munck, 1964a,b; Eggum, 1970a).

The amount of the crude glutelin fraction from 1508 and Hiproly barleys is not significantly changed from that in normal barley. Differences from normal barley in the amino acid composition of the crude Osborn fractions do occur in both mutants, indicating shifts in the relative abundance of individual proteins with different amino acid compositions within the fractions. This is especially conspicuous with regard to an increase in lysine content in the hordein and glutelin fractions of 1508 (Ingversen *et al.*, 1973) and a decrease of cysteine as well as an increase of the methionine and the albumin, globulin, and glutelin fractions of Hiproly (Munck, 1972).

Storage proteins are deposited in the aleurone grains of the cells in the aleurone layer and in the protein bodies of the cells in the endosperm. The mature aleurone cells contain extensive amounts of storage fat but virtually no starch. The endosperm cells—comprising the bulk of the storage tissue—contain huge starch grains in their amyloplasts, while the protein bodies are located in the vacuolar compartment of the cell.

Observations with light microscopy (Munck *et al.*, 1970) and scanning electron microscopy (Munck, 1972) revealed no significant difference between the protein bodies of Hiproly and an otherwise isogenic sister line. The hard endosperm character of the original Hiproly line segregated from the lysine trait with a frequency of about 5 percent, suggesting crossing over between two genes on the same chromosome (Munck, 1972). Tallberg (1973) compared the mutant 1508 with its mother variety Bomi in scanning electron micrographs and suggested an impairment of protein body formation due to the inhibition of prolamin synthesis in analogy to the observations made on opaque-2 maize (Wolf *et al.*, 1967; Christianson, 1970).

Using electron microscopy of thin sections after fixation in glutaraldehyde and osmium tetroxide, we have studied the formation of the protein bodies and aleurone grains during kernel development, especially in Bomi barley and its high-lysine mutant 1508. Aleurone grains and protein bodies can readily be distinguished by their ultrastructural morphology at all stages of development. The ultrastructure of aleurone cells was found to be very similar if not identical in Bomi barley and the two high-lysine lines 1508 and Hiproly. Pronounced differences, however, are seen in the ultrastructure and development of the protein bodies in the endosperm when these lines are compared.

In agreement with the classical light microscopic studies, the following mode of reserve protein formation can be inferred from the ultrastructural analyses: The proteins are synthesized on the ribosomes of the extensively developed endoplasmic reticulum in the cytoplasm between the amyloplasts, when these are actively engaged in starch synthesis [portions of the endoplasmic reticulum (*ER*) are seen in Figure 2]. The proteins are then transported into vacuoles bounded by

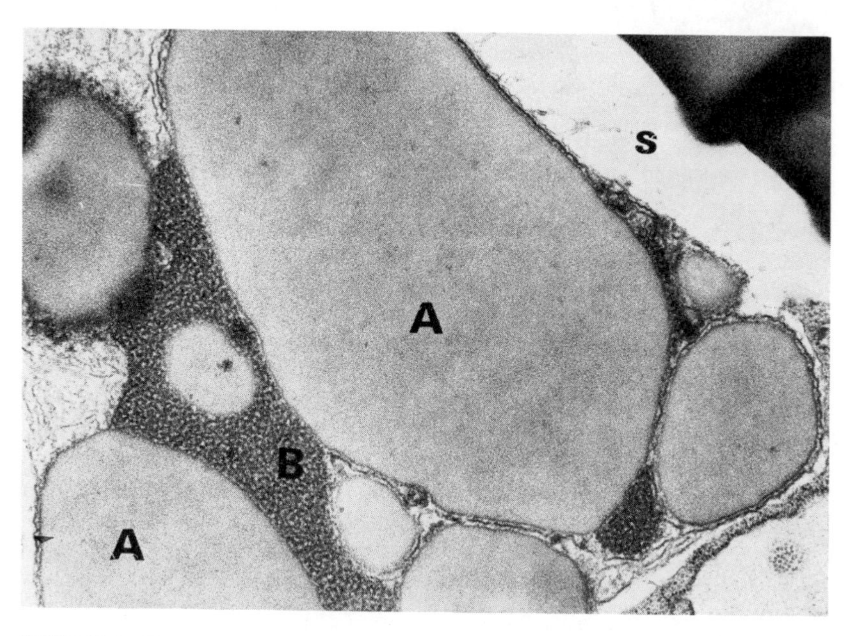

FIGURE 1 Electron micrograph of section through endosperm cell in developing barley kernel of the variety of Bomi 3 weeks after fertilization. Protein bodies are formed in vacuoles located in the cytoplasm between the large starch grains (*S*) deposited in the amyloplasts. A starch grain is present in the upper right corner. The protein body consists at this stage of a dominating homogeneous component (*A*) and a granular matrix (*B*). × 20,100.

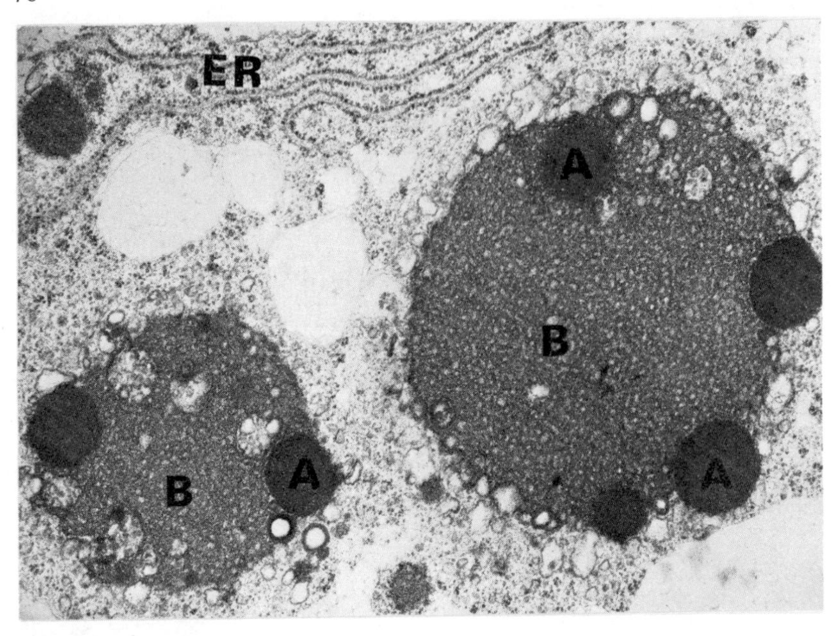

FIGURE 2 As Figure 1, but for high-lysine mutant 1508. Portion of the extensive ribosome-coated endoplasmic reticulum is seen at the upper edge of the picture (*ER*). In contrast to Bomi, the protein bodies of the mutant consist at this stage of a highly developed granular matrix (*B*) containing only a few osmiophilic globules of the homogeneous component. × 20,100.

the tonoplast. There dehydration takes place and the proteins precipitate according to solubility, the least soluble components and peptides precipitating first. The transport presumably involves the lumen of the endoplasmic reticulum, as this is frequently dilated and filled with electron-scattering material. The successive precipitation and dehydration of the proteins in the vacuoles leads to a characteristic and complex ultrastructural morphology of the protein bodies, especially at early stages of development (compare Figures 1 and 2).

Three weeks after fertilization, the protein bodies in the endosperm of Bomi barley contain homogeneous grains (*A*) embedded in a granular matrix (*B*), as shown in Figure 1. Frequently a third crystalline component, presumably consisting of sodium phytates (not shown in the micrograph), is associated with the developing protein bodies. In contrast to Bomi, the protein bodies of its high-lysine mutant 1508 contain at this developmental stage the granular matrix (*B*) as the dominating component, whereas the homogeneous component (*A*) is restricted to small globules, as seen in Figure 2. These globules are more highly osmiophilic than the homogeneous component of the wild type.

The round compact grains of the protein bodies (*A*) in 4-week-old endosperm cells of Bomi are shown in Figure 3 and resemble protein bodies described for other cereals such as maize (Khoo and Wolf, 1970). The extent to which the lighter fibrillar areas surrounding the protein grains (*b*) originate from the granular matrix (*B* in Figure 1) and contain other types of peptides is being investigated.

A loose fibrillar structure is characteristic of the protein aggregations (*C*) of 4-week-old endosperm in the high-lysine mutant 1508 (Figure 4). Their developmental relationship with the structures of the protein bodies at the 3-week stage is being investigated.

The protein aggregations of the mutant contain tubular inclusions (*t*), which are shown in cross section (Figure 4). Changes in the ultrastructure of mutant 1508 include an aberrant organization of the endoplasmic reticulum (*ER*) in the endosperm cells. We consider it likely that the morphology of the protein bodies in the gene mutant is the result of

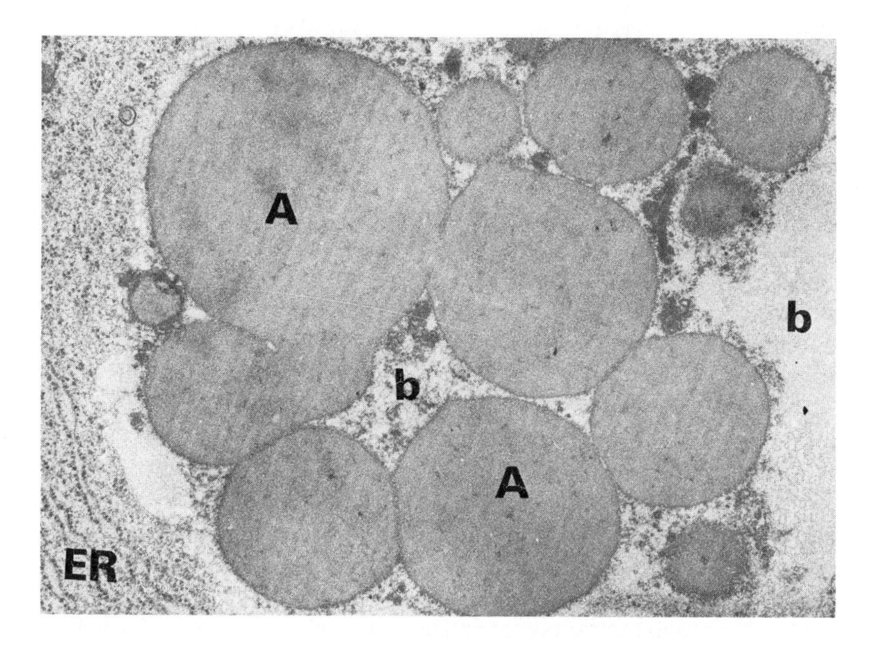

FIGURE 3 Almost mature protein body in 4-week-old endosperm cell of Bomi. The round homogeneous and compact protein grains (*A*) are conspicuous. They are surrounded by fibrillar material (*b*).

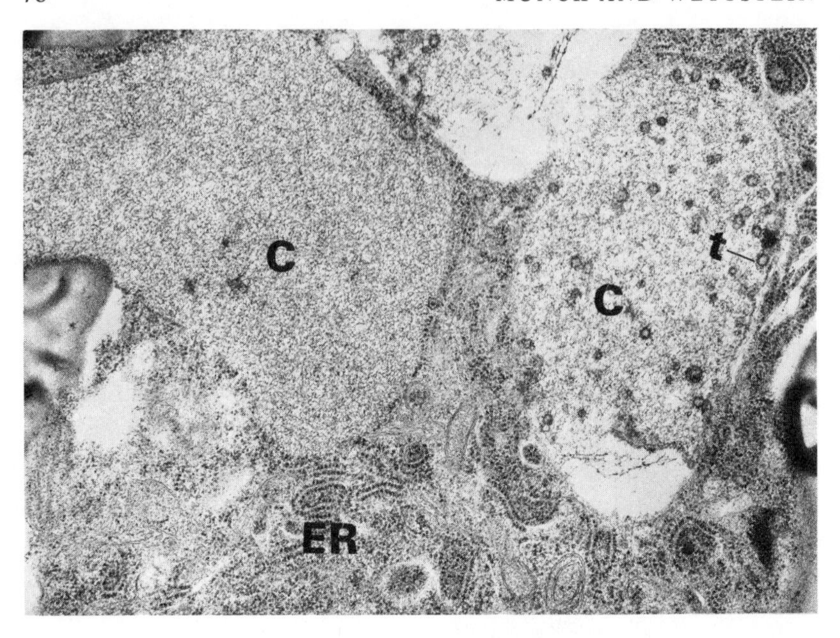

FIGURE 4 Protein aggregates in 4-week-old endosperm cells of the high-lysine mutant 1508 are inhomogeneous and appear loose in structure (*C*). Some of the aggregates contain tubular inclusions (*t*). Edges of starch grains are seen at left and at right. × 20,100.

the change in amino acid metabolism of the endosperm. Whether the absence of a compact homogeneous component is a direct consequence of the partial inhibition in 1508 of the conversion of lysine into glutamic acid (Brandt, personal communication) or of the inability of the mutant to form large amounts of the lysine-poor prolamins (hordein) is being investigated.

Different sets of genes govern the formation of the aleurone grains in the aleurone layer and the formation of the protein bodies in the endosperm. It is thus possible to change the composition of these two types of protein stores independently of each other. The recognition of ultrastructurally distinct components in developing protein bodies provides encouragement for the feasibility of isolating and separating these components. The availability of gene mutants that selectively reduce or increase certain protein fractions will be helpful in assigning these fractions to the morphologically recognizable components of the protein bodies. Such characterization of the protein bodies should allow more sophisticated selection procedures in the genetic improvement of cereal proteins, as well as a deeper understanding of the structure and composition of the indigestible fraction of barley protein.

Data collected in this review point out important similarities between the barley mutant 1508 and the mutant opaque-2 in maize. These include an increased amount of free amino acids in endosperm, reduction of prolamin synthesis, impairment of protein body formation, and the partial inhibition of the lysine-to-glutamic-acid pathway (for information on opaque-2 see the paper by Mertz, p. 57). The barley mutant Hiproly differs definitely from this pattern. Its high lysine content is primarily due to enrichment in four water-soluble protein components.

REFERENCES

Christianson, D. D. 1970. *In* Proc. 3rd Scanning Electron Microscope Symp., IIT Research Inst., Chicago. p. 161.

Doll, H. 1973. Heriditas 74:293.

Eggum, B. O. 1970a. Z. Tierphysiol. Tierernähr. Futtermittelk. 26:65.

Eggum, B. O. 1970b. *In* Improving Plant Proteins by Nuclear Techniques. IAEA/FAO STI/PUB 258, Vienna, p. 289.

Eggum, B. O. 1973. Tolvmandsbladet 45(2):7

Hagberg, A., K. E. Karlsson, and L. Munck. 1970. *In* Improving Plant Proteins by Nuclear Techniques. IAEA/FAO STI/PUB 258, Vienna, p. 121.

Ingversen, J., and B. Køie. 1973. Phytochemistry 12:1107.

Ingversen, J., B. Køie, and H. Doll. 1973. Experientia 29:1151.

Karlsson, K. E. 1972. Barley Genet. Newslett. 2:34.

Khoo, U., and M. J. Wolf. 1970. Am. J. Bot. 57:1042.

Mertz, E. T., P. S. Misra, and J. Jambunathan. 1974. Cereal Chem. 51:304.

Munck, L. 1964a. Hereditas 52:1.

Munck, L. 1974b. Hereditas 52:151.

Munck, L. 1972. Hereditas 72:1.

Munck, L., K. E. Karlsson, A. Hagberg, and B. O. Eggum. 1970. Science 168:985.

Tallberg, A. 1973. Hereditas 75:195.

Toft-Viuf, B. 1972. Roy Vet. Agr. Coll. Copenhagen Yearb., p. 37.

Wolf, M. J., U. Khoo, and H. L. Seckinger. 1967. Science 157:556.

DISCUSSION

DR. JENSEN: Dr. Mertz, I presume the data that we have seen is from inbreds. Perhaps data are also available from single crosses and double-cross hybrids. The problem for geneticists and plant breeders is to translate this high-lysine bonanza into corn that will produce a balanced product. Has any consideration been given to the formulation of a seed composite of hybrids that will put together in the final product a multiline balanced zein–lysine, and all the other components that you want?

DR. MERTZ: I do not understand exactly what you mean by "multiline." For example, there are a dozen hybrid corn seed companies in the United States that are offering opaque-2 mutants where they have simply taken their inbred lines and backcrossed the opaque-2 gene in to get. . . .

DR. JENSEN: Yes. My understanding is that the constraints of this give you an economic product yield that is somewhat lower at the present time. Is this not true?

DR. MERTZ: Yes, it is true. Some companies have done better than others on that. There are supposed to be a few hybrid opaque-2's that are competitive with the best normal hybrids. Unfortunately, I think in some cases the hybrid corn breeders have been so busy trying to raise the yield and bring it up to that of the normal that they have not spent enough money in the laboratory to follow the lysine levels, and so these, in some cases, have not come up to lysine levels that we would like to see in the best-yielding hybrids of opaque-2.

DR. JENSEN: One cannot escape this economic factor. Growers simply will not grow something if they are asked to take too much of a penalty. The same thing could be accomplished by blending varieties that have, say, high zein and high lysine and so on at the end product, but it would be so much simpler, and I am talking only about a near-term thing until the plant breeders work this out, it would be so much simpler to put together the hybrids in the seed composite that would then be grown and would provide a balanced product.

DR. MUNRO: I want to ask some orienting questions. First, I assume that you are not getting any changes in the basic composition of the zeins and glutinins, that you are, in fact, getting these changes exclusively by alterations in proportion. This links with what Dr. Dieckert was saying about amino acid homology and nonhomology. Second, you do have a free amino acid component. How much does this contribute to the important amino acids, lysine and tryptophan? In other words, is it a trivial proportion or not? For leucine, which for the human is nutritionally undesirable in excess, is it possible that with further genetic manipulation you may be able to reduce the biosynthetic pathway for leucine, which presumably piles up when your zein output is reduced? Finally, are you getting a reduction in the capacity to form storage granules, which is on a time-shift basis? In other words, if the seed is allowed to grow longer, does it develop these at a later time, and what effect does this whole pattern of change in storage granules have on subsequent generations of the plant? How far can you go with shifting the basic structures inside the grain and still have a viable plant?

DR. MERTZ: As I understand the first question, it was relating to the actual composition of the zein, for example, in normal versus opaque-2 where you still have appreciable amounts of zein. We have analyzed the Landry-Moureaux fraction from the normal counterpart and from opaque-2, floury-2, brittle-2 and the double mutant. We were very surprised at the similarity of the amino acid pattern of the fractions in the normal and in the mutant strains. This suggested to us that there was a shift in the proportions of the proteins without any major changes in the composition. No new proteins were being produced in large quantity that were changing the amino acid pattern.

For example, in the fifth fraction, the glutelin fraction with the Landry-Moureaux procedure, in the normal the lysine content is between 7 and 8 percent. It is the same in the opaque-2. Therefore, it seems that what we have done is to increase the amount of glutelin with the 7–8 percent lysine and reduce the zein with the 0.1 percent lysine. So what we are doing is pushing down the zein and raising the albumin, globulin, and glutelin.

I understand that Dr. Munck pointed out some very high lysine-containing fractions in the albumin-globulin fraction of barley. The question that I had to him was, Are those same ones present in the normal? I mean, they could be there but in lesser amounts, and there might still be a shift in the proportion. Maybe you could answer that question.

DR. MUNCK: It is a question of proportion.

DR. MERTZ: In other words, you would have those same proteins in the normal.

DR. MUNCK: Yes.

DR. MERTZ: That is what we found. What was the next?

DR. MUNRO: The question was, When you suppress zein, you stimulate the accumulation of amino acids: what is the spectrum of these? Are they high in leucine and low in lysine and tryptophan?

DR. MERTZ: Yes, in the one that we studied first, the W64A, where we had really big differences, we isolated the amino acids and ran patterns, and found that in both the normal and the spontaneous mutant the pattern was very similar. About half of the amino acids could be accounted for as glutamic acid, proline, and serine, and the level of lysine was relatively low, around 3 percent of the total in both. Therefore, this fraction does not contribute lots of lysine and tryptophan or nutritious amino acids that are needed. Even in the extreme case of W64A this still amounts to only 5 percent of the total protein. It certainly does not have much effect on the nutritional value.

DR. MUNRO: The final point was whether the plant as a plant is changed in any way as a result of zein deficiency.

DR. MERTZ: There are some differences. For example, on germination the opaque-2 plant is a little more sensitive to cold injury than the normal, but this happens the very first time you produce the plant. I do not think it is something that gets worse. It just stays at the level of being sensitive to cold, as far as I know.

DR. DIECKERT: I think that this may be an indication that these plants are not so fastidious with respect to the nature of their proteins that make up the aleurone class.

DR. WILSON: I am interested in the composition of the seed in terms of dry weight and, more important, percentage total nitrogen. A single inbred that Dr. Sodek looked at in my lab had the same endosperm weight but only two-thirds as much nitrogen as the normal. It did not make zein, but it did not do anything beneficial with the excess nitrogen. It happened to be the one in which there was a very high level of free amino acids, as you have, for example, in the brittle-2. Is this high free amino acid perhaps a symptom that it lacks the ability to incorporate the amino acids into some other protein?

DR. MERTZ: I think that what has happened here is that, in the amino acid pool, you get a damming up of the building blocks. The plant is used to making zein and a particular pattern of other proteins. You have distributed this process by this mutation or combination of mutations. As a result, you continue to have these amino acids flowing into the cob, and from the cob into the kernel. But the kernel cannot handle them, and so the amino acids pile up. They are very high at the start in the developing kernel, and then they fall off toward the end. In the normal kernel they almost disappear. You have just a fraction of a percent of free amino acids at the end in the normal, but you have a higher level in these mutants. You just cannot dispose of these building blocks, and it is true that you do have this disturbance of metabolism. I think this is why the yield is so poor in some of these. What the breeder has to do is to bring in other genes to try to counteract the effect of this disturbed metabolism on the productivity of the plant.

DR. MIFLIN: About the distribution within groups, you had one mutant—I believe it was sugary-1—that showed the same drop in zein and glutelin and yet only a 12 percent increase in lysine. It would seem to me that there has been a drastic shift in the nature of the glutelin fraction, which is, after all, made up of a large number of proteins that presumably have a range of lysine contents. Have you been able to analyze this? This would be in distinction to the facts you gave previously.

DR. MERTZ: When I looked this material over last night, I felt somebody was going to ask me that question, and I wish I had brought along the actual data I had for fractions 3 and 4. I think the answer lies in that data; unfortunately, I did not bring it along. Fraction 3 is zeinlike, and fraction 4 is glutelinlike. It is possible that we had a big increase in 3 which counterbalanced the drop in 2, and this could account for the fact that the lysine did not go up so much.

DR. MIFLIN: But you still have an increase of up to 22 percent of glutelin?

DR. MERTZ: Yes, that increased, but if we had too much of an increase in the zeinlike fraction, this might have tended to pull down the lysine. I think that a study of the fractions, individual fractions, would show where this happened. I think that the answer lies in the levels that we had for fractions 3 and 4.

G. F. SPRAGUE

Plant Breeding, Molecular Genetics, and Biology

Plant breeding as a discipline draws heavily upon genetics for its operational procedures. Other fields, such as entomology, plant pathology, and experimental statistics, have also contributed heavily to plant breeding practices and accomplishments. Plant breeding reliance on genetics has been limited to the pre-1950 developments, which I shall call classical genetics. Molecular genetics, as yet, has had no impact on plant breeding practices. This situation prevails not because of lack of possible relevance but rather because of limitations imposed by lack of suitable screening techniques and economic restrictions on population sizes. In this paper I propose to show the relation of plant breeding to classical genetics, provide a brief review of plant breeding practices and accomplishments, and, finally, speculate on research development that will be required before molecular genetics and biology can have any important impact on breeding procedures.

The genetic principles laid down by Mendel and extended during the first half of the twentieth century provide the basis for plant breeding procedures. Among these principles are the following: (a) Inheritance is particulate rather than blending, (b) genes are arranged in linear order on the chromosomes, (c) linkages may be broken through exchanges involving homologous chromosome segments, and (d) non-linked genes assort randomly at meiosis.

These same developments provide the framework for developments in quantitative genetics that permit estimation of genetic variances and covariances. These, in turn, have provided valuable information on types of gene action in polygenic traits and also measures of the relative effectiveness of alternative breeding systems.

In spite of dependence upon a common group of postulates, genetics and plant breeding involve several important differences that should be clarified at this point: The geneticist, by choice, works with traits for which genotypes can be assigned or developed, in which segregations are clear-cut, and that are relatively little influenced by environment. The plant breeder, of necessity, works with polygenic traits for which the number of loci and their interaction effects are unknown or cannot be precisely established, and for which environmental effects may be as large as or larger than the genetic effects. Under such conditions, genetic differences can be inferred from differences in mean performance measured in replicated experiments. Such tests must be conducted in the varied environments to which the new type may be exposed under conditions of commercial production. The plant breeder operates under one more very important constraint: The products of his research efforts must have sufficient economic value to win farmer acceptance. In the absence of farmer acceptance, the developmental research must be considered unproductive.

The crops useful to man may be divided roughly into two distinct groups: the self-pollinators and the cross-pollinators. In the first group the commercial product is usually a highly inbred line, and in the second group it is an F_1 hybrid or an improved random-mating population. Historically, different breeding procedures have been used for these two groups. For both groups, attention has been given to yielding ability and to disease and insect resistance as these influence farmer acceptance. In addition to these necessary performance characteristics, each crop may have other specific quality attributes that are required to achieve consumer acceptance, e.g., milling and baking properties in the hard wheats. Selection for such quality factors becomes a necessary part of the evaluation process to which all new developments must be subjected. It would be impossible to detail the breeding procedures and evaluation techniques used for each crop. I shall, therefore, present a brief résumé for only two: corn and wheat. Procedures vary widely among breeders, so this résumé can be little more than a general outline of the types of operations performed and the time required.

CORN

Corn is a cross-pollinated crop. Current U.S. production depends almost entirely upon controlled hybrids involving two (single cross), three (three-way), or four (double-cross) inbred parents.

In the 1920's the parental sources used for the development of inbred lines were open-pollinated varieties. After hybrids came into commercial use, the better hybrids generally replaced open-pollinated varieties as parental sources. Now, populations improved through some form of recurrent selection are extensively used as foundation materials. Inbreeding and evaluation procedures are roughly similar and largely independent of the foundation materials used.

Since corn is a cross-pollinating species, any form of inbreeding requires artificial controls. The most common form of inbreeding is self-fertilization. This requires the covering of ear-shoots before the silks (style and stigma) appear. When both silks and pollen are available, the tassel is enclosed in a paper bag. The following day the accumulated pollen is transferred to the silks of the same plant and then protected from pollen from other sources by enclosing the developing ear in the same tassel bag. This bag remains in place until harvest time.

After the decision as to foundation populations, selfing must necessarily be carried out on an extensive scale. Many of the self-pollinated ears are discarded at harvest due to faults (e.g., insect or disease susceptibility) that were not apparent at the time of pollination. Seed from the selected ears is planted on an ear-row progeny basis for continued inbreeding and selection. Intensive visual selection is relatively ineffective in modifying combining ability but may be highly effective in attaining required levels of inbred productivity, insect and disease resistance, and kernel sizing properties.

After three or more generations of selective inbreeding, the resulting lines have acquired a sufficient degree of genetic stability to permit tests for combining ability, i.e., performance in hybrid combinations.

The evaluation of lines in hybrid combinations is a herculean task; 100 lines can be combined to produce 4,950 different single-crosses and nearly 12 million double-crosses. At either hybrid level, facilities preclude adequate replicated tests involving numbers of these magnitudes. Time will not permit detailing the various developments that have led to a reasonable solution to this numbers game. Regardless of the system of evaluation, specific combinations having potential value must be subjected to further detailed evaluations under the array of environments characteristic of the region in which the hybrid may be used commercially.

The time required for this whole process of inbreeding and hybrid evaluation will average about 10 generations. Cumulative mortality during the entire process through inbreeding and hybrid evaluation is high: Considerably less than 1%—probably less than 0.1%—of the initial selfed ears will result in inbred lines having extensive commercial usage.

Thousands of inbred lines have been developed in public breeding programs within the United States, and hundreds have been shown to have sufficient merit to justify release for possible commercial use. In a recent survey, however, six lines were involved in the parentage of 70 percent of all hybrid seed produced in 1970. Performance under commercial conditions provides a stringent selection criterion in addition to those routinely used by the breeder.

HARD RED WINTER WHEAT

Wheat is a self-pollinating crop. The commercial varieties are roughly equivalent in genetic uniformity to inbred lines of corn. Improvements are effected by crossing selected parents and searching for still better types among the segregating progeny.

The strains selected as parents will normally represent the best materials currently available. Each strain, however, will not possess all desired attributes to equal degrees, so crosses are made to maximize the opportunity for complementarity and for transgressive inheritance.

The time at which the various selective criteria are imposed will vary with the genetic complexity of the trait and the quantity of seed required for a given evaluation. Selection for single-gene effects (e.g., resistance to certain races of rust or biotypes of the Hessian fly) may be performed as early as the F_2 generation. Milling and baking qualities and yield evaluations cannot be performed adequately on a single-plant basis.

Early in the selection and evaluation process, therefore, individual head selections would be made and then grown in head-row progenies. Grain produced from such plots would be examined for plumpness, kernel size, and other visual quality factors. Progenies surviving this culling level would be subjected to protein analyses and to micromilling and baking tests. Strains judged superior as a result of these evaluations would be moved into replicated nursery yield trials where further eliminations would be made on the basis of differences in yielding ability, resistance to lodging or other environmental hazards. The best material from such tests would then be advanced into more extensive regional trials for further yield evaluations and the magnitude of the effects of environment on milling and baking properties. After two or more years of such testing, a very few strains may be judged adequate for release and commercial production.

This entire process—from choice of parents to production of segregating populations to the sequential evaluation processes—may require as many as 10 generations. Aside from disease and insect resistance screening, often done under controlled greenhouse conditions, and the laboratory screening for quality, all evaluations are done under field conditions, where environmental effects tend to obscure real genetic differences. The general superiority of the parents chosen, in terms of field performance and quality characteristics, will be reflected in the average merit of the progeny produced and the level of superiority of the best genotypes finally chosen for release and commercial production.

Plant breeding progress has been achieved by small but significant improvements with each cycle of selection. These small stepwise improvements are due to changes in gene frequency that are reflected in improved performance at either the inbred or hybrid level.

This very abbreviated outline of procedures used in wheat and corn improvement emphasizes the fact that although numerous screening techniques have been developed to simplify genotypic evaluations, the process retains an undesirably high degree of trial and error. In spite of this limitation, the practical accomplishments have been impressive. In the last 40 years wheat yields have doubled and corn yields have tripled. A part of these increases is due to various improvements in production practices, but the genetic inputs have been substantial. Yields per acre for other major crops would exhibit roughly comparable improvement rates.

Progress in plant breeding with any crop is dependent upon the efficiency of the screening and evaluation techniques used to identify superior genotypes. Tremendous advances have been made in both molecular genetics and cell biology, again largely dependent upon the effectiveness of the selection techniques used. It appears logical therefore to assume that if some of these new techniques were employed by plant breeders, the gains in efficiency of operation would be substantial. Popular speculations along these lines have tended to minimize both the nature of the problem and the extent and specificity of research needed to apply those techniques in a plant breeding program.

Unfortunately, little current research is aimed at the solution of problems related to simplification of assay techniques. Such research will not be done by the plant breeder. Work in this area would be an add-on to his current operations, since yield and quality considerations must remain his primary concern. Research in this intermediate area has not been looked upon with favor by the several federal granting agencies. Research proposals that can be interpreted as having any applied flavor are usually not funded. This attitude, of long standing, reflects a peculiar philosophy: Work on the C_3 photosynthetic pathway in spinach or chlorella is basic research of the purest sort, while similar work on the C_4 types, corn for instance, smacks of application and therefore does not warrant grant support. A more realistic philosophy will have to be developed if any substantial progress is to be achieved in making current and future developments in molecular biology relevant to either plant or animal improvement.

TRANSDUCTION AND DNA HYBRIDIZATION

For some years it has been possible to transfer chromosome segments from a donor bacterium to a recipient cell through use of a DNA phage as the transfer agent. This transfer is a rare event, but selective screening techniques permit the elimination of all or essentially all cells lacking the desired transfer. Such a transfer can be accomplished at low cost—a few pennies per petri plate.

A few of the factors that currently preclude the use of this technique in plant breeding may be of some interest. The great majority of the plant viruses have RNA rather than DNA as the carrier of genetic information. In no instance has transduction been accomplished with RNA viruses in higher plants. Even if appropriate techniques were developed, the important problem of identification of the desired transduced genotype remains. A turbid bacterial population may contain 10^9 individuals. If a culture with this population density is used in transduction studies, essentially all of the individuals will be killed before cell division is completed. A corn population of equivalent size would occupy roughly 50,000 acres. Unless a selective screening technique, applicable to dry or germinating seed, could be devised, one such transfer attempt could cost, in terms of land preparation, seed and loss of crop, as much as 10 million dollars. This sum is greater than the total annual expenditure for public corn breeding research in the United States. Until these and other obstacles can be overcome, the plant breeder will continue to use the backcross procedure for simple inherited gene substitutions, as he has for the past 50 years.

Technical ability now exists to join DNA molecules from diverse sources. Thus DNA from viruses could be combined with bacterial DNA to form new genotypes. Whether such techniques, still in developmental stages, can be extended to higher organisms remains to be established. The use of DNA manipulations in cell culture has been suggested as offering great potential for various types of genetic engineering.

CELL CULTURES

Great progress has been made in developing techniques for producing cell cultures of several species. This technique may well have more potential for plant breeding than either transduction or somatic hybridization. Further improvements in techniques should permit the development of cellular genetics at a rapid rate. Many mutants of potential interest might be more readily identified and their influence on biochemical synthesis more readily investigated than would be possible at the sporophytic or seed level. Sporophytes and seed progeny could then be obtained for mutants of special interest.

Such mutant forms could then be incorporated into inbred lines having commercial value. It must be recognized that an inbred line of corn or its genetic equivalent in a self-pollinating species represents a physiologically integrated genotype. Even single-gene substitutions may upset this balance to an extent requiring additional genetic modifications to maintain its commercial value. Thus, even simply inherited traits must be considered as a component of a polygenic system in any subsequent evaluation procedures.

Current information on the inheritance of polygenic traits is meager. On the basis of several studies, we know that the number of loci involved is large and that they are distributed more or less randomly throughout the genome. Information on the existence and distribution of such loci has been obtained from studies involving translocations. Such studies identify chromosome arms that carry loci conditioning differences between the translocation stock and the stock under study. Information on the effects of the individual alleles involved is almost completely lacking. Much more precise chromosome mapping followed by characterization of gene products and interaction systems will be required before any massive genetic engineering becomes feasible. This type of massive genotypic restructuring has always been the concern of the plant breeder. His approach, of necessity, has been a gradual one through small sequential changes in gene frequency that must be verified and evaluated at each step in the progression. Alternative procedures for effecting genotypic changes would require the same sort of evaluation under a wide array of environments.

SOMATIC HYBRIDIZATION

Great strides have been made in the ability to produce fusion between naked protoplasts and, in some instances, to obtain subsequent development of a mature plant with seed production capabilities. The argument has been advanced that with further development and refinements of techniques new opportunities will become available for increasing genetic variability opportunities now restricted by the difficulty or impossibility of producing hybrids through the sexual route. Somatic hybridization as a technique for the production of new polyploids will provide a valuable mechanism for study of a broad array of genetic and biochemical problems. It appears reasonable to expect, however, a very appreciable time lag between technical capabilities in producing new polyploids and the use of this approach in plant breeding.

If somatic hybridization is to be an effective plant-breeding tool, it must produce new types with a greater seed yield potential or yield equivalent than species currently grown or with special properties that would lead to increased economic value. Species or varieties currently grown have been selected for their adaptation to particular ecological niches and for specific properties that are compatible with current production practices. New polyploids must satisfy these same requirements or, if possessing altered characteristics, must competitively replace existing types.

A number of questions must be posed and resolved. The potential combinations are so large as to preclude a detailed examination. Therefore, judgment, however poorly based, must be used to reduce the potential number to some manageable figure. For instance, the first judgment is, what other small-seeded cereal might be combined with a hard red winter wheat type to produce a new economic crop. Almost certainly the new type would not possess the milling and baking properties required for acceptable quality bread. These properties might be recovered through subsequent selection, but even if this could be accomplished, many generations would be required. If one is willing to forgo the special properties that determine the current production and use of different classes of wheat, of malting barley, and so on and is prepared to accept the new product solely as a feed grain, the problem is not materially simplified. Possible combinants with wheat would include rye, oats, and barley. When grown under optimum ecological conditions, the yield levels of these three crops are roughly comparable. Differences in protein value of these three cereals exist, but these differences seem hardly great enough to make the somatic hybridization route highly attractive. Opportunities for increases in protein quality or quantity or both are known for each species and have not been fully exploited by conventional means.

Corn, our major feed grain, is one of the more efficient species in the utilization of solar energy. Sexual combinations of corn with its American relatives, *Euchlaena* and *Tripsacum*, have not led to viable commercial products. Relatively little work has been done with *Zea* × *Tripsacum* hybrids. Extensive studies have been conducted with *Zea* × *Euchlaena* hybrids over the past 20 years without development of a superior commercial product. If still wider somatic hybrids are produced, it appears likely that a proportionately greater time span will be required between initiation and ultimate commercial usefulness.

All somatic hybrids would represent new polyploid combinations. Results to date with artificial polyploids obtained through conventional means have not been impressive. Corn tetraploids, obtained through use of the elongate gene, have initially been characterized by unacceptable levels of sterility. This sterility has been rather easily overcome by selection, but the fertile product has no known advantage over diploid corn.

Much more extensive work has been done with Triticale, a polyploid product of *Triticum* × *Secale*. No selections from this sexually developed polyploid have equaled the best adapted wheats in yield. Processing quality is reduced, as judged by current standards, although there may be a nutritional advantage in increased protein and lysine content, and the product may best be considered as a feed grain. Eventual use in this category also poses extensive problems. The Triticales abound in various toxicants that adversely affect rates of gain and biological efficiency of protein.

The existence of production problems among relatives sufficiently close to permit sexual hybridization suggest that still greater problems may be encountered with progeny of somatic hybrids involving components having lesser degrees of taxonomic relationships.

MOLECULAR BIOLOGY

Molecular genetics has made its greatest advances in the prokaryotes, permitting a great increase in our understanding of the mechanics of DNA transcription and chromosome replication, the details of the structure and function of individual genes, and the nature of mutation. Techniques used in such studies are, since they cannot be used for handling tremendous numbers, not generally available for studies in the genetics of eukaryotes and still less available to the plant breeder, who is forced to deal with relative differences rather than precisely determined genotypes. Given strains differing in average performance, the plant breeder, using quantitative genetic techniques, can characterize such differences in terms of types of gene action involved. Little progress has been made, however, in relating such differences to the biochemical systems involved and their component enzymes. Developments in this area may be more immediately useful to the breeder than developments in molecular genetics have been.

A single example, relevant to the topic of this volume, may serve to illustrate this point. Genotypes differ markedly in their ability to produce storage proteins when grown under similar field conditions. The Illinois high- and low-protein strains of maize are a classic example. Smaller differences, however, have been reported for other cereals and the legumes. Where studies have been conducted, differences between such strains appear to be influenced by many loci. Many steps are also involved in the biosynthesis of protein. One may speculate that the quantitative geneticist's evidence for many loci might be resolved into a more tractable system if suitable assays were available for each of the steps in protein biosynthesis.

The general features of protein synthesis are known. Aspects currently limiting maximum efficiency are unknown, but certain steps appear to be of particular importance. Three enzyme systems appear to be of particular interest and importance: nitrate reductase, which initiates the reduction of NO_3 to NH_3; glutamic dehydrogenase, which plays a key role in the *de novo* synthesis of amino acids; and the as yet unresolved systems of enzymes that determine the composition of the storage proteins.

The enzyme nitrate reductase, the rate-limiting enzyme in NO_3 assimilation, has been studied in some detail in both corn and wheat. Evidence has been obtained by Hageman and co-workers (p. 103) that variation in NRA (nitrate reductase activity) is positively correlated with grain protein, grain yield, and total reduced nitrogen in above-ground plant parts at maturity. Evidence has also been presented indicating that NRA activity is under genetic control. Assay techniques have been developed with the capacity for handling large numbers and the degree of simplicity necessary for routine use in a breeding program. Selection studies are now in the planning stage to establish whether more efficient genotypes may be isolated. Such genotypes might lead to more efficient use of applied nitrogen or greater protein production or both.

Little work has been done in glutamic dehydrogenase beyond establishing its role in combining NH_3 and keto-organic acids to form amino acids. The degree of limitation on protein synthesis imposed by this reaction remains unknown, and assay techniques suitable for plant breeding use are not available.

Other enzyme systems playing a major role in crop production that would appear to be useful as selection criteria include CO_2 fixation (chloroplasts), energy generation (mitochondria), and lipid and amino acid synthesizing complexes. The first step, for each system, would be to identify the rate-limiting process. Next, an *in vivo* assay must be developed for this enzyme that requires a minimum of time, expensive auxiliary equipment or training for its effective use. Only when such techniques become available will their effective use by the plant breeder become feasible.

Several simply inherited traits are now known in corn that affect protein quality: opaque-2, opaque-7, and floury-2. In each case the improvement in protein quality appears to be related to the reduced production of zein, a low-quality protein. Wilson and Sodek (1970) have shown that availability of lysine is not the limiting factor: Free lysine equal to a 2- or 3-day supply is present in the developing corn kernel. Knowledge of the biosynthetic pathways and limiting factors might permit much more effective selection criteria in a breeding program.

COMMENT

The efficiency of plant breeding is dependent upon two factors: the choice of parental material and the efficiency of selection practiced among subsequently derived populations. Most traits of economic importance have a polygenic base. For any single locus, selection efficiency is related to both genetic frequency and degree of dominance. Genetic advance may be slow initially because of the low frequency of the desired allele. Gains from selection reach a maximum at intermediate gene frequencies and decrease again as one approaches a frequency of 1.0 for the favored allele. The same general principles apply when dealing with a polygenic trait, except that gene frequency shifts are expected to occur at even slower rates.

The problem becomes even more complex because the plant breeder is seldom interested in a single trait, but rather in a combination of traits of essentially equal economic value. The univariate distribution of genotypes affecting a single trait is, in practice, replaced by a multivariate distribution of genotypes affecting the several traits. The presence of linkage and the degree of dominance of the favored alleles and of interactions modifying mean effects add to the complexity of the problem.

In the plant breeding field, little use has been made of selection indices that permit weighting several traits in terms of their individual contributions to economic value. Under this system, accurate estimation of the pertinent genetic variances and covariances requires very extensive and well-designed experiments. If selection is effective, these population parameters may change rapidly, thus requiring allocation of a sizable fraction of available resources to maintaining a currently adequate index. A "levels of culling" system of selection is used much more extensively. Under this system the first selection is concentrated on traits with high heritability. Subsequent selections are concentrated on traits having a lower heritability (e.g., yield) in which repeated tests under a range of environments is required to obtain valid estimates of mean performance.

With this very abbreviated and simplistic outline of selection procedures, we may turn to the possible relevance of developments in molecular genetics. We have noted earlier that transduction techniques or mutant types from cell cultures, when they have become available in higher organisms, are best suited to transfer of traits controlled by a single locus. As long as transduction remains a largely random process, its utility in modifying a polygenic trait will be quite limited. If the use of this method is limited to single-locus effects, any advantage over the long-used backcross technique will be dependent upon the reduction in numbers of generations required for transfer. The efficiency of either procedure is limited to traits with high heritability, i.e., when classification of genotypes has a high degree of accuracy.

When somatic hybridization becomes available to plant breeders, the effect may be to broaden choice of parents. The problems relating to modification of progeny through selection will remain unchanged or may even be intensified. Sterility problems have been common in newly formed polyploids, thus adding another trait to the selection complex. Interactions between genomes, of unpredictable extent and importance, are also to be expected.

We have stressed that plant breeding progress is limited by the efficiency of selection procedures. When superior genotypes are developed and identified, the biological basis for superiority remains largely unknown. Use of such material as parents for further improvement operations is based on the presumption of a higher frequency of the desired alleles and the probability of further gene frequency modifications within their progenies rather than on any understanding of the biological functions that account for the increased productivity. It is precisely an increase in this understanding that offers the greatest possibility for increased efficiency of selection.

Each of the loci involved in some one or more aspects of production efficiency exerts its control through the production of specific peptide chains that, in combination, become functional proteins. Extensive studies have been conducted with bacteria and *Neurospora* in the identification of loci, their enzyme systems and their role in biosynthesis. Comparable developments for higher plants are quite limited. The development of an extensive body of information will require a greatly expanded research program. The relevance of such developments to plant breeding will depend on the utility of enzyme assay techniques.

Plant breeding progress has been closely related to developments in related fields. First came developments in classical genetics that provided the necessary theoretical foundation and understanding of some practices already in use and the development of new and more efficient techniques. The second major development involved improvements in statistical methods and experimental design. These new designs permitted testing large numbers of entries with acceptable levels of precision. The same type of design permitted the expansion and evaluation of the newer developments in quantitative genetics.

In the last 20 years two new tools have become available to the plant breeder interested in modifying chemical composition. The nuclear magnetic resonance technique is being used for the nondestructive estimation of fats, and the amino acid analyzer is being used for establishing differences in protein quality. Both pieces of equipment are expensive and beyond the reach of most breeding programs.

The improvements in protein quality that have been achieved with opaque-2 and similar genes in corn and the high-lysine gene in sorghum were made possible because lysine, the most commonly limiting amino acid in cereals, is relatively easily determined with an amino acid analyzer. The first limiting amino acid in soybeans and several of the edible legumes is methionine. The amino acid analyzer has less utility in measuring variation for this amino acid. Colorimetric techniques have been developed for both lysine and tryptophans that permit handling a large number of samples, but these appear to have a lower order of reliability than the amino acid analyzer. It is to be hoped that adequate colorimetric techniques will be developed for other essential amino acids.

Even though several loci have been identified as affecting protein quality, information is generally lacking on the underlying changes in biosynthetic pathways. If the enzyme systems involved in these protein quality changes were identified and characterized, still better analytical tools might become available.

The development of improved analytical tools is not likely to result from plant breeders' efforts. At present levels of staffing and funding, the main concern will continue to be increased efficiency in total production. If improved analytical tools are to be developed, they will most likely have to come from the research efforts of the molecular biologist. The efficiency of such investigations, however, can be maximized only if close cooperation is established between the two groups of scientists. If major inputs from the molecular biologist are not forthcoming, the plant breeder's major attention will continue to be directed toward those aspects of productivity for which he has adequate evaluation criteria.

SUMMARY

Plant breeders have played a dominant role in the increases in agricultural production achieved during the last 50 years. Their gains have been achieved by capitalizing on improvements in statistical design that permit accurate comparisons among large numbers of genotypes and improvement in production practices. Gains have come through small increments, each new development representing an improvement in yield and disease and insect resistance over the varieties replaced. The genetic base has been drastically reduced, but there has as yet been no decrease in rate of progress achieved.

Opportunities for the immediate utilization of molecular genetics or biology in plant breeding appear minimal. However, opportunities for long-term technology transfer exist if such developments can contribute to increases in selection efficiency.

REFERENCE

Wilson, D. M., and L. Sodek. 1970. Arch. Biochem. Biophys. 140:29.

DISCUSSION

DR. DOY: As the person who later in this workshop has the task of talking about some of the new approaches to plant genetics, I would like to say straightaway I agree with much of what we have just heard.

At present, we do not know how to select for improvement of plant seed proteins at the level of cells in culture. It is relatively simple to select for overproduction of certain amino acids, but this does not automatically imply improvement of protein quality, especially in the seed. What is needed is more fundamental information at the cellular level, and therefore it is necessary to fund relatively simple studies, including development of simple enzyme assays and methods of selection in culture.

DR. INGLETT: It has been over 10 years since the discovery of high-lysine corn by Mertz and his colleagues. Each year we dry-mill over 120 million bushels of corn. A lot of it goes into food and some into brewing. The primary purpose of much of the cereal breeding for nutrient content is for human food. I am wondering why we have not made greater progress in getting high-lysine corn in agronomically successful form.

DR. SPRAGUE: In my talk I told you that this whole evaluation, production, and evaluation cycle requires about 10 years. So, in that light we have made rather remarkable progress in getting high-lysine corn types into commercial production. Now, these are not as high yielding as we have every reason to believe they will be eventually, but plant breeding is a slow process. It is geared to the generation time of the particular species involved. There is no crash program you can initiate to reduce generation time, and, furthermore, up until now the development has been largely an attempt at a single-gene transfer. It made very limited use of additional genetic variability that undoubtedly exists in these populations. In order to do this, the plant breeders are going to have to have access to an amino acid analyzer, and obviously most of them do not have such access. They have had to depend on the work on opaque-2 and very largely on a phenotypic classification of these two types of kernels. So, again you come down to exactly the thing that I hoped that I stressed, and that is that if we are going to make continued progress in any of these areas, we have got to have better techniques than we have at the present time.

An amino acid analyzer is an expensive bit of machinery. It requires a technician to operate it, and when you consider this as an add-on to most of the plant-breeding operations, it is just completely out of the question. They are not that well financed.

DR. DOY: No one has said anything about the possibility of cutting down the breeding program time as a way one might be able to get at this. One way is haploid cell lines, which by chromosome doubling immediately give a homozygous situation that would normally take years and years of breeding. There are many difficulties involved, but I would like to make this point and ask Dr. Sprague to comment.

DR. SPRAGUE: This would depend entirely on the origin of the haploid → diploid lines. If the group of diploids to be supplied had been screened for some important rate-limiting enzyme activity, they would be of considerable interest. If the diploids were merely a random sample of possible homozygous lines, my interest would be rather limited. The major problem in breeding has never been the production of lines but rather the evaluation of lines. The evaluation of lines is a task that is time-consuming. Whether the lines must first be combined into hybrids or whether they may be evaluated directly, they must be subjected to yield evaluations under the range of environments to which they will be subjected in actual commercial production. It will likely also be necessary that such lines be evaluated for insect and disease reactions, quality factors, or other characteristics that make for commercial acceptability.

DR. MARCUS: I have two comments. I think that screening for specific characteristics at the level of an enzyme assay is probably the right thing to do in order to pick out your selection to begin with. The problem is twofold. One is for the person who has a vested interest when he starts the screening to learn when he has had enough. That is, just because he particularly likes an enzyme that he learned how to assay, to know when to stop assaying for it.

For high-lysine strains, another problem is testing in breeding programs to be sure that the hybrids keep the high-lysine characteristic. I gather a limiting feature of this is the availability of amino acid analyzers to look at the hydrolysates of the proteins. It seems funny to me that people have not developed a simpler assay for lysine. For instance, there is a lysine decarboxylase that one could purify from a bacterial source and perhaps make available as a kit. I just wonder why this kind of thing has not been looked at. It is not even high-powered molecular biology. Really, it is plant physiology.

DR. SPRAGUE: There are simple tests for both lysine and tryptophan. As far as I know, there is not an equally simple one for methionine or some of the limiting amino acids for the legumes.

DR. BOULTER: I would suggest that we now have assays for methionine and cysteine that, although not perfect from the viewpoint of an analytical chemist, are satisfactory as primary screens in breeding programs, so that we now have tests for four of the essential amino acids, including lysine and tryptophan.

II

BIOCHEMICAL AND BIOPHYSICAL LIMITATIONS

ROBERT RABSON
Session Chairman

Biochemical and Biophysical Limitations: Introduction

In this session we will begin to consider some of the physiological parameters influencing the yield of protein in seeds. Clearly, it is possible to alter the synthesis of protein in a specific genotype by manipulating the environment, for example, by supplying nitrogen fertilizer.

Ultimately, the farmer and the breeder need to know more about the physiological–genetic interactions to obtain the optimal yields of protein desired. Thus, we shall be examining the probable physiological and biochemical determinants of protein synthesis in the next two sessions.

As already mentioned by Dr. Sprague, it is hoped that another very important aspect of breeding will be considered: the possibility of obtaining better indicators of protein yield in both a quantitative and qualitative way that can be used as a tool by the breeder.

It is obvious that, unlike most characters that the breeder is searching for, there appear to be no good morphological features that the breeder may use in his quest for protein improvement. The problem is a chemical as well as a genetic one, and this adds a new and difficult dimension to the problem.

Environmental effects on protein quantity are great, and to pick out the genetic differences is, at times, a difficult task. To put it in terms of electronics, the genetic signal is weak, and the environmental noise is strong. Thus, it is appropriate to be thinking about ways of either amplifying the signal or diminishing the noise or both in order to assist the breeder in his analyses.

First we shall hear about the well-known enzyme nitrate reductase, which is responsible for the initial conversion of nitrate to nitrite in the many-stepped complex process of making protein from inorganic nitrogen. Dr. Hageman, our speaker, will examine the evidence bearing on whether the levels of the enzyme nitrate reductase in the plant are limiting in the overall process of protein synthesis in the seed. He will also discuss the possibility of using the analysis of nitrate reductase as an indicator of the production of high protein in the grain.

Our second paper will deal with the ideas of how amino acids are synthesized and how these syntheses are regulated. Conventional wisdom of molecular biology suggests that protein synthesis is controlled by the availability of ribosomes, messenger RNA, transfer RNAS, and other factors. Are there other levels of control of protein synthesis that may be associated with the regulation of synthesis of the constituent amino acids into protein, such as in the storage proteins? Dr. Miflin's talk will cover this possibility of additional control mechanisms. Recently, there has been renewed interest in the prospects of identifying induced chemical autotrophic mutants of plants in cell cultures that can then be grown into mature plants carrying higher levels of specific amino acids. Peter Carlson (1973. Science 180:1366) of Brookhaven National Laboratory has made an elegant demonstration of this in tobacco plants where a high-methionine line was picked out. Carlson's experiment may serve as a model for other experiments.

The interactions between protein synthesis and the stress conditions of temperature and moisture will be discussed by Dr. Boyer. It is a common observation that drought conditions will cause the plant to produce shrunken seeds, oftentimes at the expense of polysaccharide synthesis. I am also told by Paul Mattern (personal communication) that adverse environmental conditions may substantially alter the milling and baking qualities of wheat which are certainly reflections of the protein content.

R. H. HAGEMAN, R. J. LAMBERT,
DALE LOUSSAERT,
M. DALLING, *and* L. A. KLEPPER

Nitrate and Nitrate Reductase as Factors Limiting Protein Synthesis

NITRATE

The importance of nitrogen as a factor in limiting production of corn (*Zea mays* L.) (yield and percent protein) is illustrated in a gross way by the parallelism of nitrogen fertilizer usage and corn production in Illinois from 1950 to 1973 (Figure 1). The average production of soybeans (*Glycine max* L. Merr) (bu/acre) in Illinois is included for comparison because nitrogen is not applied directly to this crop. Since approximately 90 percent of the nitrogen used in Illinois is for corn production, the average application of nitrogen per acre in 1971 (year of maximum application) can be estimated at approximately 65 pounds. However, the rates of application of nitrogen are not uniform statewide. Because a bushel of corn grain contains approximately 1 pound of nitrogen, producers use a rule of thumb of applying 1 pound of nitrogen annually for every bushel of grain anticipated.

Although the increased application of fertilizer nitrogen is associated with increased grain yields, there has been little change in the percentage of crude protein (N) in the corn and wheat grain available for industry (Figure 1 and Table 1).

The concept that grain yields and grain protein content (percent) are negatively correlated, especially for wheat (*Triticum aestivum* L.), is widely held and supported by experimental evidence (Terman *et al.*, 1969). This negative correlation results primarily from management practices (amount and time of fertilizer applications, genotypes used,

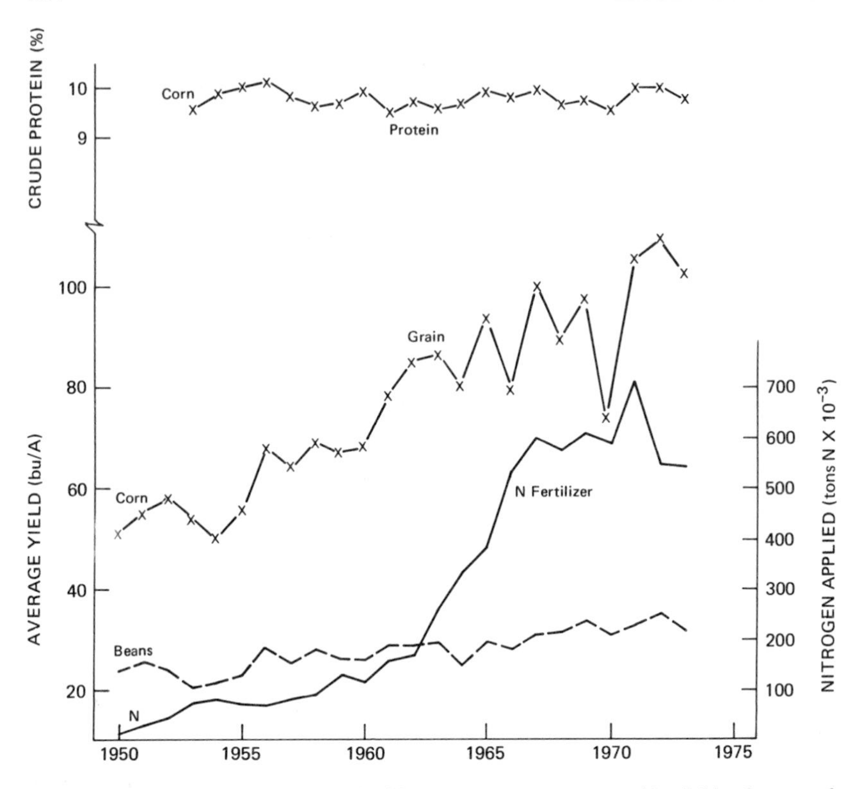

FIGURE 1 The trends in nitrogen fertilizer usage, average statewide yields of corn and soybeans, and percentage of corn grain over 23 years in Illinois. Approximately 90 percent of the fertilizer was applied to corn. The protein data represent the average protein content (dry basis) of corn processed by mills at Argo and Pekin, Illinois, and were supplied by J. Freeman, CPC International Inc., Argo, Illinois.

and environmental conditions); however, it is not obligatory. The prime cause of the negative correlation is the unavailability to the plant of soil nitrogen just before and during the reproductive phase. The lack of availability of nitrogen can be direct (exhaustion of soil nitrogen by the plant or competing systems) or indirect (lack of water, which limits availability and uptake and decreases nitrate assimilation).

Although the managerial practice of partitioning the spring applications of nitrogen required for maximum production of small grains has been developed in Europe (Van Dobben, 1966), applications of nitrogen made at late tillering and boot stages of development have not been used extensively in the United States. In the United States, nitrogen is

TABLE 1 Average Statewide Yields and Percentage Protein Content of Wheats Produced in Kansas, Nebraska, and Ohio

	Ohio		Nebraska		Kansas	
Year	bu/A	%	bu/A	%	bu/A	%
1952	25	10.0	22	—	14	11.1
1953	29	10.6	22	—	16	13.5
1954	27	10.8	19	—	17	12.3
1955	29	9.9	25	—	15	12.5
1956	26	10.3	20	—	19	14.1
1957	22	10.7	27	—	19	12.0
1958	31	9.8	33	—	27	11.8
1959	25	10.3	22	—	20	12.5
1960	35	9.9	29	—	28	11.5
1961	31	10.0	25	—	27	10.7
1962	32	10.8	20	—	24	11.7
1963	38	—	22	—	22	12.1
1964	33	—	25	—	23	12.2
1965	32	11.5	20	10.7	24	11.3
1966	39	11.0	35	11.7	20	12.2
1967	34	10.1	27	11.9	20	12.9
1968	37	10.4	32	11.3	25	11.7
1969	37	10.3	32	11.3	31	10.9
1970	38	10.1	38	11.6	33	11.5
1971	44	10.4	42	11.5	35	12.0
1972	45	9.9	37	11.1	34	11.5
1973	32	11.0	—	—	37	11.0

normally applied to wheat in the fall or early spring and in minimal amounts (50 lb/acre), because larger amounts of nitrogen applied at these times to many varieties currently used promotes excessive vegetative growth and increases susceptibility to lodging.

Production of wheat protein in the United States can be increased by late spring applications of nitrogen as shown by the data of Table 2 (Hucklesby *et al.*, 1971). The minimum increase in protein production from supplemental nitrogen treatments was 30 percent over control (Parker, 56 kg N/ha, applied April 2), and the maximum increase was 118 percent (Arthur, 224 kg N/ha, applied April 23). The average increase in protein over the respective controls, for all treatments and varieties, was approximately 65 percent or nearly 300 kg crude protein per hectare. The quality of the protein, as judged by amino acid composition, was unaffected by the nitrogen treatments. The nitrogen

TABLE 2 Effect of Spring Nitrogen Applications on Yield of Grain and Grain Crude Protein of Wheat[a]

Varieties and Date of Treatment	Rates of N (kg/ha)											
	0	56	112	224	0		56		112		224	
	Grain Yield (kg/ha)				Grain Protein %	kg/ha	%	kg/ha	%	kg/ha	%	kg/ha
Blueboy	4,700				10.0	470						
April 2		5,460	5,760				11.3	617	12.9	743		
April 23		6,140	5,760	5,320			11.5	706	13.8	795	15.3	814
May 9		5,700	6,400				12.0	684	14.3	915		
Arthur	3,820				10.4	397						
April 2		4,450	4,510				12.0	534	14.7	662		
April 23		4,450	4,950	5,140			13.2	587	15.5	767	16.9	869
May 9		3,960	4,200				14.6	578	16.9	710		
Parker	3,820				11.0	420						
April 2		4,450	4,510				12.2	553	12.2	640		
April 23		4,880	5,200	4,320			12.9	630	12.9	811	18.6	803
May 9		4,260	4,510				14.5	618	14.5	780		

[a]Hucklesby *et al.* (1971).

treatments did not alter the baking quality (cookie test) of the two soft red winter wheats (Blueboy and Arthur).*

In comparable experiments, average increases in protein production were 70 percent over controls with Ponca and Monon wheats (Croy and Hageman, 1970) and 40 percent over controls with Gage (Eilrich and Hageman, 1973). Thus, the treatment effect is reproducible.

Assuming that late spring applications of nitrogen would enhance protein production by the minimum value observed (Table 2), such treatments, when applied on a commercial scale, would increase wheat protein production in Illinois by 8×10^7 kg annually. The cost of producing this extra protein was minimal (approximately $0.15/kg N/ha, applied); however, the cost currently is about twice as high.

As shown by the data in Figure 1, there has been no appreciable change in the percentage protein content of commercial corn grain, in spite of the marked increase in nitrogen usage. Because the percentage composition has remained constant, the total protein production has increased in direct proportion with grain yield. It is also possible to increase the percentage of protein and total protein of corn grain by supplemental nitrogen applied 2 to 4 weeks after tassel initiation (Table 3). The average increase in protein production for all hybrids and nitrogen treatments was 21 percent over control, or 141 kg protein/ha. Hybrid 5 showed the greatest increases in protein production per hectare (a maximum of 296 kg/ha) and percentage composition (from 8.9 percent to 11.3 percent). Similar results were obtained with these same six hybrids when grown at a higher plant density (79,000 plants/ha). No significant increases in grain yield were obtained from any of the supplemental nitrogen treatments with any hybrid or plant population, because nitrogen adequate for the grain yields achieved was applied before planting. Thus, the increases in protein production resulted from changes in percentage composition. No evaluation of protein quality was made.

The increases in percent protein in corn after the massive applications of supplemental nitrogen are not as marked as those achieved in wheat with smaller amounts of supplemental nitrogen. Calculations based upon differences in total nitrogen between control and nitrogen-treated plants show that approximately 16 percent (54 kg/ha) of the supplemental nitrogen was taken up by the plants. It is not known whether the minimal utilization of supplemental nitrogen by corn is due to a decrease in uptake efficiency with the onset of the reproductive phase; senescence of the roots in the surface soil zone where nitrogen content

* Evaluation, courtesy of Dr. W. T. Yamazaki, Soft Wheat Quality Lab., Wooster, Ohio.

TABLE 3 Crude Protein Production and Percent Composition of Grain from Six Corn Hybrids in Response to Supplemental Nitrogen Treatments

	Time of Nitrogen[a] Application (days after planting)							
Hybrid	Protein (kg/ha)				Protein (%)			
Number	Control	44	59	72	Control	44	59	72
1	661	781	787	826	7.7	9.0	9.1	9.3
2	686	833	851	817	8.8	10.3	10.4	11.2
3	631	588	617	627	7.9	8.1	8.1	8.4
4	663	845	826	815	7.6	9.4	9.4	8.9
5	668	888	964	865	8.9	11.1	11.2	11.3
6	667	843	869	822	9.1	10.8	11.3	10.8

[a] Nitrogen (337 kg N/ha) was applied to separate plots on dates indicated. All plots received 5 cm irrigation water at each date of nitrogen application. Plant density was 59,300 plants/ha.

is highest; or higher soil temperatures that affect ion absorption or stimulate soil microbial competition for nitrate.

The use of 427 kg N/ha (90 kg N/ha applied before planting plus 337 kg N/ha supplemental) to achieve the corn protein production (Table 3) was excessive and most uneconomical. Although additional research is required to establish the fact, it is most likely that 150 kg N/ha (90 kg N before planting plus 60 kg N supplemental applied 60 days after planting) would have achieved comparable results.

It is concluded that management practices that would increase the availability of soil nitrogen to the plant at the onset and during the reproductive phase would provide substantial increases in corn and wheat protein production, especially the latter. Work with rice (*Oryza sativa* L.) (Matsushima, 1965) suggests that partitioning of nitrogen applications is probably the most effective way of increasing grain yields of higher protein content. Multiple applications of relatively small amounts of nitrogen rather than a single preplanting application of a large amount would also minimize loss to the environment. However, multiple applications of nitrogen are not used extensively in the United States because the cost of nitrogen has been low and the application costs high. The use of supplemental nitrogen applications is not considered practical under dry-land conditions where soil moisture levels would limit or prevent uptake. In reality, the grain producers will not be interested in extra protein production achieved by management practices that add to operational expense until grain is marketed on a protein rather than a volume basis. However, in the case of wheat, considering current management practices, the use of supplemental nitrogen in conjunction with improved varieties is considered economically feasible.

Production of wheat proteins per acre can also be increased by use of improved genetic types. For example, the short stiff-strawed types (e.g., Gaines and Mexican, developed by Vogel and Borlaug) can tolerate more nitrogen without excessive lodging. In contrast, varieties produced by crossing Atlas 66 with standard hard red winter wheats have a marked increase in actual protein content (Johnson *et al.*, 1963). In the initial crosses with Atlas 66, the higher protein content was associated with lower grain yields; however, varieties have been developed that retain the higher protein content and have the yield potential of the best commercial varieties (Johnson, 1972).

Corn genotypes with the potential for producing grain of high protein content are available (Dudley and Lambert, 1969). However, hybrids that produce grain with high protein content are not currently available for release for commercial production.

NITRATE REDUCTASE

The concept that nitrate reductase is a factor that limits production of cereal grain and grain protein is based upon the facts that nitrate is the primary form of soil nitrogen available to the plant and that in the assimilation of nitrate by the plant, nitrate reductase is the rate-limiting step between nitrate and amino acids (Beevers and Hageman, 1969). The level of nitrate reductase is a measure of the amount of available catalyst and reflects the genetic potential for synthesis and degradation of the enzyme. While the amount of enzyme is also a reflection of the genetic potential to supply nitrate to the tissue, the correlation between enzyme and nitrate content of the tissue is not close throughout most of the life cycle of the plant. The flux of nitrate to the enzyme induction site is probably the major factor regulating the level of nitrate reductase activity. It can be argued that the mechanism of protein synthesis, rather than amino acid supply, is the major factor regulating protein production. However, the pronounced effect of increased and supplemental nitrogen supply on vegetative growth, grain yield, and grain protein content suggests that nitrogen supply is the primary limiting factor under commercial field conditions. Because nitrate reductase is induced by substrate (NO_3^-), some investigators believe that nitrate reductase activity is not strongly related to nitrogen content of the grain and is only a reflection of availability of nitrate. While we are aware of the strong relationship between nitrate supply and nitrate reductase activity, we have been more successful in using nitrate reductase activity as a productivity index of grain and grain protein production than tissue nitrate content (Deckard, 1970). Genotypic differences in levels of nitrate reductase also have been found to be independent of nitrate content of the tissue (Purvis, 1972). Nitrate reductase activity can be correlated with nitrate content of leaf tissue only in the grain development phase of corn, when leaf nitrate content is extremely low (Deckard *et al.*, 1973).Nitrate reductase is also a much better criterion of plant response to drought stress than nitrate content of the tissue (Morilla *et al.,* 1972). For these reasons we have attempted to establish nitrate reductase as a biochemical criterion useful for the selection of superior cultivars.

A biochemical criterion should meet the following minimal requirements:

1. The method of evaluation of the criterion should be simple, rapid, and reproducible.
2. The criterion should exhibit variability among genotypes within a species.
3. The criterion should be highly heritable.
4. The criterion should show a reasonable relationship to productivity or quality of the product.

The *in vitro* assay for nitrate reductase (extraction and assay under optimal conditions) is only relatively simple and rapid; however, it is reproducible. A team of 3 or 4 persons can assay 100 to 120 samples per 8-hr day. The reproducibility of the assay is excellent, with the major variability resulting from the tissue-sampling procedure (plant-to-plant variability). An *in vivo* assay that utilizes excised tissue sections (Klepper *et al.*, 1971) can also be used to assay nitrate reductase. This assay is simpler than the *in vitro* assay in that it requires less equipment; it is as rapid, but it exhibits more variability with duplicate samples. Although the *in vitro* assays show 2 to 5 times more activity than the *in vivo* assays, correlations between the assay values obtained by the two methods are high ($r = +.85^{**}$ to $+.99^{**}$).* Even with the lower levels of activity measured by the *in vivo* assay, the estimated input of reduced nitrogen based on enzymatic assay exceeds the actual accumulation of reduced nitrogen by the plant. Thus, neither the *in vivo* nor the *in vitro* assay is a true reflection of the *in situ* activity.

* Two asterisks following a correlation coefficient indicate that the value is significant at the 0.01 level. A single asterisk indicates significance at the 0.05 level.

Nitrate reductase activity varies markedly among genotypes within a species. Reports show that activities vary up to fivefold among corn inbreds (Zeiserl and Hageman, 1962), and twofold for corn hybrids, wheat varieties, and Sudan grass [Sorghum *vulgare sudensis* (Piper) Stapf] varieties (Zeiserl *et al.*, 1963; Eck and Hageman, 1974).

The level of nitrate reductase activity is highly heritable. Schrader *et al.* (1966) observed that F_1 hybrids obtained by crossing two inbreds with relatively high nitrate reductase activity were high in activity but no higher than the parental inbreds. Crosses between parents of high and low nitrate reductase activity produced hybrids with intermediate activity. Hybrids from inbreds low in activity possessed either a low level of activity or heterotic levels of activity. This superior ability (heterosis) to reduce nitrate existed from seedling stage through maturity. Studies of this heterotic effect with two corn inbreds and their progeny (F_1, F_2, F_3, and F_4 generations) indicated that the level of activity was under the control of two loci (Warner *et al.*, 1969). One locus appeared to affect the rate of enzyme synthesis while the other locus altered stability or decay rate. It should be emphasized that these studies were based on inheritance of levels of activity and were not concerned with structural genes. Work with *Neurospora* (Sorger and Giles, 1965) and *Aspergillus nidulans* (Pateman *et al.*, 1964) mutants indicate that four and six loci, respectively, are involved in the synthesis of nitrate reductase.

General combining ability (GCA) and specific combining ability (SCA) effects were estimated for a diallel set of crosses using 10 parents and 45 hybrids. Statistical analyses developed by Gardner and Eberhart (1966) for estimating parameters of this type were used. Table 4

TABLE 4 The General and Specific Combining Ability Mean Squares of Nitrate Reductase Activity in a 10-Parent Diallel with 45 Hybrids Grown at DeKalb, Illinois, 1973

Source of Variation	Degrees of Freedom	Mean Square
Entries	54	55.70[a]
Parents	9	48.61[a]
Parents versus hybrids	1	195.27[a]
Hybrids	44	53.97[a]
General combining ability	9	111.67[a]
Specific combining ability	35	39.14[a]
Error	108	11.01

[a] Significant at .05 level.

illustrates that both GCA and SCA effects are important in this set of crosses. However, the GCA mean square is about three times greater than the SCA. The magnitude of GCA effects relative to SCA indicates that selection for increased nitrate reductase activity should be effective in a synthetic variety made up of these inbred lines. In addition, the difference between the parent and hybrid mean square was also large, indicating the importance of heterosis for nitrate reductase activity in this set of crosses. Table 5 illustrates the GCA effects and heterosis effects for each parent line in the diallel set. The data indicate that line 2 and line 8 could be used as testers in a selection program for nitrate reductase activity.

Correlations between nitrate reductase activity and production of grain and grain protein have been variable. It is not known whether it is difficult to establish such a correlation because nitrate reductase is not a good criterion or because the actual production of a given variety cannot be accurately evaluated with respect to time and environment.

Nitrate reductase was first tested as a predictive criterion of yield with four corn hybrids (Illinois 1996 and Hy2 × Oh7, reported as high yielders; and WF9 × Oh7 and WF9 × C103, reported as low yielders, especially at high plant populations). While nitrate reductase assays supported the reported yield evaluation for all plant populations tested, there was no difference among the four varieties in both 1960 and 1961 (Zeiserl *et al.*, 1963). However, in 1964 trials with

TABLE 5 General Combining Ability and Heterosis Effects of Nitrate Reductase Activity in a 10-Parent Diallel with 45 Hybrids

Line	$\overline{H} = 2.82$	GCA[a]
1	+0.31	+2.22
2	+1.65	+5.18
3	−0.33	+0.45
4	−1.10	−4.08
5	−0.51	−2.72
6	−2.69	−3.11
7	−0.50	−0.71
8	+3.24	+3.50
9	−0.09	+1.23
10	+0.02	−1.96

[a] General combining effects for each line.

Hy2 × Oh7 and WF9 × C103, nitrate reductase activity, grain yields, and grain protein production were significantly related (Schrader and Hageman, 1965).

In experiments with wheat, Monon and Ponca were selected because they were reported to be high and low protein producers, respectively. Again, nitrate reductase assays agreed with the predicted performance, but actual protein production in 1965 and 1966 did not (Croy and Hageman, 1970). Monon appeared to have a more efficient enzyme than Ponca, in that it accumulated a larger amount of grain and grain protein per unit of nitrate reductase activity. However, for each genotype the seasonal input of reduced nitrogen (kg/ha) estimated from enzyme assays was significantly correlated with grain protein when nitrate fertilizer was the variable. (Seasonal input was estimated by measuring nitrate reductase in samples representative of the entire canopy taken at weekly intervals throughout the season.) This work also suggested that transport of vegetative nitrogen to the grain was an independent genotypic variable.

Eilrich and Hageman (1973) found that with field-grown Arthur wheat, the input of reduced nitrogen estimated from nitrate reductase activity of the entire canopy was significantly related to the amount of reduced nitrogen (determined by Kjeldahl assay) of the total shoot tissue (r values of +.89** for March 30 to April 20, +.81** for April 20 to 28, +.82** for April 28 to May 9, +.75** for May 9 to 24, and +.53* for May 29 to June 7). In this same study, the total seasonal nitrate reductase activity correlated significantly with grain yield and grain protein (kg N/ha) production (r values = +.94** for grain yield and + .87** for grain protein) when nitrogen fertilizer was the variable. Similar results were obtained with Ottawa wheat.

These studies are open to criticism because nitrogen supply was a variable and because the relationship between nitrate reductase levels and grain protein production did not hold among genotypes [e.g., Monon and Ponca (Croy and Hageman, 1970)]. Perhaps it is not surprising that the relationship did not hold for Ponca and Monon (soft and hard red winter wheats), not only because many factors (disease, lodging, insects, weather) play a role in the production of grain and protein under field conditions but because the two varieties differed in genetic composition and, presumably, in metabolic functions. To examine the effect of genetic diversity, Eilrich (1968) compared the relationship between enzyme activity and nitrogen components of 14 wheat varieties, 10 of which were closely related genotypes (derived from Atlas 66 × Comanche cross, supplied by V. A. Johnson, ARS, University of Nebraska, Lincoln). The other four varieties were of diverse origin. The correlation coefficient between seasonal enzyme activity and grain nitrogen (kg/ha) for the 10 related varieties was $r = +.57^*$, in contrast to a nonsignificant $r = +.38$ when all 14 varieties were compared.

This study also revealed two other factors that would interfere with the relationship between enzyme activity and grain protein. The first factor is illustrated by the variation in the percentage of total nitrogen retained by the straw as a function of genotype (Table 6). The relative proportion of nitrogen ultimately "transported" to the grain was less variable among the 10 related genotypes than among the diverse genotypes. The effect of this "translocation efficiency" on the relationship between enzyme activity and grain protein is reflected by the higher correlation ($r = +.84^{**}$) obtained by omitting from the calculations three of the related genotypes (4, 7, 11) that showed low

TABLE 6 Estimated Cumulative Seasonal Input of Reduced Nitrogen, Compared with Actual Nitrogen in Straw and Grain at Maturity for 14 Wheat Varieties

Variety No.	Seasonal[a] Input (kg/ha)	Reduced-N (kg/ha)[b]		Input-N/ Actual-N Ratio	Straw-N (%)	Total Plant-N in Straw (%)
		Grain	Total			
Related Varieties—Atlas 66 × Commanche						
6	837	86	113	7.4	0.35	24
2	798	93	128	6.2	0.42	27
7	687	80	116	5.9	0.45	31
1	680	85	115	5.9	0.37	26
9	644	76	107	6.0	0.41	29
11	641	61	98	6.6	0.54	38
4	638	70	103	6.2	0.41	32
15	616	81	103	6.0	0.31	22
14	553	79	101	5.5	0.31	22
5	512	74	103	4.9	0.39	29
Nonrelated Varieties						
10	736	56	98	7.5	0.55	43
8	689	93	122	5.6	0.37	24
12	433	57	110	3.9	0.60	48
3	332	74	113	2.9	0.57	35

[a] Estimated from nitrate reductase assays.
[b] Determined by Kjeldahl assay; the total represents the nitrogen content of both grain and straw.

"transport efficiency." As can be deduced from the data in Table 6, there are two components involved in this transport efficiency: the total amount of straw and the percentage of nitrogen. Lodging causes an increase in the amount of nitrogen retained by the straw and thus can be a major factor in the relationship between enzyme activity and grain protein.

The second factor is illustrated by the variation in the ratio of cumulative seasonal input of reduced nitrogen estimated from enzyme activity to the actual reduced nitrogen in grain and straw at maturity as a function of genotype (Table 6). In a general way, this ratio value can be used as a measure of "enzyme efficiency." The degree of variation in enzyme efficiency was greater for the diverse varieties than for the related varieties. This overestimation (higher input/actual values) was expected because the *in vitro* assays were made under optimal conditions (temperature, substrate, and cofactor). Such conditions would rarely be obtained *in situ*. In some instances, correction for temperature differences alone would reduce the ratio values by half. The variability in enzyme efficiency among the genotypes is much more disturbing than the overestimation. Such factors as availability of nitrate (specifically, the rate of influx to the induction and assimilation sites) and the availability of energy sources for reduction could account for the higher ratio values. The presence of inhibitors of nitrate reductase that are released on homogenization of the tissue (Schrader *et al.*, 1974) or instability of the enzyme after homogenization could also cause the low values.

Data obtained with five Australian wheat varieties that differed in date of maturity, vegetative mass, and grain protein showed a significant relationship between input of reduced nitrogen estimated from nitrate reductase assays and actual vegetative nitrogen from seedling to booting stage. The individual r values were +.92** for Olympic, +.95** for Argentine IX, +.98** for Timgalen, +.98** for Gatcher, and +.97** for Petit Rojo. A significant relationship was also noted between seasonal input and actual nitrogen in the straw and grain at maturity with all five varieties (Figure 2*A*). However, a significant relationship between enzyme activity and grain nitrogen was not observed without allowing for transport efficiency. The difference in translocation efficiency among the genotypes is shown in Table 7. When the varieties were grouped according to the ratio of grain nitrogen to total plant nitrogen (translocation efficiency), a significant relationship was observed between enzyme activity and grain protein production for both groups (Figure 2*B*).

The data (Table 7) also show that the seasonal input of nitrogen

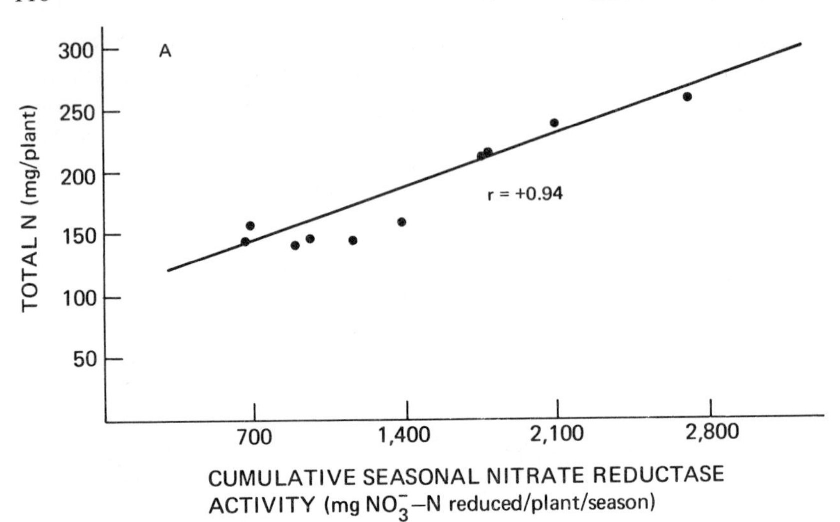

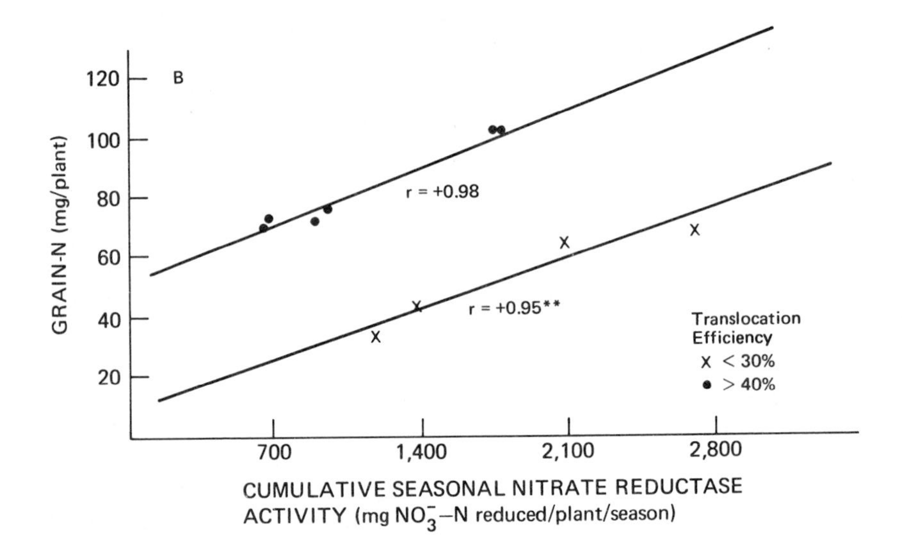

FIGURE 2 Relationship between input of reduced nitrogen estimated from nitrate reductase activity and total nitrogen content of the shoot and grain at maturity (*A*) and total grain nitrogen (*B*), for five varieties of Australian wheat. The varieties were grouped according to translocation efficiencies, estimated by ratio of grain nitrogen to total nitrogen in straw and grain in ***B***.

(enzyme) exceeded actual nitrogen accumulation and that the ratio values (enzyme efficiency) varied as a function of genotype.

Deckard *et al.* (1973) determined the relationship between nitrate reductase activity and grain yields and grain protein of high-performance commercial corn hybrids. The 48 variables were 6 hybrids, 2 plant populations (59,300 and 79,500 plants/ha), and 4 levels of supplemental nitrogen. The amount of nitrogen initially applied to all plots was high, as supplemental nitrogen applications had no effect on grain yield. The correlation coefficients between seasonal average nitrate reductase activity and grain yields and grain nitrogen, obtained by pooling all data, were $r = +.60^{**}$ and $r = +.62^{**}$, respectively. When enzyme activity was expressed as the cumulative seasonal input (to correct for differences in amount of canopy) the r values were $+.58^{**}$ for yield and $+.51^{**}$ for nitrogen. When the data were grouped by nitrogen treatments, correlations between seasonal averages of enzyme activity and (a) grain yields of the 6 hybrids at both populations and (b) grain nitrogen of 5 of the hybrids were as shown in Table 8. The correlations with grain nitrogen were not significant when computed with all 6 hybrids because one of the hybrids (No. 6, Table 9) exhibited a low ratio value of input N/actual N (high enzyme efficiency).

The enzyme efficiencies (estimated input N/actual N) for the six corn hybrids ranged from 2.2 to 4.1 (Table 9) and thus are generally lower and less variable than the enzyme efficiencies for wheat. In spite of the marked differences in enzyme efficiency of hybrids 5 and 6, increases in enzyme activity (obtained with supplemental nitrogen treatments) were closely associated with increases in grain protein for both hybrids 5 and 6 ($r = +.90^{**}$ and $+.83^{**}$, respectively). Since there was no marked difference in total residual nitrogen in the vegetation at maturity, or in nitrate content in leaf blade or stalk throughout the season, for these two hybrids (Deckard, 1970) neither translocation or nitrate availability can be cited as causal factors for the differences in enzyme efficiency.

If a single hybrid from the six commercial hybrids were to be chosen on the basis of actual yield of grain and protein obtained in a single trial at one location (Table 6), hybrid 1 would have been selected for yield and hybrid 5 for protein production. Both of these hybrids had high nitrate reductase activity.

These studies with corn and wheat (Croy and Hageman, 1970; Deckard, 1970; Eilrich, 1968) were undertaken only to show the degree of relationship between nitrate reductase activity and grain yields and grain protein production. The determination of seasonal input of

TABLE 7 Estimated Seasonal Input of Reduced Nitrogen from Enzyme Assay, Total Reduced Nitrogen Content in Shoot Material plus Grain at Maturity, Grain Nitrogen and Calculated Translocation Efficiency for 5 Australian Wheats

Genotype	Seasonal input (mg/plant/season)	Input-N/Actual-N Ratio	Total-N (mg/plant)	Grain-N (mg/plant)	Efficiency (Grain-N ÷ Total-N)
Timgalen	1302	8.6	151	38	0.25
Olympic	2408	9.7	248	67	0.27
Petit Rojo[a]	1778	8.3	213	102	0.48
Gatcher	672	4.5	149	72	0.49
Argentine IX[a]	927	6.5	143	74	0.52
LSD (0.05)	—	—	14	7	0.26

[a] Classified, agronomically, as high-protein varieties.

TABLE 8 Seasonal Average of Enzyme Activity: Correlation with Grain Yield and Grain Nitrogen[a]

	Correlation (*r*) with Seasonal Average			
		Supplemental Nitrogen Treatment (days after planting)		
	Control	44	59	72
Grain yield (6 hybrids)	+.46	+.72**	+.73**	+.55
Grain nitrogen (5 hybrids)	+.76	+.89**	+.86**	+.94

[a] Data from Deckard *et al.* (1973).

reduced nitrogen or even seasonal average levels of activity is far too difficult to be of use as a plant breeding tool. However, the finding by Deckard *et al.* (1973) that enzyme activity (expressed as activity per gram fresh weight or as daily input of reduced nitrogen) measured during the reproductive period was more closely correlated with grain yields and grain protein than activity measured during the vegetative phase drastically reduces the amount of sampling required to permit genetic selection. These studies also suggest that the precision of the assay as a selection tool is also improved by the use of related genotypes. Further, the development of the *in vivo* assay (Klepper *et al.*, 1971) should eliminate some of the problems (inhibitors, enzyme instability) encountered in the *in vitro* assay.

Two experiments with partial diallel sets of corn grown at the University of Illinois, Urbana, and at DeKalb Agricultural Research Inc., DeKalb, Illinois, were made in 1973 to test these new developments. In both experiments plants were sampled twice, with a 7- or 10-day interval between samplings during the postsilking period. The ear leaves from five plants were composited per sample, and duplicate samples were taken from each plot. Nitrate reductase activity was measured by the *in vivo* assay procedure and expressed as relative units of activity obtained by integrating the area delineated by the two sets of assay values.

The data from the Urbana experiment is summarized in Tables 10 and 11. The correlation coefficients between enyzme activity and grain yield and protein production for all 36 hybrids of +.41* and +.33 are low; however, they are probably as high as can be expected between a single metabolic factor and such complex processes as grain and

TABLE 9 Nitrate Reductase Activity as Cumulative Seasonal Input, Grain Yields, Total Nitrogen Content of the Vegetation and Grain at Maturity, Ratio Values of Input to Actual Nitrogen, and Estimated Translocation for Six Corn Hybrids[a]

Variety	Grain Yields (kg/ha)	Grain-N	Estimated Input (kg N/ha/season)	Input Actual (kg N/ha)	Input-N/Actual-N Ratio	Translocation (Grain-N ÷ Actual-N)
5	9,166	132	745	182	4.1	0.72
4	10,370	125	667	169	4.0	0.74
1	10,337	123	615	167	3.7	0.74
2	9,258	125	515	163	3.1	0.77
3	8,426	93	416	146	2.8	0.64
6	8,697	124	364	170	2.2	0.73

[a] Values are the average of two populations and four treatments.

TABLE 10 Relative Nitrate Reductase Activity, Grain Yield, and Grain Protein for 36 Corn Hybrids (Urbana, Illinois, 1973)

Genotype[a]	Relative NR Activity	Grain Yield (bu/A)	Grain Protein (kg/ha)
1 P-338[b]	16.0	171	955
2 FRB7 × B73[b]	14.3	189	949
3 Mo17 × H96	12.4	178	994
4 P 3517[b]	12.0	181	1011
5 B14 × B37	12.0	136	802
6 A 19775B[b]	11.6	144	732
7 B37 × Oh43	11.3	161	909
8 B37 × N28	11.1	164	937
9 P 3369A[b]	11.1	179	831
10 B37 × R177	10.7	166	854
11 B14 × Oh43	10.5	180	1107
12 B14 × N28	10.2	143	861
13 WF9 × Oh43	10.0	154	880
14 WF9 × C123	9.9	153	873
15 B37 × C123	9.7	167	860
16 A 19793[b]	9.5	176	972
17 Mo17 × B73	9.4	193	1005
18 A 19792A[b]	9.4	158	912
19 WF9 × N28	9.3	154	889
20 B37 × H95	9.3	145	765
21 B37 × Mo17	9.1	158	916
22 Mo17 × N28	8.8	173	956
23 DXL81[b]	8.8	154	899
24 B14 × C123[b]	8.6	157	917
25 DXL64[b]	8.5	177	978
26 B14 × Mo17	8.4	155	901
27 WF9 × C103	8.3	142	838
28 FR4A × Mo17[b]	8.2	173	923
29 Mo17 × H95	7.8	173	955
30 D805A[b]	7.0	141	779
31 WF9 × R177	6.8	152	887
32 Mo17 × R177	6.7	146	797
33 Mo17 × Oh43	6.6	162	911
34 Vu26 × FR632[b]	6.6	161	869
35 WF9 × H95	6.5	140	755
36 B14 × WF9	6.2	128	803
Average	9.5	161	893

[a] Listed in order of decreasing nitrate reductase activity.
[b] Commercial hybrid, parentage is proprietary.

TABLE 11 Relationship between Relative Nitrate Reductase Activity and Grain Yield and Grain Protein of 36 Corn Hybrids Grown at Urbana, Illinois[a]

Hybrid Group and No.		Regression Equation ($y = ax + b$)	Correlation Coefficient r
		Grain Yield	
All	36	$y = 2.9^{*} x + 133$	+.41*
Commercial	12	$y = 2.3\ x + 144$	+.43
B14[b]	5	$y = 2.8\ x + 124$	+.31
B37[b]	7	$y = 2.6^{*} x + 130$	+.88*
Mo17	9	$y = 3.9\ x + 135$	+.48
WF9	7	$y = 4.9^{*} x + 105$	+.74
		Grain Protein	
All	36	$y = 11.9^{*} x + 781$	+.33
Commercial	12	$y = 8.7\ x + 815$	+.29
B14+	5	$y = 15.1\ x + 754$	+.27
B37+	7	$y = 21.1\ x + 662$	+.59
Mo17	9	$y = 23.7\ x + 722$	+.65*
WF9	7	$y = 20.8\ x + 836$	+.67

[a]The hybrids were also grouped into five groups, the commercial hybrids and four groups with one parent in common. The specific hybrids in each group are given in Table 9.
[b]B14 × B37 was arbitrarily omitted from the calculations.

protein production. When the hybrids were divided into groups, each with one parent in common, some increase in the degree of association was observed in a majority of the groups (Table 11). Additional data are needed to determine if the higher coefficients observed within some groups (e.g., WF9, Mo17, and B37) are reproducible. If reproducible, this would indicate a dominant effect, probably an enzyme efficiency, of the common parent.

The variation in the values of the regression equation calculated for the different groups of hybrids (Table 11) indicate a twofold-to-threefold difference in enzyme efficiency. This difference in efficiency is probably more apparent than real and suggests that the *in vivo* assay as currently used (NO_3^- added to the assay) is not superior to the *in vitro* assay in predicting *in situ* nitrate reduction, especially among diverse genotypes.

If nitrate reductase assays had been used as the sole criterion to select the 12 best hybrids from this group of 36, the 12 selected would have had an average yield of 166 bu/acre and an average protein production of 911 kg protein/ha. Such a selection would have included three of the four hybrids with the highest grain yield and two of the best protein producers, but would have included three genotypes with yield and protein production lower than the average for the 36 hybrids. In comparison, the 12 commercial hybrids have an average grain yield of 167 bu/acre and an average protein production of 903 kg/ha. The commercial group included 5 of the top 10 hybrids with respect to yield and protein production but also included 5 hybrids with yields and protein production lower than the average for the 36 hybrids.

The data for the 10 corn inbreds and the partial diallel (45 hybrids) from the DeKalb experiment are summarized in Tables 12 and 13. The correlation coefficients between enzyme activity and grain yields and protein production for the 45 hybrids were +.35 and +.04, respectively. Reasons for this lack of relationship, especially with protein, could be due to one or more of the following reasons:

The plants were exposed to a period of water stress (drought) during the reproductive phase that could have interfered with accumulation of protein in the grain.

The first sampling was made too early in the reproductive phase. The enzyme activity values were higher in some instances at the second sampling than at the first sampling. [The significant correlations between enzyme activity and grain yield and protein on single sampling dates occurred when nitrate reductase activity was in the declining seasonal phase (Deckard *et al.*, 1973)].

The genetic diversity among the 45 hybrids.

When the hybrids were separated into groups, each with one parent in common, the degree of relationship between enzyme activity and grain yield and grain protein increased markedly for most groups (Table 13). Enzyme efficiency varied markedly among the 10 groups for both grain and protein production. In general, the data show that increases in nitrate reductase activity are associated with increases in yield and protein production and that grouping the hybrids to minimize genetic variability increases this association.

If nitrate reductase had been used as the sole selection criterion to select the 10 best and 10 poorest hybrids, the following results would have been obtained: The top 10 hybrids had an average yield of 118 bu/acre and produced 743 kg protein/ha. This group would have included 2 of the best 10 hybrids with respect to grain yield and 3 of the best 10 hybrids with respect to protein. The poorest 10 hybrids had an

TABLE 12 Relative Nitrate Reductase Activity, Grain Yield, and Grain Protein for 10 Inbreds and a Partial Diallel of 45 Hybrids[a]

Genotype	Relative NRA	Yield (bu/A)	Protein (kg/ha)	Genotype	Relative NRA	Yield (bu/A)	Protein (kg/ha)
2	22.4	61	451	1 × 3	19.1	131	823
1	19.2	69	515	3 × 9	18.7	138	790
9	18.0	84	557	7 × 8	18.5	80	500
3	16.9	99	638	3 × 5	18.2	131	833
8	15.8	92	701	9 × 10	18.2	113	697
7	14.9	90	614	5 × 8	17.6	111	684
6	14.5	84	588	3 × 6	17.2	103	643
10	11.4	70	593	3 × 7	17.0	122	728
5	10.9	95	663	2 × 5	16.8	100	680
4	9.4	87	649	1 × 6	16.8	112	755

2 × 7	27.3	139	844
1 × 8	25.8	118	779
1 × 2	23.6	124	798
2 × 9	23.5	118	709
1 × 9	22.9	130	738
3 × 8	22.6	122	736
8 × 10	22.1	104	698
2 × 10	21.9	102	691
2 × 3	21.8	118	756
2 × 4	21.0	126	836
2 × 6	20.8	103	667
8 × 9	20.6	97	594
1 × 7	20.2	135	868
7 × 9	19.9	121	725
1 × 5	19.6	119	812
6 × 8	19.6	90	601
5 × 9	19.2	137	812
3 × 4	16.6	129	792
1 × 10	16.4	110	757
7 × 10	15.8	102	701
3 × 10	15.7	100	664
6 × 9	15.7	111	692
4 × 9	14.4	120	700
5 × 10	14.0	105	703
6 × 7	13.8	92	575
5 × 7	13.1	115	754
5 × 6	12.6	87	564
4 × 10	12.1	100	663
4 × 7	12.0	128	797
6 × 10	11.5	98	640
4 × 5	10.6	95	663
4 × 6	10.6	101	640
Average	18.1	113	720

[a] Data arranged in order of decreasing nitrate reductase activity (NRA). DeKalb, Illinois, 1973.

TABLE 13 Relationship between Nitrate Reductase Activity and Grain Yield and Grain Protein of 45 Corn Hybrids Grown at DeKalb, Illinois[a]

Hybrid Group	and No.	Regression Equation	Correlation Coefficient r
		Grain Yield	
All	45	$y = 1.2^{*}\ x + 91$	+.35
1	9	$y = 0.3\ x + 118$	+.01
9	9	$y = 0.4\ x + 113$	+.08
10	9	$y = 0.5\ x + 96$	+.36
6	9	$y = 0.5\ x + 92$	+.21
8	9	$y = 1.1\ x + 80$	+.32
3	9	$y = 1.5\ x + 94$	+.27
7	9	$y = 1.7\ x + 85$	+.40
2	9	$y = 1.9\ x + 72$	+.48
4	9	$y = 2.8^{**}\ x + 75$	+.67*
5	9	$y = 3.8^{*}\ x + 52$	+.75*
		Grain Protein	
All	45	$y = 6.2\ x + 608$	+.04
1	9	$y = -4.2\ x + 887$	+.25
9	9	$y = 0.4\ x + 710$	+.02
10	9	$y = 3.5\ x + 633$	+.40
6	9	$y = 4.8\ x + 568$	+.29
2	9	$y = 6.3\ x + 597$	+.31
3	9	$y = 7.4\ x + 615$	+.26
7	9	$y = 7.9\ x + 583$	+.31
8	9	$y = 10.1\ x + 450$	+.44
4	9	$y = 17.8^{*}\ x + 487$	+.69*
5	9	$y = 18.7^{*}\ x + 429$	+.69*

[a] The hybrids are also grouped into 10 groups, each group with one parent in common.

average yield of 104 bu/acre and produced 670 kg protein/ha. This group would have had only one of the top 10 grain producers and none of the top 10 protein producers. A group of 22 hybrids with the highest enzyme activity would have included 7 of the best-yielding hybrids and 8 of the top 10 protein producers.

In both experiments (Urbana and DeKalb), the lower correlations between enzyme activity and grain protein than between activity and grain yields were not expected and are puzzling. However, the data in Figure 1 show that protein composition in corn has remained relatively constant throughout the past 20 years, although fertilizer nitrogen usage and grain yields have increased. In general farming practice, nitrogen is applied either before planting or during the early vegetative stage. The work of Deckard *et al.* (1973) shows that, although supplemental nitrogen applied to highly fertile plots at 44, 59, and 72 days after planting did not increase grain yields, such applications did significantly increase grain protein (both percentage and amount). In this experiment, irrigation water was applied to all plots each time fertilizer was applied. It is also known that a considerable proportion of the grain protein is deposited in the late stages of grain development. The early application of nitrogen at both DeKalb and Urbana coupled with the droughty conditions in August and September at DeKalb could have limited the amount of nitrate available to the plant. Unfortunately, tissue nitrate was not assayed in the experiments; however, other work (Deckard *et al.*, 1973; Zeiserl *et al.*, 1963) has shown leaf blade nitrate to be extremely low throughout the reproductive period. Thus, it is believed that in these two experiments and under general farming conditions, high-nitrate-reductase corn plants are unable to express their full genetic potential for protein production because of lack of substrate (nitrate).

The observation that the 10 DeKalb corn hybrids with lowest enzyme activity were low producers of grain and protein suggests that normal selection procedures also discard the low-nitrate-reductase lines, albeit unknowingly. Additional support for this concept is provided by a study with wheat that compared the class distribution (based on levels of nitrate reductase activity) of 97 F_2 genotypes with the class distribution of 87 parental varieties. Nitrate reductase assays (*in vivo*) were made with 3,546 individual F_2 seedlings and with replicated bulked samples of the 87 varieties. The 87 varieties were derived from the world wheat collection and from Mexican dwarf wheats. A total of 896 individual F_2 seedlings (25 percent) had activities lower than any variety (Figure 3). The greatest number of F_2 seedlings fell in the 2–3 class range (μmoles NO_2^- reduced/g fresh wt/hr), while the greatest number of the varieties fell in the 3–4 class range. The absence of the low (1–2) class range among the 87 varieties shows that the normal selection procedures (yield, morphology, baking quality) used in selecting the varieties also discarded the low-nitrate-reductase lines. This suggests that high nitrate reductase activity is a more desirable trait than low nitrate reductase activity for grain yield as well as protein production. In contrast, only 24 of the 3,546 individual seedlings (0.67 percent) had activities higher than the highest parental varieties. This indicates that varieties could be selected and developed from this genetic material with higher levels of nitrate reductase activity.

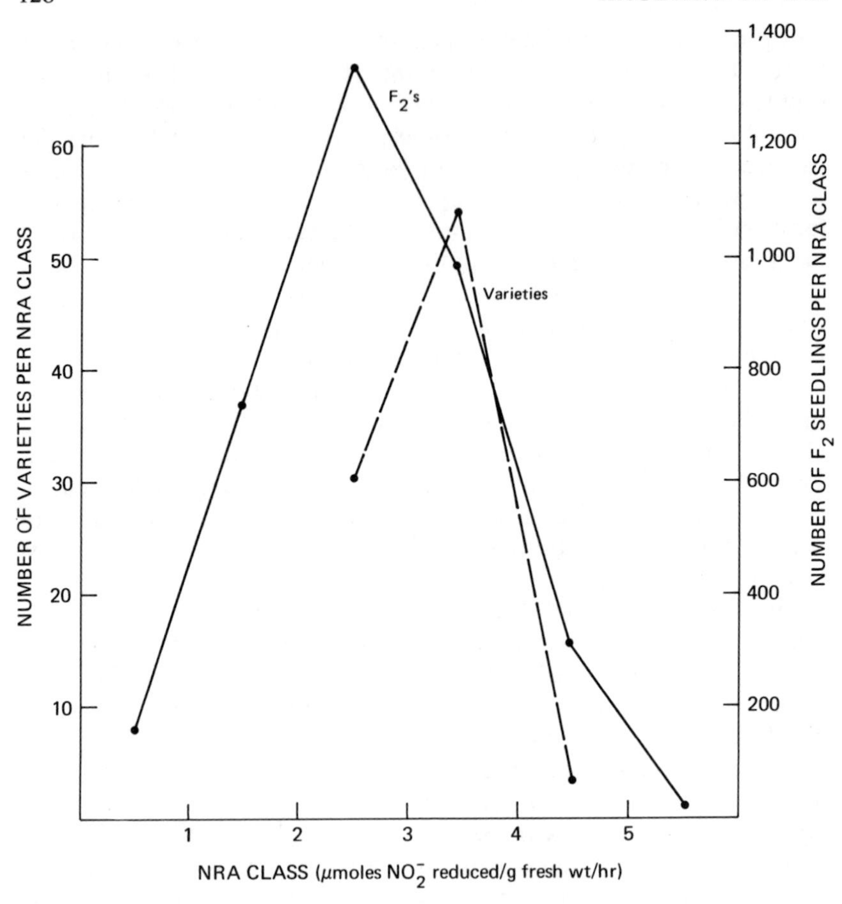

FIGURE 3 A comparison of the class distribution based on nitrate reductase activity (NRA) of 97 F_2 wheat genotypes with the 87 parental varieties from which the F_2 genotypes were derived.

Nitrate reductase as a selection criterion would be of greatest use if nondestructive evaluations could be made on seedling material. That this might be possible was first suggested by the findings of Zeiserl *et al*. (1963) and Warner (1968; Warner *et al*., 1969), who observed that the level of activity in corn seedlings reflected the level of activity throughout the season. Croy and Hageman (1970) compared the level of nitrate reductase activity at the seedling stage for 32 winter wheat varieties with their grain protein production the following summer. While no significant correlations were obtained, selection of the 16 varieties with highest nitrate reductase activity would have included 9 of the highest-protein-producing varieties. Eilrich (1968) conducted a similar study with 14 wheat varieties, 10 of which were related varieties derived from Atlas 66 × Comanche cross. The *in vitro* assays were made on growth-chamber-grown seedlings, and the grain nitrogen values were obtained from field-grown plants. Correlation coefficients between nitrate reductase activity (μmoles NO_2^- reduced/g fresh wt/hr) and percent and amount grain nitrogen were $r = +.60^*$ and +.43, respectively, for all 14 varieties.

The relationship between nitrate reductase activity in wheat seedlings grown in the field at Lincoln, Nebraska, and the percentage grain protein of comparable genotypes grown in the field at Yuma, Arizona, are shown in Figure 4. The following procedure was used to select the 53 genotypes tested. Each lot of F_2 seed in the original group was made from individual plants that were selected primarily for high grain yield and desirable morphological traits. [This original selection was made at the Centro Internacional de Mejoramiento de Maiz y Trigo (CIMMYT), Mexico.] From this large group of genotypes, a second selection was made for F_2 seeds with high protein. This selection process was repeated for the F_3 generation, with selection being made first for yield and morphology and second for protein, to provide the 50 genotypes (F_4 seed) used in the experiment. The other three entries were Atlas 66, Inia 66 and Nap Hal.

Although the correlation is nonsignificant, the general trend shown by the data in Figure 4, in conjunction with the high levels of nitrate reductase activity in Atlas 66 and Nap Hal, known progenitors of high-protein wheats, show that nitrate reductase has potential as a selection tool for developing superior wheats. These data also suggest the validity of screening genotypes of the world collection as an initial step in a program designed to develop wheats of high protein content.

Currently, investigations are under way to determine whether relative performance of corn genotypes can be assessed by measurement of nitrate reductase activity at the seedling stage. Successful development of the seedling assay could drastically reduce field evaluations. In view of the high general combining ability of nitrate reductase, evaluation of the breeding stock for nitrate reductase provides a predictive basis for making crosses. Alternatively, the results obtained from the 1973 experiments at Urbana and DeKalb are sufficiently simple and valid to warrant a program of selection of individual plants of high nitrate reductase levels in a synthetic variety. In such selections the enzyme assay should be used as an adjunct to the normal selection practices for good agronomic characteristics.

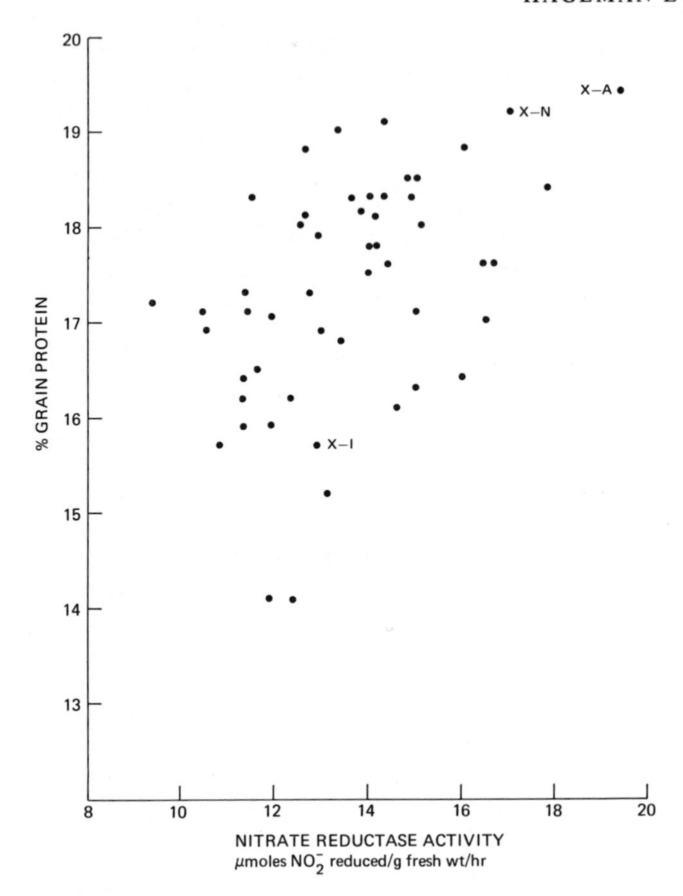

FIGURE 4 The correlation between nitrate reductase activity of wheat seedlings grown in the field at Lincoln, Nebraska, with the percentage grain protein produced in Yuma, Arizona (r = +.65*). Of the 53 entries, 50 were F_4 genotypes derived from the worldwide collection and Mexican dwarf wheats. The other entries were varieties A (Atlas 66), I (Inia 66), and N (Nap Hal). Atlas 66 and Nap Hal have been used as progenitors of high-protein wheat varieties.

REFERENCES

Beevers, L., and R. H. Hageman. 1969. Annu. Rev. Plant Physiol. 20:495.

Croy, L. I., and R. H. Hageman. 1970. Crop Sci. 10:280.

Deckard, E. L. 1970. Ph.D. Thesis, University of Illinois, Urbana, Illinois.

Deckard, E. L., R. J. Lambert, and R. H. Hageman. 1973. Crop Sci. 13:343.

Dudley, J. W., and R. J. Lambert. 1969. Crop Sci. 9:179.

Eck, H. V., and R. H. Hageman. 1974. Crop Sci. 14:283.

Eilrich, G. L. 1968. Ph.D. Thesis, University of Illinois, Urbana, Illinois.
Eilrich, G. L., and R. H. Hageman. 1973. Crop Sci. 13:59.
Gardner, C. D., and S. A. Eberhart. 1966. Biometrics 22:439.
Hucklesby, D. P., C. M. Brown, S. E. Howell, and R. H. Hageman. 1971. Agron. J. 63:274.
Johnson, V. A. 1972. *In* Seed Protein (G. England, ed.). Avi Publishing Co., Westport, Conn., p. 183.
Johnson, V. A., J. W. Schmidt, P. J. Mattern, and A. Haunold. 1963. Crop Sci. 3:7.
Klepper, L. A., D. Flesher, and R. H. Hageman. 1971. Plant Physiol. 48:580.
Matsushima, S. 1965. *In* The Mineral Nutrition of the Rice Plant. Johns Hopkins Press, Baltimore, p. 219.
Morrilla, C. A., J. S. Boyer, and R. H. Hageman. 1972. Plant Physiol. 51:817.
Mulder, E. G., R. Boxma, and W. L. Van Veen. 1959. Plant Soil 10:335.
Pateman, J. A., D. J. Cove, B. M. Rever, and D. B. Roberts. 1964. Nature 201:58.
Purvis, A. C. 1972. Ph.D. Thesis, University of Illinois, Urbana, Illinois.
Schrader, L. E., and R. H. Hageman. 1965. Agron. Abstr. 30.
Schrader, L. E., D. M. Peterson, E. R. Leng, and R. H. Hageman. 1966. Crop Sci. 6:169.
Schrader, L. E., D. A. Cataldo, and D. M. Peterson. 1974. Plant Physiol. 53:688.
Sorger, G. L., and N. H. Giles. 1965. Genetics 52:777.
Terman, G. L., R. E. Ramig, A. F. Dreier, and R. A. Olson. 1969. Agron. J. 61:755.
Van Dobben, W. H. 1966. *In* The growth of cereals and grasses. (F. W. Milthorpe and J. D. Ivins, ed.). Butterworth, London, p. 320.
Warner, R. L. 1968. Ph.D. Thesis, University of Illinois, Urbana.
Warner, R. L., R. H. Hageman, J. W. Dudley, and R. J. Lambert. 1969. Proc. Nat. Acad. Sci. U.S.A. 62:785.
Zeiserl, J. F., and R. H. Hageman. 1962. Crop Sci. 2:512.
Zeiserl, J. F., W. L. Rivenbark, and R. H. Hageman. 1963. Crop Sci. 3:27.

DISCUSSION

DR. JENSEN: I just want to point out that the breeding program that we are involved in in the Northeast is concerned with white wheats, and the objective there, insofar as the industrial use of this product, is low protein. We therefore developed, I am sure, the lowest-protein wheats in the world. We also have another program to develop higher-protein wheats for feed production. There is something very interesting there, in that these are not low-yielding wheats. Have you looked at the differences between the reds' and the whites' nitrate reductase? You did show the Genesee variety as low in nitrate reductase.

DR. HAGEMAN: Genesee is low in nitrate reductase.

DR. JENSEN: Yes.

DR. HAGEMAN: It is not particularly low in yield.

DR. JENSEN: No, it is not. These are high-yielding wheats. So there is a conflict there.

DR. RABSON: It needs resolution.

DR. HAGEMAN: I cannot resolve the conflict at this time. What is needed is an explanation for the variation in B values. For example, Monon, which is a soft winter wheat, had a low nitrate reductase activity, and yet it was very effective in accumulating the total amount of protein. Now, percentage protein usually is quite low, although in our hands our percentage protein in Monon was not low. I have attributed this higher percentage of protein to increased application of nitrogen fertilizer.

DR. JOHNSON: Dick, I can confirm what you say about the two varieties, Atlas 66 and Nap Hal. Both of them are very high in protein, and they are, at least in part, high in protein because of elevated nitrate reductase activity. You show data, Dick, scattergrams that suggest to me that in some of the genotypes or varieties you are looking at the protein produced may not be involved exclusively with level of nitrate reductase activity. They appear to fall outside of the trend that you have established.

From a genetic point of view, this suggests to me that there are genes operating there that are affecting some other part of the nitrogen metabolism system, not necessarily nitrate reductase. So, first of all, I would like to have you comment about this. Second, how does this make you feel about substituting a level of nitrate reductase activity for measurement of protein as a test if in doing so you would run the risk of losing genes that are affecting another part of the system?

DR. HAGEMAN: To answer your last question first, I would simply say that my approach is not to make any change in the normal selection procedure. You simply add a nitrate reductase assay to the normal selection procedure.

The major question is why one genotype has a low nitrate reductase activity and a relatively high accumulation of grain nitrogen, whereas in another genotype it takes more enzymes to produce the same amount of protein. While I can glibly say that this is a difference in enzyme efficiency, this does not solve the problem.

It is conceivable that there is genetic variation in induction potential. Thus, certain genotypes would require less nitrate to induce a high level of enzyme than other genotypes. Then, if the rate of influx of nitrate to the leaves is low, the enzyme of the genotype with high induction potential would not be saturated with respect to substrate. This would cause variation in the B value. What is needed is an assay that would measure *in situ* reduction of nitrate. Neither the *in vitro* or *in vivo* assay for nitrate reductase does this. Until we can correct this flaw in methodology, the best we can do is to use genetically related lines to show the relationship between enzyme activity and grain and protein production. I am not discouraged, because even with unrelated genotypes we find correlations between enzyme and yield and protein production.

There are many reasons why the correlation between nitrate reductase activity and yields of grain and grain protein are low and variable. One reason is that yields of grain and protein vary markedly with environmental factors, even for a given genotype. A second reason is that there may be inherent errors in the assay, especially among genotypes. Assuming that the errors in the assay procedures among genotypes are small, the difference observed in "enzyme efficiency" constitutes a third reason for the variation. Such differences in enzyme efficiency support the belief that genes other than those for level of nitrate reductase activity are operative. A fourth reason is the observed variation in "transport efficiency" or harvest index. This variation also supports the concept that other genes are involved.

Therefore, it is not reasonable to substitute the nitrate reductase assay for protein assays in the selection procedure. However, the use of both assays should provide a better approach than either assay used separately.

We are both in agreement that other genes are involved. Unless these genes are closely linked, which seems unlikely, selection based on protein assays above would not separate or identify those genotypes that have high potential for nitrate uptake and nitrate reductase, "enzyme efficiency," or "Harvest Index." In fact, the protein assays are made after the fact and do not measure maximum genetic potential under the ideal environment. This separation and identification of the various genetic components involved in protein production should provide the information needed for "genetic engineering."

In my view the nitrate reductase is responsible for the protein. The enzyme is the causal agent. We have compared nitrate reductase activity with the total nitrogen accumulated by the seedling stage, during vegetative development and by maturity. We have obtained correlation between the enzyme activity and the accumulation of total nitrogen in the plant that are highly significant (r values = +.99).

We have tried induction. We have not yet been able to develop a workable system.

DR. MUNRO: Is there not a question of statistical confounding in a sense that when you have more protein total yield, you also have more enzyme? Is this excluded from the correlations you showed, and would it not be better if you could demonstrate a short-period induction response to nitrate, say, of 24 hours differing in different varieties?

DR. ELLIOTT: You are expecting unadapted genotypes to respond normally in your tests. If someone develops a series of lines adapted to Illinois for your studies, that is one possibility. If you are going to make these studies in one year, you need several locations in Illinois so that means of lines are available rather than depending upon one location. I would like you to comment on adaptation because you now have a whole series of genotypes unadapted to Illinois conditions.

DR. HAGEMAN: Thank you very much. I will speak to the corn breeders.

DR. DOY: Do you think this system could be adapted to tissue culture? You could have nitrate as your sole source of nitrogen going to protein, and you could positively select for clones at that time and then grow plants from them. You would have thereby selected on the criterion you like to apply. Then you could hand it over to the breeders and say, "I have done my part. Now, you breed in anything else that you want." Second, why won't they let you select from natural populations? Aren't they pleased to get anything, no matter where it comes, natural or unnatural?

DR. HAGEMAN: The answer to the last question is an emphatic "yes." I have been playing the devil's advocate; attempting to get the geneticists to react. In fact, the plant breeders at Illinois have been very tolerant of me, and we are attempting to select the best individuals from natural populations. With respect to the cell culture approach, I do not feel qualified to respond.

DR. DOY: I would think that this is a beautiful system where the new ideas could be applied extraordinarily simply.

DR. HAGEMAN: The cell culture method should be very useful. The assay is adaptable to cell culture and has been used extensively by P. Filner (Biochim. Biophys Acta 118:299 (1966); 215:152 (1970); 230:362 (1971).

DR. HARDY: We have been talking so far about the genetic side. I think the chemical side may be worth thinking about, too. From time to time one sees in the literature some chemicals that have been used to stimulate nitrate reductase activity. Has anyone taken any of these chemicals out to the field to see if you can make one of the low-nitrate-reductase plants into a high-nitrate-reductase plant and at the same time increase its yield?

DR. HAGEMAN: I have not tried that.

DR. HARDY: There are a variety of things that are increasing measurable nitrate reductase activity among the factors you are measuring here. Can you use them as your agent to increase activities?

DR. HAGEMAN: We have recently attempted to use benzyl adenine treatments to increase nitrate reductase in growth-chamber-grown corn seedlings, but without success. About 10 years ago, I attempted to use 2-4D, indole, and napthalene acetic acids and gibberellin to increase nitrate reductase activity in greenhouse-grown corn seedlings, again without success. I am not saying that chemicals will not work, only that I have not been successful with the compounds and conditions that I have employed.

B. J. MIFLIN

Metabolic Control of Biosynthesis of Nutritionally Essential Amino Acids

The general problem of protein quality in plant products has already been outlined by previous speakers. However, I would like to re-emphasize a few points. First, the nutritional quality of plant proteins depends on their amino acid composition. Thus, a potentially limiting factor in protein quality is the supply of constituent amino acids, particularly those that are nutritionally essential. Second, the work of Hageman's group (Croy and Hageman, 1970; Eilrich and Hageman, 1973; Deckard *et al.*, 1973; and Hageman, p. 103, this volume) has shown that increased nitrate feeding means increased amino acid production and increased protein. However, the increases do not occur across the whole spectrum of amino acids. Table 1, taken from some work done in my laboratory by Dr. G. C. Blackwood (1973), illustrates this point. When nitrate is added to nitrogen-starved maize plants, the α-amino N level in the roots increases approximately twofold in the first 20 hours and the protein levels also increase. When the soluble α-amino N is analyzed, it is seen that, whereas amino acids such as glutamate and aspartate increase sixfold to sevenfold, the nutritionally essential ones remain at about their original level. Similarly, when the level of nitrogen fertilizer is increased, cereal plants respond by producing grain of higher protein content, but this protein has a lower percentage of lysine and some other essential amino acids.

This is no new concept. Bishop (1928, 1929) in his pioneering studies with barley in the late 1920's showed how this could arise through changes in the relative amounts of hordein, glutelin, and salt-soluble

TABLE 1 Increase of Amino Acids Following Addition of Nitrate to Nitrogen-Starved Maize Roots

Amino Acid	Amino Acids in Root (μmole/g fresh wt)		Amino Acids in Exudate $+NO_3^-$ (μmole/g fresh wt/hr)
	NO_3^- Starved	$+NO_3^-$ for 20 hr	
Aspartate	0.14	1.00	0.014
Asparagine	0.95	1.08	0.088
Glutamate	0.80	3.32	—
Glutamine	1.68	5.90	0.388
Serine	0.08	0.15	0.01
Glycine	0.32	0.52	0.01
Alanine	1.00	4.52	0.19
Tyrosine	0.14	0.28	—
Phenylalanine	.19	.19	—
Threonine	.06	.06	—
Lysine	.03	.03	—
Methionine	.03	.03	—
Isoleucine	.07	.04	—
Leucine	.06	.03	—
Valine	.07	.11	—
Total α amino N	8.6	15.5	0.58
NO_3 levels	11.0	590.0	—

nitrogen in the grain. As the protein content increased, the total amount of the low-lysine hordein (barley prolamin) increased much more rapidly than the other fractions, so that it formed an ever-increasing proportion of the total seed protein, thereby depressing the overall protein quality. Although Bishop proposed an internal regulation of the synthesis of reserve proteins nearly 50 years ago, we have yet to understand how this regulation might operate.

Genetic studies are only a little more hopeful. Although one or two varieties combining both high protein content with above average lysine levels have been found, the results of screening 7,000 varieties indicate an overall negative correlation between lysine and protein content (Johnson *et al.*, 1970).

To summarize this brief introduction, it would seem that the following constraints operate, even under conditions where the plant is plentifully supplied with nitrogen:

1. The production of high-quality protein depends on the presence of high-quality amino acids.

2. There is a constraint regulating the synthesis of the individual reserve proteins and classes of proteins.

3. There is a constraint regulating the synthesis of certain amino acids.

I should like to consider this last point and to discuss briefly how it might be related to the first two points.

THE REGULATION OF AMINO ACID BIOSYNTHESIS

Although the field is relatively new, and until a short while ago only a few laboratories were working in it, there is now good circumstantial evidence that regulation of amino acid biosynthesis occurs *in vivo*. This evidence has been gathered in a number of ways (Miflin, 1973; Bryan, 1974). The most usual approach has been to feed a ^{14}C-labeled precursor, e.g., acetate or glucose, and determine its incorporation into the soluble and protein pool of a given amino acid in the presence and absence of that amino acid. Using this approach, Oaks (1965b; Oaks *et al.*, 1970) has shown that lysine, threonine, proline, arginine, leucine, isoleucine, valine, and tryptophan regulate their own synthesis. Other studies (Miflin, 1969a; Dunham and Bryan, 1969; and Furuhashi and Yatazawa, 1970) suggest that methionine, phenylalanine, and tyrosine should probably be added to this list. However, this work does not tell us the mechanisms by which these controls operate. In bacteria there are two main systems controlling biosynthetic pathways—those of induction and repression of enzyme synthesis and those of allosteric feedback modification of enzyme activity. Although the classic definition of induction and repression involves changes in the rate of synthesis of enzymes (Jacob and Monod, 1961), the terms are now often used in a wider context to reflect changes in the extractable level of enzymes.

CHANGES IN THE LEVEL OF EXTRACTABLE ENZYME OF AMINO ACID BIOSYNTHESIS

The production of many amino acids in bacteria is regulated chiefly by repressing the formation of the enzymes that synthesize them. The represser molecule is the amino acid either by itself or in combination with another molecule. Despite the overenthusiastic acceptance of the original Jacob and Monod model by plant biochemists and physiologists and its application far outside its original context, no good hard evidence for the operation of this control mechanism in anything approaching its original concept has been obtained in higher plants. This is particularly true in relation to amino acid biosynthesis.

Table 2 lists a number of studies in organisms ranging from blue green algae to barley in which no evidence could be obtained for any effect of the level of amino acids on the amount of their biosynthetic enzymes in the tissue. This is likely to be a genuine negative result, since in most studies it is known that an exogenous supply of these amino acids does turn off their internal synthesis.

Although there still remains the outside possibility that derepression may occur under extreme conditions of nitrogen starvation, it seems reasonable to conclude that repression and induction mechanisms have little or no part to play in regulating amino acid biosynthesis under normal conditions in higher plants.

ALLOSTERIC FEEDBACK CONTROL OF ENZYME ACTIVITY

In this system the presence of an end product of a biosynthetic pathway modifies the activity of a key enzyme involved in its synthesis. If we consider the pathway $A \rightarrow B \rightarrow C \rightarrow D$, then, in general terms, the activity of the enzyme $A \rightarrow B$ will be inhibited by D so that as the concentration of D increases, it shuts off its own synthesis.

TABLE 2 Studies Showing an Absence of Repression of the Enzymes Involved in the Synthesis of Amino Acids

Enzyme	Organism	References
Threonine deaminase	Plant tissue culture	Dougall (1970)
Threonine deaminase	Blue green algae	Hood and Carr (1972)
Acetolactate synthetase	Barley seedlings	Miflin and Cave (1972)
Acetolactate synthetase	Blue green algae	Hood and Carr (1972)
Isopropyl malate synthetase	Blue green algae	Hood and Carr (1972)
Isopropyl malate isomerase	Maize embryos	Oaks (1965a)
Isopropyl malate dehydrogenase	Maize embryos	Oaks (1965a)
Leucine aminotransferase	Maize embryos	Oaks (1965a)
Leucine aminotransferase	Blue green algae	Hood and Carr (1972)
DAHP synthetase	Blue green algae and Euglena	Weber and Bock (1968)
Anthranilate synthetase	Plant tissue culture	Widholm (1971)
Tryptophan synthetase	Plant tissue culture	Widholm (1971), Belser *et al.* (1971)
Chorismate mutase	Plant tissue culture	Chu and Widholm (1972)
Ornithine transcarbamoylase	Algae	Holden and Morris (1970)
Arginosuccinase	Algae	Hood and Carr (1971)
α-*N*-acetylornithine transacetylase	Blue green algae	Hood and Carr (1971)
N-acetylglutamate phosphokinase	Blue green algae	Hood and Carr (1971)

The relationships become more complex when we consider a branched pathway:

$$A \rightarrow B \rightarrow C \begin{matrix} \nearrow D \rightarrow E \\ \searrow F \rightarrow G \end{matrix}$$

A number of ways have evolved to deal with this situation and, based on results with bacterial systems, it is generally found that E will modify the activity of the first enzyme unique to its biosynthesis (i.e., $C \rightarrow D$) and, likewise, G will modify the activity of the enzyme carrying out $C \rightarrow F$. When both are present in sufficient quantities, they then cooperate to inhibit $A \rightarrow B$. Modification can also be positive, in that an excess of E stimulates the production of G by activating the enzyme $C \rightarrow F$. For further discussion of this type of regulation in bacterial and mammalian systems a number of reviews are available (Umbarger, 1969; Stadtman, 1970). Examples of this type of regulation are now well known in plants, and I shall discuss some of these.

CONTROL OF LEUCINE, ISOLEUCINE, AND VALINE BIOSYNTHESIS

The pathway leading to the formation of these amino acids is outlined in Figure 1. This pathway is unusual in that, although leucine and valine use different carbon precursors from isoleucine, a common set of enzymes is utilized for many of the steps. Thus the enzyme carrying out the second step in isoleucine biosynthesis and the first step in leucine and valine biosynthesis is the same enzyme and this also holds true for the subsequent steps in the pathway until the branch point for leucine biosynthesis (Kanamori and Wexom, 1963; Greenberg, 1969; Miflin, 1971). This, in effect, gives a pathway with three branches.

From the general principles outlined briefly above, we can predict that there are three key enzymes. First, it might be expected that leucine would regulate the first enzyme unique to its own biosynthesis—isopropylmalate synthetase. Second, the enzyme acetolactate synthetase is the first one unique to all three of the amino acids. Third, the enzyme threonine deaminase is the first enzyme unique to isoleucine synthesis, even though it comes before the common enzyme acetolactate synthetase.

Isopropylmalate synthetase has been studied by Oaks (1965a). The enzyme from maize is inhibited 75 percent by leucine at concentrations

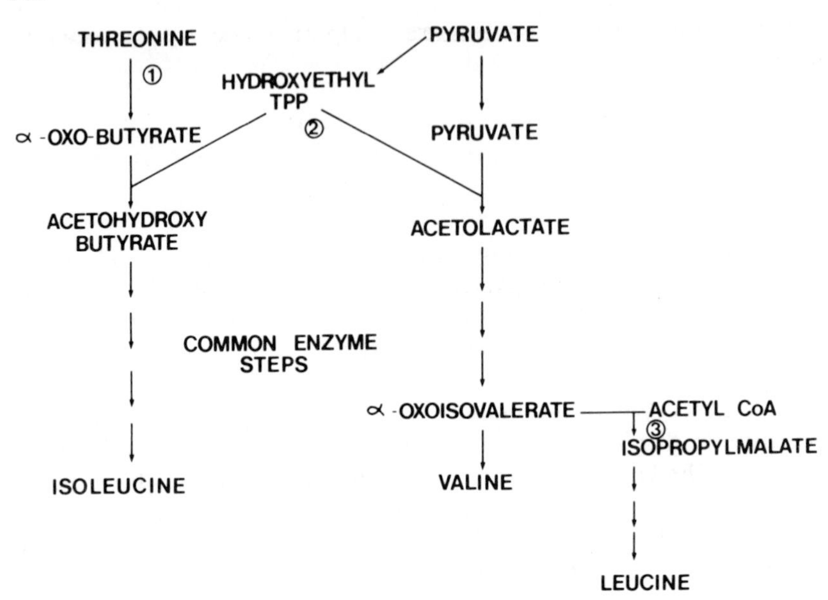

FIGURE 1 Pathway of leucine, isoleucine, and valine biosynthesis. (*1*) Threonine deaminase; (*2*) acetolactate synthetase; (*3*) isopropylmalate synthetase.

of 5×10^{-4} *M*. We have looked at the regulation of acetolactate synthetase in barley and in a range of other species (Miflin, 1969b, 1971; Miflin and Cave, 1972). It is inhibited to some extent by leucine and to a lesser extent valine but very little by isoleucine and other amino acids, except at high concentrations (Table 3). However, when leucine and valine are present in equal concentrations, only one-tenth of the total concentration of either amino acid supplied singly is required to cut the rate of enzyme activity in half. This cooperative effect is not further enhanced by the addition of isoleucine, although the pairs isoleucine and leucine do show enhanced inhibition. This cooperation has been discussed in detail elsewhere (Miflin, 1971); the main point to emphasize is that only when two amino acids are present is it likely that the enzyme is significantly inhibited *in vivo*.

Threonine deaminase has been studied by Dougall (1970). The results of his studies show that the enzyme is inhibited by 90 percent by 5×10^{-4} *M* isoleucine. Threonine deaminase also provides an example of positive as well as negative control operating to regulate amino acid biosynthesis. When valine is added in the absence of isoleucine, it has

TABLE 3 Effect of Various Amino Acids on the Rate of Acetolactate Synthetase

Amino Acid	Concentration (m*M*)	Rate
None	—	100
Leucine	0.5	59
Valine	0.5	70
Norvaline	0.5	80
Isoleucine	0.5	84
Norleucine	0.5	86
Alanine	0.5	98
Homoserine	10	88
Aspartate	10	88
Methionine	10	85
Leucine + valine[a]	0.05	51
Leucine + valine	0.5	18

[a] Added in equal proportion to give the total amino acid concentration shown.

little effect on activity; however, when the enzyme is inhibited by isoleucine, the further addition of valine substantially removes the inhibition. Similar results have been obtained by Sharma and Mazumder (1970), and Blekhman *et al.* (1971) have found that threonine deaminase from peas is also activated by L-aspartate and DL-norvaline.

The control of the branched chain pathway by allosteric feedback mechanisms can thus be summarized as in Figure 2. There are a number of regulatory loops in operation, both positive and negative. Together these controls provide the possibility of fine regulation of the synthesis of the branched chain amino acids. They can be envisaged as operating as follows. Under conditions where surplus leucine is produced, it will inhibit its own synthesis at the isopropylmalate synthetase step. If isoleucine alone is overproduced, then threonine deaminase will be inhibited. When both leucine and valine (or leucine and isoleucine) are in excess of requirements, they will shut off the whole pathway at the acetolactate synthetase step. Conceivably, this could lead to a shortage of isoleucine, but this is probably prevented by valine nullifying any feedback inhibition of threonine deaminase. These studies provide evidence that plants have highly sophisticated mechanisms to control their amino acid biosynthesis. The operation of these controls should enable the plant to keep a reasonable balance between the amino acids during protein synthesis.

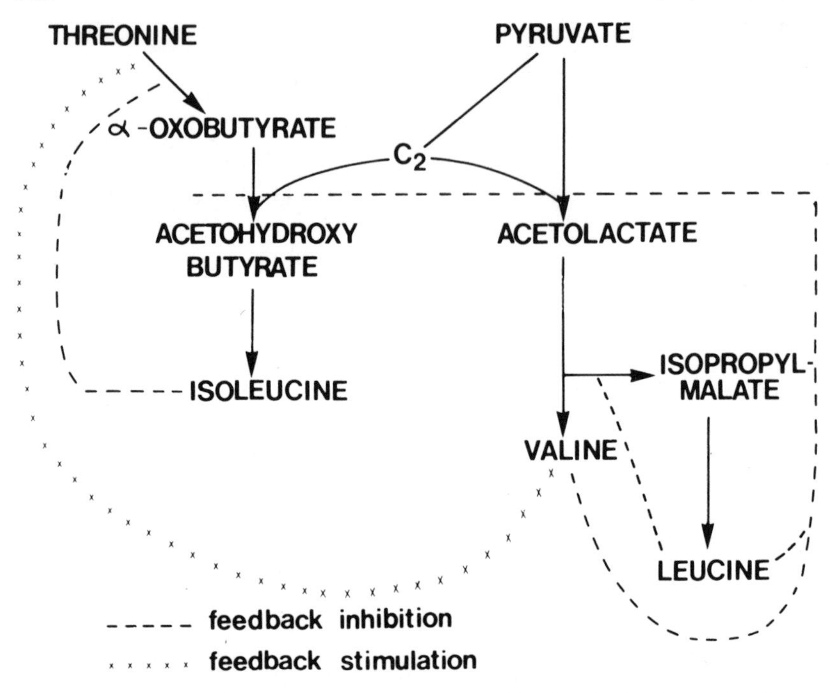

FIGURE 2 Regulatory loops in the biosynthesis of leucine, isoleucine, and valine.

REGULATION OF AROMATIC AMINO ACID BIOSYNTHESIS

Further support for this hypothesis is found from studies on the biosynthesis of the aromatic amino acids. The pathway for their synthesis is shown in Figure 3. Again we can identify a number of potential regulatory enzymes: the initial enzyme of the pathway catalyzing the condensation of erythrose-4-phosphate and phosphoenolpyruvate to give 3-deoxy-D-arabino-heptulosonic-7-phosphate (DAHP); the first enzyme unique to tryptophan biosynthesis, anthranilate synthetase; the first enzyme unique to phenylalanine and tyrosine biosynthesis, chorismate mutase; and the initial enzyme leading solely to either phenylalanine or tyrosine (prephenate dehydratase and prephenate dehydrogenase, respectively).

There have been few studies of DAHP in higher plants (Minamikawa, 1967), but the enzyme is subject to regulation by tyrosine and to a lesser extent by phenylalanine in algae (Weber and Bock, 1968). Studies by Belser *et al.* (1971) and by Widholm (1972a, b) have shown that tryptophan regulates its biosynthesis by feedback control of anthranilate synthetase. The enzyme from tobacco cells is completely inhibited by 5×10^{-5} *M* tryptophan. Phenylalanine and tyrosine inhibit chorismate mutase, but tryptophan activates the enzyme in the presence and absence of the other two amino acids (Cotton and Gibson, 1969; Gilchrist *et al.*, 1972). There is no evidence of any cooperative effect between the two amino acids. Interestingly, the enzyme exists in two forms, only one of which is subject to regulation. The insensitive form does not appear to be an artifact of isolation and is apparently not interconvertible with the regulatory form. It may be that this unregulated enzyme is concerned with the production of phenylalanine and tyrosine as intermediates in the biosynthesis of flavanoids, phenolics, and lignins rather than with protein synthesis. The enzymes metabolizing prephenate do not appear to be subject to regulation (Gamborg and Keeley, 1966).

THE REGULATION OF THE BIOSYNTHESIS OF THE ASPARTATE FAMILY OF AMINO ACIDS

The pathway of the synthesis of amino acids derived from aspartate is outlined in Figure 4. Our knowledge of the enzymology of this pathway is relatively scarce and as yet incomplete. Similarly, information on the regulatory mechanisms operating in this pathway is also still limited.

As in the previous pathways, we can suggest a number of key enzymes. The first is aspartokinase, which is one of the classical enzymes for study of bacterial regulatory systems. A number of elegant ways of controlling this step have evolved in bacteria, all of which involve some means whereby the various end products lysine,

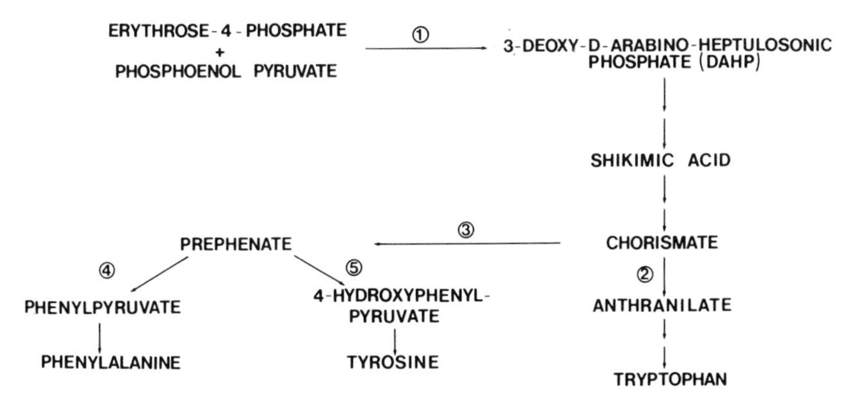

FIGURE 3 Pathway of aromatic amino acid biosynthesis.

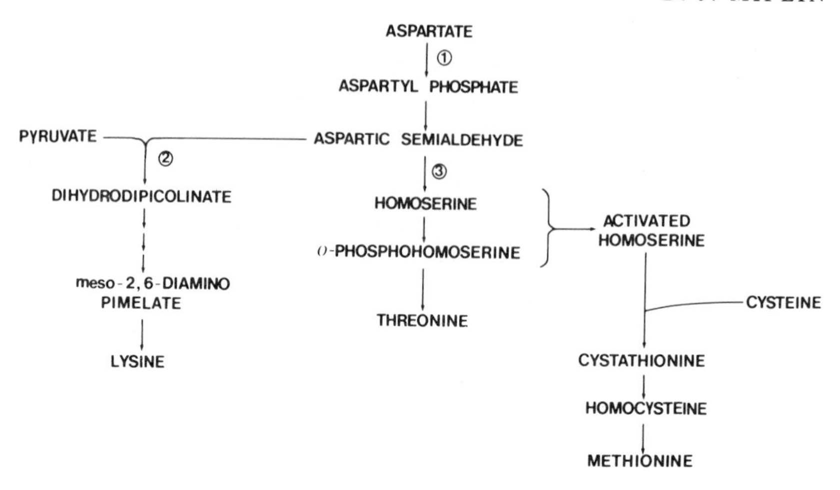

FIGURE 4 Amino acids derived from aspartate. (*1*) Aspartokinase; (*2* dihydrodipicolinate synthetase; (*3*) homoserine dehydrogenase.

threonine, and methionine can each, separately or together, regulate the flow of carbon through this initial step of the pathway (Datta, 1969).

The enzyme was first isolated in higher plants by Bryan *et al.* (1970), who obtained an active preparation from maize shoots. Surprisingly, the maize enzyme was found to be significantly affected only by lysine. Threonine inhibited the rate of aspartokinase activity to a small degree at very high concentrations (10 m*M*). We have repeated these studies using a more sensitive assay procedure and have confirmed Bryan's findings (Cheshire and Miflin, 1973). Lysine at low concentration (0.1 m*M*) causes 80 percent inhibition of aspartokinase. The approximate K_i for lysine is 20 μM. The inhibition is to some extent competitive with respect to aspartate concentration. The effect of other amino acids on their own or in combination with lysine is minimal (Table 4). In most cases, the effects are of the order of 20 percent or less at 10 m*M* concentrations and are additive to that of lysine. The physiological significance of effects at this concentration is doubtful.

Wong and Dennis (1973) have investigated aspartokinase in wheat germ. Unfortunately, their preparation had very low specific activity and most of their work was done with very low optical density changes, which limits the significance that can be placed on their conclusions. However, they too found that lysine was the most effective single amino acid, causing 50 percent inhibition at 0.5 m*M*. Addition of tenfold higher concentration of threonine to low concentrations of lysine caused a slight synergistic inhibition, indicating a possible cooperative control mechanism involving lysine and threonine. According to the tenets of the conventional wisdom, such a cooperative effect is theoretically more likely, but the evidence so far in maize is that it does not occur and the evidence in wheat is as yet inconclusive.

The disadvantage of a control mechanism dictated solely by the lysine concentration is that if this amino acid were to build up, it would tend to prevent the synthesis of methionine and threonine as well as lysine and would inhibit growth. However, the above work with extracted enzymes may reflect the controls *in vivo* or may merely indicate that our ability to extract and handle the enzymes is not yet sufficiently good to provide us with the right answers. Unfortunately, the evidence on *in vivo* regulation of the pathway has been obtained on different species from those used in the enzymological studies. We hope to be able to rectify this deficiency in the near future.

The next key enzyme in lysine biosynthesis is the one past the branch point of lysine and homoserine. This enzyme catalyzes the condensation of aspartic semialdehyde with pyruvate to form dihydrodipicolinic acid. As far as I know, there is no published work on this enzyme in higher plants. However, Dr. R. M. Cheshire, working in my laboratory at Newcastle, has demonstrated that this enzyme is present in maize. The levels of activity are reasonable in relation to aspartokinase and appear to be genuine. The partially purified extract is completely inhibited by 1 m*M* lysine.

The first enzyme on the threonine and methionine branch of the pathway is homoserine dehydrogenase. This enzyme has been carefully studied by Bryan (1969) and shown to be subject to feedback control by threonine. Although the enzyme does show some response

TABLE 4 Effect of Various Amino Acids on the Inhibition of Aspartokinase by Lysine

Amino Acids Present (10 m*M*)	Rate as % of Control	
	Alone	+Lysine (0.02 m*M*)
None	100	65
Homoserine	101	65
Methionine	90	55
Isoleucine	124	89
Valine	115	78
Alanine	119	85
Leucine	105	70
Threonine	87	52

to other amino acids, it is likely that this response to threonine is the only physiologically significant control on the enzyme.

The enzymology of the early steps of methionine biosynthesis had not been studied in plants until recently. In microorganisms two possible routes to cystathionine formation have been established in which homoserine is activated to form either the *o*-acetyl or *o*-succinyl derivative. The activated homoserine then combines with cysteine to give cystathionine. Recent work by Giovanelli *et al.* (1973; Datko *et al.,* 1973) has shown that pea extracts contain an active *o*-acetyl-CoA homoserine acetyl transferase and that *Lathyrus sativum* contains a corresponding oxalyl transferase. This is correlated with the known occurrence of *o*-acetyl and *o*-oxalyl homoserine in these plants (Grobbelaar and Steward, 1969; Przybylska and Pawelkiewicz, 1965). However, a range of other plant extracts were unable to catalyze the transfer of acetyl, malonyl, succinyl, or oxalyl groups to homoserine. They were, however, capable of forming *o*-phosphohomoserine. When cystathionine synthetase was studied, it was found that besides using *o*-acetyl homoserines as substrates, it could also use *o*-phosphohomoserine as an α-amino-butyryl donor. A survey of a range of plants showed that it was the only α-amino-butyryl donor present in most plants, with the exception of pea and *Lathyrus*. This capability of using *o*-phosphohomoserine is apparently unique to plant enzymes and suggests that plants may have a unique route to methionine biosynthesis.

There is an alternative route to homocysteine not involving cystathionine. In this, the activated homoserine reacts directly with H_2S to give homocysteine (Giovanelli and Mudd, 1967). The physiological significance of this direct sulfhydration pathway is not known.

If *o*-phosphohomoserine is the true intermediate in methionine biosynthesis, it means that the branch point between methionine and threonine is at the level of *o*-phosphohomoserine and not homoserine as indicated for the bacterial pathway. This would mean that a key enzyme in the regulation of methionine biosynthesis might well be cystathionine synthetase. This possibility is at present under investigation in our laboratories.

As far as the biosynthesis of threonine is concerned, no enzymology has been done in higher plants on the steps after homoserine production. It might be expected that the enzyme after the branch point to methionine could be regulated. However, much work remains to be done to clarify this part of the pathway.

In summary, therefore, it must be stated that our knowledge of both the enzymology and the control of the aspartate family of amino acids is fragmentary. The lack of knowledge is probably due to the relatively small number of people who are active in this field and to the difficulty of handling these particular enzymes in plants. This deficiency of knowledge is particularly unfortunate, as the amino acids of this pathway are those of most nutritional importance. A ready understanding of their synthesis and the way in which it is controlled should be of value, both to the plant physiologist and breeder. However, the progress in the last few years has been encouraging, and it should not be too long before a clear picture of this pathway emerges. At present we can state that there is good evidence that the synthesis of lysine, methionine, and threonine is subject to feedback control *in vivo* and that those key enzymes that have been studied show regulatory properties.

GENERAL COMMENTS

The studies mentioned above are test tube experiments carried out on isolated enzymes from, for the most part, either young seedlings or cells in tissue culture. There are many questions to be raised and answered before we know their exact relevance to the deposition of high-quality protein in plant seeds. Among these questions are the following:

1. Do the test-tube observations correlate with the situation in the intact plant?
2. Are the concentrations of the amino acids used to produce these effects relevant to the situation *in vivo*?
3. Do the controls remain the same throughout the development of the plant?
4. Given that such controls exist, what can be done to circumvent them?

I will try to answer these questions below. The first point has been raised previously and been discussed elsewhere (Miflin, 1973; Bryan, 1974) in some detail. From our studies of acetolactate synthetase and the effects of amino acids on the growth in barley (Miflin, 1969a), we have been able to show good circumstantial evidence for the operation of the controls *in vivo*. Similar evidence has been obtained for other enzymes.

INTERNAL REGULATORY CONCENTRATIONS OF AMINO ACIDS

I have already mentioned the problem of the concentrations of amino acids that are physiologically relevant. It is not easy to get an answer, but it is conceivable that the levels are lower, rather than higher, than the gross levels of the amino acids in the tissue. For example, in carrot tissue culture cells, the level of the tryptophan pool in normal cells is 80 μM, whereas only 10 μM tryptophan is required to inhibit anthranilate synthetase by over 90 percent (Widholm, 1972c). The simple conclusion might be made that the enzyme is not therefore subject to control *in vivo*. However, if we take a line of cells where the anthranilate synthetase is relatively insensitive to tryptophan, then the cellular tryptophan rises to 2,000 μM.

The reason tryptophan can accumulate to levels eight times greater than that required to turn off enzyme activity is probably due to the high degree of compartmentation in plant cells (Oaks and Bidwell, 1970). This is a refinement in metabolic control not present, to any degree, in bacterial systems.

We can consider that amino acids in plant cells exist in at least four pools (Figure 5). In addition, we can postulate a regulatory pool that probably overlaps with one or more of the metabolic pools. Some indication of the relevant pool can be gleaned from the knowledge of the subcellular location of the enzyme in the tissue. If the biosynthetic enzymes are present within a membrane-bound organelle, then it can be presumed to be sensitive to the end-product amino acid concentration within that organelle and to be unaffected by the cytoplasmic pool or pools in other organelles. Also, the rate of synthesis will be a

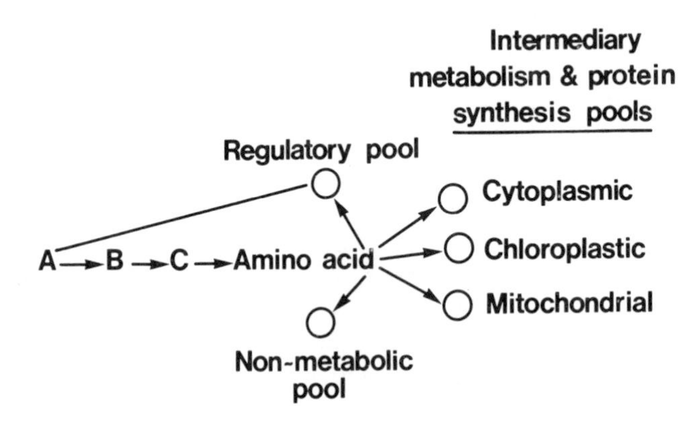

FIGURE 5 Compartmentation of amino acids.

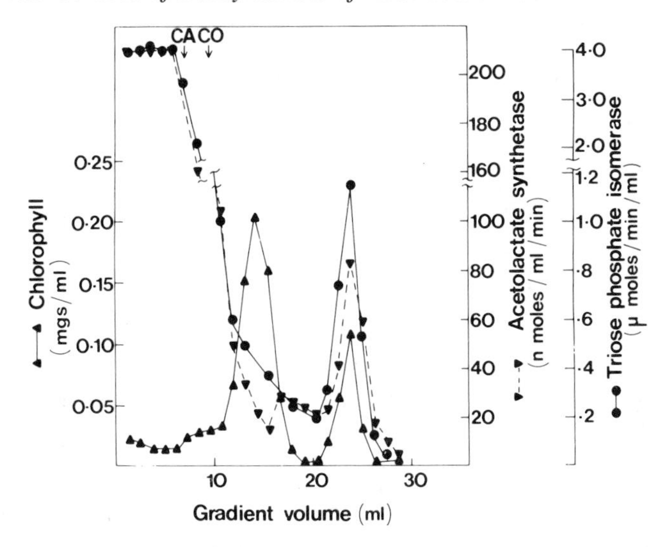

FIGURE 6 Distribution of acetolactate synthetase and organelle marker enzymes from pea leaves after centrifugation of tissue homogenates on sucrose density gradients.

function of transport of the amino acids out of the organelle. Recently, I have been investigating the location of certain enzymes within cells (Miflin and Beevers, 1974), and I have looked at the distribution of acetolactate synthetase in peas (Miflin, 1974).

The tissue is homogenized in isotonic medium and debris filtered off, and the resulting filtrate is layered on top of a sucrose density gradient and centrifuged for 15 min at up to 10,000 *g*. The gradient is then fractionated and various enzymes assayed in the different fractions. The results of such a density gradient separation of a homogenate of leaf tissue is shown in Figure 6. The presence of organelles in the gradient is measured by marker enzymes: cytochrome oxidase for mitochondria, catalase for the microbodies, and triosephosphate isomerase for intact plastids. A peak of acetolactate synthetase activity is observed in the gradient coincidental with the peak of triosephosphate isomerase and the lower chlorophyll band. This indicates that at least part of the enzyme is present in the plastids.

A similar separation of root tissue is shown in Figure 7; again the peak of acetolactate synthetase activity correlates with that of triosephosphate isomerase. In this experiment about one-third of the total acetolactate synthetase activity is found in the plastids. When the almost inevitably high degree of plastid breakage is taken into account,

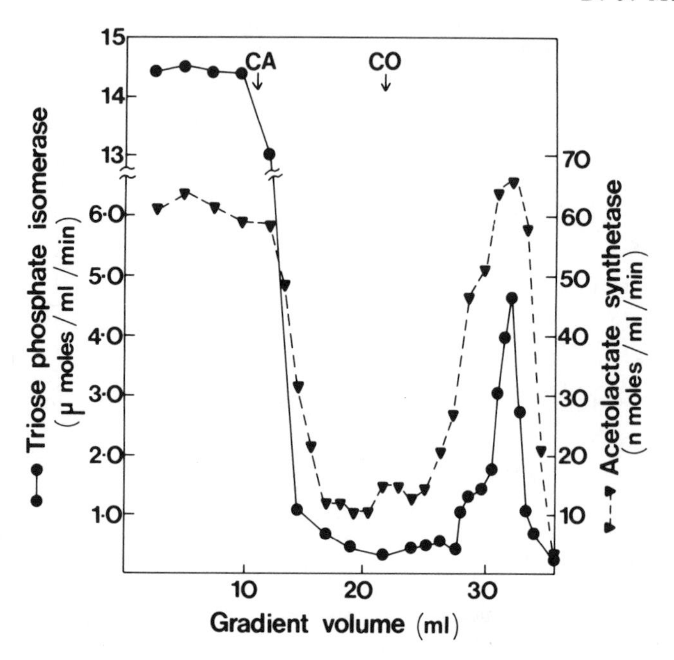

FIGURE 7 Distribution of acetolactate synthetase and organelle marker enzymes from pea roots after centrifugation of tissue homogenates on sucrose density gradients.

this result would suggest that probably all of this enzyme is present in plastids, but it is not possible to exclude a cytoplasmic location. There is no evidence to suggest that the enzyme is present in mitochondria or microbodies, and the degree of contamination of the plastids by these organelles is extremely small. It would thus appear that for this enzyme, at least, the pool of leucine and valine in the chloroplasts is an important regulatory pool. As yet, nothing is known of the subcellular location of other key enzymes of amino acid biosynthesis.

From the studies mentioned above, it would seem likely that the effective regulatory level of amino acid is likely to be smaller than the total concentration in the cells. This means that experiments in which high concentrations (greater than 1 m*M*) have been used to obtain effects must be treated with caution. In general it would seem that biosynthetic enzymes isolated from plant systems are more sensitive to low levels of amino acids than those from bacterial systems.

REGULATION IN A MATURE PLANT

Although we have obtained evidence that acetolactate synthetase is strongly feedback-regulated in mature clover leaves, most studies have utilized tissue derived either from rapidly growing tissue culture cells or from young seedlings. This raises the problem of whether the phenomenon of regulation is limited to these tissues or whether it occurs throughout the growth phases of the plant. In a study of the regulation of proline biosynthesis *in vivo*, Oaks *et al.* (1970) obtained evidence that suggested that as the plant ages, the degree of regulation diminishes. Studies with homoserine dehydrogenase in maize have suggested that the enzyme is progressively desensitized (rendered insensitive to feedback inhibition by homoserine) as detached and intact shoots age (J. K. Bryan, private communication). Although there is more work to be done, the results suggest that the final system of control, operating during the development of a highly differentiated multicellular plant, is more complex and subject to more variables than the simple approach, modeled on bacterial systems, would indicate.

The regulation of amino acid biosynthesis in the context of the whole plant is worthy of further discussion. I have already considered the complications involved in moving from a bacterial to a plant cell. Superimposed on this is the cooperation that occurs between plant cells and plant organs. Nitrogen appearing in seed protein is derived from leaves and roots (Pate, 1968; Hay *et al.*, 1953). Apart from a small amount of still photosynthetically active leaf tissue that, presumably, is still reducing nitrate and forming amino acids, the leaves contribute nitrogen to the seed via the amino acids derived from protein breakdown during their senescence. This source is outside the controls mentioned above. However, the nitrogen derived from the active leaf tissue and from the roots via the transport stream is likely to be subject to feedback control.

The majority of uptake of nitrate and its subsequent reduction by roots is carried out in the young growing root tips (Van Die, 1962; Grasmanis, 1969). This suggests that the phenomenon of age desensitization may not be important in this case, and we might expect the synthesis to be subject to feedback regulation. Further, either because of these controls or because of selective translocation, the range of amino acids transported up the plant in the xylem is limited—chiefly aspartate, asparagine, glutamate, glutamine, proline, and possibly alanine. This results in the developing seed being supplied with an unbalanced set of amino acids, particularly when heavy or late nitrogen fertilization stimulates the reduction in the root during seed maturation (Folkes, 1970). Under such conditions the only way the balance can be restored in the seed is if it is capable of massive interconversions of amino acids. As yet we have no information on the complement of the amino acid biosynthetic enzymes present in developing seeds. However, some feeding experiments (Finlayson and McConnell, 1969) suggest that there is little interconversion of amino acids occurring during the later stages of protein deposition in the wheat grain. Unless these interconversions occur, it is conceivable that the synthesis of the higher-quality reserve protein is limited by the supply of its constituent amino acids. This limitation can probably be related to metabolic controls operating on their synthesis.

SELECTION OF REGULATORY MUTANTS

The knowledge of the biosynthetic pathways leading to amino acid formation and the way in which they are regulated is important in widening our understanding of the plant and providing a sounder basis for work in plant breeding and plant physiology. It is also important because it enables us to locate a potential blockage in the production of high-quality protein. The obvious next question is whether, having found such a blockage, we can remove it. The answer, fortunately, is yes.

Now that techniques of plant cell tissue culture are well established, it is possible to treat plant cells much as we treat microorganisms. In bacteria, regulatory mutants have been selected by exposing cells to growth-inhibitory analogues of metabolites and harvesting those that survive. Some of these are found to be regulatory mutants in which the feedback controls are no longer operating.

Widholm has used this approach to select mutants resistant to 5-methyl-tryptophan (Widholm, 1972a,b,c). This analogue inhibits the activity of anthranilate synthetase almost as effectively as tryptophan, but it is unable to supply tryptophan for growth. When tobacco or carrot tissue culture cells were exposed to the analogue, a small proportion survived (of the order of 1 in 10^6); these were then cultured and reselected and a mutant line resistant to 5-methyl-tryptophan established. This line of cells was able to retain this resistance even after being subcultured for some time in the absence of the analogue. Anthranilate synthetase extracted from these cells was found to be much less sensitive to both tryptophan and 5-methyl-tryptophan. When the tryptophan pool within the cells was measured, it was found that the mutant lines overproduced tryptophan, producing 27 times that produced by the normal cells.

Other selection programs are under way in a number of other laboratories, including ours at Rothamsted, and a range of such mutants should soon be selected and characterized. Once the mutant cells have been selected, it should prove possible in a number of cases to regenerate plants from them. Indeed, Carlson (1973) has already selected and regenerated a mutant tobacco resistant to methionine sulfoximine, the toxic principle of *Pseudomonas tabacii*. Exactly what effect this mutation will have on the physiology, growth, and protein content of the plant remains to be seen. Exciting prospects for this type of genetic and biochemical engineering in crop plants lie before us and although, inevitably, the potentialities will be oversold and the results unpredictable, there is hope that in the end something worthwhile will be achieved.

SUMMARY

Evidence exists to show that synthesis of amino acids, particularly those important in nonruminant animal nutrition, is subject to regulation in crop plants. Biochemical studies on a number of isolated enzymes have shown that the regulation is due to allosteric feedback inhibition of the activity of amino acid biosynthetic enzymes by the end-product amino acids. The actual pool of amino acids affecting this regulation is likely to be that present in the same subcellular compartment as the enzyme. Studies with acetolactate synthetase have shown that this enzyme at least is located chiefly, and possibly solely, in the plastids. These regulatory controls could limit protein quality by limiting the production of the amino acids required for the synthesis of high-quality protein. Ways in which mutants lacking such controls can be selected are available. These mutants are likely to be able to produce nutritionally essential amino acids in large quantities.

REFERENCES

Belser, W. L., J. B. Murphy, D. P. Delmer, and S. E. Mills. 1971. Biochim. Biophys. Acta 237:1.

Bishop, L. R. 1928. J. Inst. Brewing 34:101.

Bishop, L. R. 1929. J. Inst. Brewing 35:316.

Blekhman, G. I., Z. S. Kagan, and W. L. Kretovich. 1971. Biokhimiya 36:1050.

Blackwood, G. C. 1973. Ph.D. Thesis, University of Newcastle upon Tyne, U.K.

Bryan, J. D. 1969. Biochim. Biophys. Acta 171:205.

Bryan, J. K. 1974. *In* Plant Biochemistry (J. Bonner and J. E. Varner, ed.). 2nd ed. Academic Press, New York.
Bryan, P. A., R. D. Cawley, C. E. Brunner, and J. K. Bryan. 1970. Biochem. Biophys. Res. Commun. 41:1211.
Carlson, P. S. 1973. Science 180:1366.
Cheshire, R. M., and B. J. Miflin. 1973. Plant Physiol. 51 (suppl.): abstr. 291.
Chu, M., and J. M. Widholm. 1972. Physiol. Plant. 26:24.
Cotton, R. G. H., and F. Gibson. 1969. Biochim. Biophys. Acta 156:187.
Croy, L. I., and R. H. Hageman. 1970. Crop Sci. 10:280.
Datko, A. H., J. Giovanelli, and S. H. Mudd. 1973. Plant Physiol. 51(suppl.):abstr. 270.
Datta, P. 1969. Science 165:556.
Deckard, E. L., R. J. Lambert, and R. H. Hageman. 1973. Crop Sci. 13:343.
Delaney, S. F., A. Dickson, and N. G. Carr. 1973. J. Gen. Microbiol. 79:89.
Dougall, D. K. 1970. Phytochemistry, 9:959.
Dunham, V. L., and J. K. Bryan. 1969. Plant Physiol. 44:1601.
Dunham, V. L., and J. K. Bryan. 1971. Plant Physiol. 47:91.
Eilrich, G. L., and R. H. Hageman. 1973. Crop Sci. 13:59.
Finlayson, A. J., and W. B. McConnell. 1969. Can. J. Biochem. 47:415.
Folkes, B. F. 1970. Proc. Nutr. Soc. 29:12.
Furuhashi, K., and M. Yatazawa. 1970. Plant Cell. Physiol. 11:569.
Gamborg, O. L., and F. W. Keeley. 1966. Biochim. Biophys. Acta 115:65.
Gilchrist, D. G., T. S. Woodin, M. S. Johnson, and T. Kosuge. 1972. Plant Physiol. 49:52.
Giovanelli, J., and S. H. Mudd. 1967. Biochem. Biophys. Res. Commun. 27:150.
Giovanelli, J., S. H. Mudd, and A. H. Datko. 1973. Plant Physiol. 51(suppl):abstr. 269.
Grasmanis, V. O. 1969. Aust. J. Biol. Sci. 22:1313.
Greenberg, D. M. 1969. Metabolic Pathways. Vol. III. Academic Press, New York.
Grobbelaar, N., and F. C. Steward. 1969. Phytochemistry 8:553.
Hay, R. E., E. B. Earley, and E. E. DeTurk. 1953. Plant Physiol. 28:606.
Holden, J., and I. Morris. 1970. Arch. Microbiol. 74:58.
Hood, W., and N. G. Carr. 1971. J. Bacteriol. 107:365.
Hood, W., and N. G. Carr. 1972. J. Gen. Microbiol. 73:417.
Jacob, F., and J. Monod. 1961. J. Mol. Biol. 3:318.
Johnson, V. A., P. J. Mattern, and J. W. Schmidt. 1970. Proc. Nutr. Soc. 29:20.
Kanamori, M., and R. L. Wexom. 1963. J. Biol. Chem. 238:998.
Miflin, B. J. 1971. Arch. Biophys. Biochem. 146:542.
Miflin, B. J. 1973. *In* Biosynthesis and Its Control in Plants (B. V. Milborrow, ed.). Academic Press, London, p. 49.
Miflin, B. J. 1969. J. Exp. Bot. 20:810.
Miflin, B. J. 1969b. Phytochemistry 8:2271.
Miflin, B. J. 1974. Plant Physiol. 54:550.
Miflin, B. J., and H. Beevers. 1974. Plant Physiol. 53:870.
Miflin, B. J., and P. R. Cave. 1972. J. Exp. Bot. 23:511.
Minamikawa, T. 1967. Plant Cell Physiol. 8:695.
Oaks, A. 1965a. Biochim. Biophys. Acta 111:79.
Oaks, A. 1965b. Plant Physiol. 40:149.
Oaks, A., and R. G. S. Bidwell. 1970. Annu. Rev. Plant Physiol. 21:43.
Oaks, A., D. J. Mitchell, R. A. Barnard and F. J. Johnson. 1970. Can. J. Bot. 48:2249.
Pate, J. S. 1968. *In* Recent Aspects of Nitrogen Metabolism (E. J. Hewitt and C. V. Cutting, ed.). Academic Press, London, p. 219.

Przybylska, J., and J. Pawelkiewicz. 1965. Bull. Acad. Pol. Sci. Ser. Sci. Biol. 13:327.
Sharma, R. K., and R. Mazumder. 1970. J. Biol. Chem. 245:3008.
Stadtman, E. R. 1970. *In* The Enzymes, Vol. 1 (P. D. Boyer, ed.). Academic Press, New York, p. 398.
Umbarger, H. E. 1969. Annu. Rev. Biochem. 38:323.
Van Die, J. 1962. Acta Bot. Neerl. 11:1.
Weber, H. L., and A. Bock. 1968. Arch. Microbiol. 61:159.
Widholm, J. M. 1972a. Biochim. Biophys. Acta 261:44.
Widholm, J. M. 1972b. Biochim. Biophys. Acta 261:52.
Widholm, J. M. 1972c. Biochim. Biophys. Acta 279:48.
Widholm, J. M. 1971. Physiol. Plant. 25:75.
Wong, K. F., and D. T. Dennis. 1973. Plant Physiol. 54:322.

DISCUSSION

DR. RABSON: I would first like to call on Dr. John Ingversen from RISØ (Danish Atomic Energy Commission Laboratory, Roskilde, Denmark) Lab to discuss a point that was brought up in regard to high lysine and the amino acid changes that go on. There has been some recent work at RISØ that Mr. Brand, who is a graduate student with Professor von Wettstein, has been doing on determining the amino acid content of the 1508 barley high-lysine mutant and contrasting it with the parental type.

DR. INGVERSEN: Yes, actually, I am going to refer to some work that even I have not seen performed. I have got letters, and I am trying to describe what he has done.

Some years ago Sodek and Wilson (1970. Arch. Biochem. Biophys. 140:29) reported some work done on segregating ears where both the opaque character and the normal character were in the ear of the corn plant. They injected lysine into this ear, and they found the opaque endosperms did not degrade lysine to the same extent as the normal endosperms did.

That indicated that there might be some difference in the catabolic processes of lysine between these two corn varieties. That was the point that Ernest Brand, who is a graduate student at RISØ, took up. He first isolated the free amino acids from our mutant 1508 and from the parent variety. He ran a thin-layer chromatography to determine semiquantitatively the content of the catabolic products of lysine degradation.

What he had was to go from lysine—and there are several steps in between to pipecolate—and further on to glutamate. That is not a fully worked out degradation procedure for lysine, but there is some indication that in yeast (and I think also in liver cells) this degradation process occurs.

He found pipecolate in very limited amounts in mutant 1508. That was a semiquantative determination. He took off the compound from the thin-layer plate, and he determined it by some reaction. Since then he has incubated some developing endosperms with ^{14}C lysine, and whereas in the normal variety 8 percent of the uptake in lysine was found, in the mutant only 2 percent of the uptake in lysine was found. This suggests that there might be a block in the degradation process. He is going to continue work on this by isolating the enzymes involved and looking at the activity.

I do not suggest that this is the reason the mutant is high in lysine. It is probably an indirect effect of the mutation, but it might give us some hints as to the direct effect of the mutated gene.

I want to make another comment. Dr. Miflin said that in general we see that the protein composition of endosperm protein is more or less determined by the supply of free amino acids from the green part of the plant and it is exemplified in the deposition of hordenines, which are rich in glutamic acid and proline. But in the mutant that we have, we have a very decreased content of these prolamins, and instead the ethylenediopamine fraction is increased greatly. That means that the endosperm has a tendency to change the supply of amino acids from glutamic acid to whatever amino acid it really needs.

DR. DURE: Dr. Miflin, I want to ask you two questions. The first is, Do you have evidence that phosphohomoserine serves as a substrate for cystathionine synthesis in addition to being on the pathway to threonine?

Second, the fact that you can pile up enormous levels of tryptophan brings up the question of whether plants actively have a degradative pathway that is distinct from the biosynthetic one. The data we just heard suggest that certainly for lysine one must exist. Do you or Dr. Oaks have any further information on degradation of amino acids and its regulation?

DR. MIFLIN: First, the work on phosphohomoserine was done by Dr. Giovanelli and co-workers [A. H. Datko, J. Giovanelli, and S. H. Mudd. 1973. Plant Physiol. 51 (suppl. abstr.):269]. As far as the further metabolism of phosphohomoserine in threonine biosynthesis is concerned, there is no evidence yet on the presence of the enzymes involved in higher plants. Also there is nothing to suggest that phosphohomoserine is not involved in threonine biosynthesis. The evidence suggests that the branch point between methionine and threonine biosynthesis is at a different place than in bacteria, which suggests that the possible regulatory enzymes may also be different. As regards tryptophan breakdown, we have little information. The pathways of amino acid biosynthesis are only just being studied and we have not yet worked out how amino acids break down. It is probable that in the carrot mutant tryptophan is accumulated away from any degradation pathways that may be present—Dr. Oaks may wish to comment on this matter.

The other point I would like to make is that I did not quite say what I was quoted by Dr. Invergsen as saying—that it was an absolute open and shut case that what happens in the transport stream determined what proteins were formed. All I was suggesting was that there may be some reason behind the correlation between the amino acids found in large amounts in the transport stream and the amino acid composition of the prolamins which are formed late in the development of the seed. Obviously, while there is still a lot of green, metabolically active tissue, there is the possibility of a lot of interconversions of amino acids, but during the later part of grain development this is less likely. I refer to some work by Finlayson and McConnell (1969. Can. J. Biochem. 47: 415–418), that showed that if they injected ^{14}C phenylalanine or ^{15}N ammonia at different times during ear development, the incorporation of the ^{14}C into a number of amino acids declined as the ear matured, and in the latest period, during the time in which gluten synthesis predominated, virtually all the phenylalanine went straight into protein. During this period there was very little incorporation of ^{15}N from ammonia into any of the amino acids. This suggests to me that this critical phase of development, which is probably not the time when albumins and globulins are formed, is the time when there is less interconversion of amino acids in the seed.

DR. OAKS: I would like to say a little bit about interconversion and the degradation of amino acids. I think we have to be very careful of the plant part that we use. For example, leucine is not metabolized extensively by the corn root tip. Valine and lysine are also not metabolized in root tip tissue. If we look at other amino acids, we find that proline, aspartic, and glutamic acid are degraded extensively in a number of different tissues in corn. I think this is something that we need to consider when we are concerned about catabolism of these substances.

DR. MUNCK: My comment is on the point that seeds, mature seeds, germinating seeds, are very often used for amino acid metabolism studies. There are so many indications that germinating seeds are quite a different system biochemically from developing seeds that I would suggest that we split those two fields and that biochemists urgently go into the problem of the developing seeds also.

DR. DOY: Some of you here will know that I have spent most of my working life as a biologist investigating the control of the multibranched aromatic pathway leading to phenylalanine, tyrosine, and tryptophan. I have worked with eukaryotic and prokaryotic microorganisms, and my main interest had been in DAHP synthetase, the first enzyme of the common pathway and a major point of control. As a result of this experience there is a whole list of points I want to make. One is that prokaryotic microorganisms are not eukaryote microorganisms, and neither are plants. Therefore I get anxious if I see people taking work with microorganisms and relating it directly to whole plants. I was very glad that you brought out in your talk that control mechanisms may vary between microorganisms and whole plants. In higher plants we can expect feedback control on the action of the preformed enzyme, but there is little evidence for control of enzyme synthesis, even in lower eukaryotic microorganisms, by feedback repression. Repression may, therefore, not be a mechanism that is very important in higher plants. In trying to relate prokaryotic microorganisms to whole plants, it should be remembered that the microorganisms depend on the close interplay of control of enzyme synthesis with control of enzyme function. Widholm's results with 5-methyl-tryptophan [see discussion, p. 152] make a very good point. It is easy to select mutants when you have a positive selection system. One is looking for resistance and can frequently pick out the mutant easily. It is a different matter altogether when one is looking for a negative selection, that is, for an auxotrophic type of mutant. It is when we look for the latter kind of mutant that we expect problems and one needs to develop selection methods.

Dr. Widholm's work illustrates several other points as well. He may have expected his selection to yield the one class of mutation, one that affected anthranilate synthetase. In actuality he found that some of his mutants had an unaltered anthranilate synthetase. The point that I want to make is that when you use such a positive selection method you will probably select every class of mutant that will be resistant. There are often several classes. Other examples are those that will not take up the amino acid or will compartmentalize it where it will not affect the enzyme.

Widholm's work can be used to illustrate a situation where haploid tissues might give cleaner results than diploid or polyploid tissue. Widholm was using diploid tissue, and therefore one would not expect to mutate the multiple genes concerned in anthranilate synthetase. As a result, the data are difficult to interpret kinetically. In a haploid situation there is at least a chance that only a single gene copy specifies the function, and therefore the effect of mutation could sometimes be absolute. I shall return to this in my own talk.

Dr. Dorothy Halsall, of our Department, has found that when rice is germinated in the presence of lysine there is an effect on both shoot and root development. At 10^{-2} M lysine, 50 percent inhibition of shoot growth occurs and development of roots is markedly affected. At 10^{-1} M lysine, no roots develop. Aspartokinase from etiolated rice shoots measured in crude extracts and in partly purified preparations is inhibited by both lysine and threonine about equally, each to 50 percent of the activity. The separate inhibitions are, however, not additive in mixtures, and, in the assays used, inhibition is only slightly enhanced over that by a single inhibitor. Aspartokinase is also inhibited by the lysine analogue *S*-(β-aminoethyl)-cysteine (AEC). The lysine path-specific enzyme, dihydrodipicolinic acid synthetase, a site of lysine (or AEC) inhibition in *Escherichia coli* (D. M. Halsall. In press. Biochem. Genet.) appears insensitive to lysine inhibition. This suggests that a single mutation in rice conveying resistance to AEC inhibition might lead to overproduction of end products in this pathway. These findings, results of a study funded by The Rockefeller Foundation, have been summarized in *Annual Report 1972*, The Research School of Biological Sciences, Australian National University, and *Genetics Report 1972*, CSIRO Division of Plant Industry, Canberra. Dr. Halsall intended to use the analogue to select mutants in tissue cultures of various crop plants but, using her work and suggestion as a basis, Chaleef and Carlson (Chaleef, personal communication) have already done this. Dr. Halsall's results also show that information gained with the prokaryote *Escherichia coli* cannot blindly be applied to plant systems.

DR. MIFLIN: I think I would agree with almost everything Dr. Doy said, and certainly Widholm has selected out some of these transport mutants, which I believe he has been able to regenerate.

J. S. BOYER

Stress Relationships in Protein Synthesis: Water and Temperature

From the point of view of total yield, the two most limiting aspects of the agricultural environment are water availability and unfavorable temperatures. Large areas of every major continent are affected by these problems, in spite of the presence of soils that would otherwise favor agriculture. When protein production is considered as well, the interaction of water and temperature with agriculture becomes even larger, because the percent protein in the crop frequently changes at the same time that total yield changes. Consequently, protein production is highly sensitive to growth conditions, and the character of the growing season bears strong implications for the nutrition of the world's population.

In the last few years, several excellent books (Slatyer, 1967; Rose, 1968; Kramer, 1969; Levitt, 1972) and a review (Hsiao, 1973) have been devoted to the effects of environment on plants, largely with emphasis on the broad biological problems of water or temperature. Unfortunately, considerations of agronomic yield are frequently missing from this literature, particularly with regard to the metabolic mechanisms that control yield. In a review of the physiological factors that control grain yield, Yoshida (1972) was able to devote a few paragraphs to the effects of unfavorable temperatures but was unable to find any literature dealing with the effects of water availability on grain yield. The effects of these factors on protein yield were similarly neglected. Perhaps it is not surprising that these kinds of experiments are rarely undertaken—they are both time-consuming and expensive. Nevertheless, if the agricultural challenges of the next few years are as large as has been predicted, there are hardly more relevant experiments that can be done.

In this review, I would like to explore a few recent ideas that bear on how temperature and water may affect protein production in crops. An exhaustive review is not intended, nor will every facet of these problems be covered. However, some recent data have come to light that may be relevant for the future direction of research in these areas, and it is probably appropriate at this point to gather perspective and comment on the directions for research that are likely to be beneficial in the future.

One of the most significant developments of recent years has been the accumulation of literature on plant enzymes whose synthesis responds to the character of the environment. Notable among these is nitrate reductase (NR), which is the first enzyme to act on nitrate in the series of biochemical steps leading to the incorporation of reduced nitrogen into amino acids and eventually into protein (Beevers and Hageman, 1969). This initial step is thought to limit the flux of reduced nitrogen for the plant, because NR has a lower activity than the other enzymes in the biochemical sequence leading to amino nitrogen (Beevers and Hageman, 1969). It is an unstable enzyme, having a half-life of about 4.5 hours at 30°C in maize (Morrilla *et al.*, 1973). NR activity is inducible by nitrate (Beevers and Hageman, 1969), limited by supraoptimal or suboptimal temperatures (Beevers *et al.*, 1965; Onwueme *et al.*, 1971), and affected by drought (Mattas and Pauli, 1965; Huffaker *et al.*, 1970; Bardzik *et al.*, 1971; Morrilla *et al.*, 1973).

In wheat (Croy and Hageman, 1970) and maize (Deckard *et al.*, 1973), Hageman and his coworkers have shown correlations between high NR and high grain yields. In many cases, the increased yields in high-NR lines were accompanied by increased protein in the grain. This work is of particular significance because it suggests that the availability of reduced nitrogen may affect grain yield, at least in some cases.

Thus, there may be a link between the total protein accumulated by the plant and the enzyme NR, because of the limiting nature of NR activity. Since rapid protein synthesis is required for the maintenance of NR activity and NR in turn appears to limit the flow of reduced nitrogen to protein, NR represents a possible point of control for protein production by plants, and a considerable portion of this review will be devoted to it.

THE EFFECTS OF TEMPERATURE ON PROTEIN SYNTHESIS

The two aspects of the temperature response of plants that are most interesting are (a) the short-term responses to high and low temperatures and (b) the adaptive responses.

The short-term responses are largely determined by the activation energies of the complex of reactions that make up the process of protein synthesis. In those studies where the response of growth can be compared with that of protein synthesis, an important generality emerges: Protein synthesis occurs at the highest rate at those temperatures that also give the highest rates of growth for the plant. Beevers *et al.* (1965) have shown that the optimum temperature for NR synthesis in leaves of maize was 35°C. This compares with an optimum for leaf growth at 35°C for leaves of maize (Watts, 1972). For radish, the optimum for protein synthesis was 31°C (Beevers *et al.*, 1965). Radish grows well in cool seasons. Leaf growth in bean, which also does well at cool temperatures, has an optimum for leaf growth at 25–30°C (Wilson and Ludlow, 1969).

In the bacteria, a similar correspondence between growth and protein synthesis holds true. Friedman *et al.* (1969) showed that free 30S and 50S ribosomal subunits accumulate when *E. coli,* a mesophile, is grown at 8°C. In this species, cell division virtually ceases at 8°C. These authors conclude that the formation of some portion of the initiation complex was inhibited by cold but that peptide chains could be completed if they had already been initiated.

Micrococcus cryophilus, a psychrophilic bacterium that grows poorly above 25°C, exhibits an inhibition of protein synthesis above this temperature (Malcolm, 1969). In this species, the block can be traced to the amino acyl transferases for glutamate and proline. These two enzymes lose activity for the cognate species of tRNA at these temperatures, in part because of changes in the tRNA's to configurations that are unfavorable for the enzymes and in part due to changes in the enzymes themselves. Somewhat similar results have been found with the psychrophilic yeast *Candida gelida* (Nash *et al.*, 1969). Several amino acyl tRNA synthetases appear to be denatured at temperatures above the growth optimum (25–30°C). Leucyl tRNA synthetase was most affected. In addition to the synthetases, the formation of ribosome-bound polypeptide chains was also inhibited at high temperatures.

It is perhaps not surprising that the temperature sensitivity of growth and protein synthesis are correlated. Cells require new protein for growth, and there must be some type of relationship between the two processes. Although the literature on higher plants includes little about the mechanism of temperature effects on protein synthesis, the bacterial literature cited above suggests that certain factors required for protein synthesis, notably particular enzymes and initiation factors, may control the temperature sensitivity of the process.

The regulation of protein synthesis is only part of the story when one is concerned with protein production, however. The protein content of a tissue is determined by the difference between the rate of protein synthesis and the rate of protein degradation. In spite of considerable work demonstrating the role of protein degradation in thermal injury at moderately high temperatures, there is little that compares the rate of degradation and synthesis of total protein or of single enzyme systems *in vivo* when the organism is exposed to unfavorable temperatures. In one of the few studies, Onwueme *et al.* (1971) have shown that the reduction in NR activity at supraoptimal temperatures is caused primarily by an inhibition of synthesis and only to a small extent by an increase in the rate of degradation. This suggests that changes in synthesis are more important than changes in degradation. Unfortunately, this valuable study is an isolated one, and others have tended not to distinguish between synthesis and degradation. Guinn (1971), for example, found that tissue protein declined when cotton plants were exposed to cool (10–15°C) temperatures for a few days. On the other hand, Chowdhury and Zubriski (1973) observed that plant protein content was high but that there was an overall decrease in plant growth. Thus, it appears that the protein content of plant tissue may increase or decrease with temperature extremes, but the roles played by synthesis and degradation remain uncertain.

The total protein content of plants also can increase or decrease when plants adapt to cool temperatures. Guinn (1971) showed that cotton exposed for several days to cool temperatures was undamaged, while "unhardened" controls were extensively damaged at low temperatures. The protein content of the hardened tissue was less than that of the control. On the other hand, Siminovitch *et al.* (1967) showed that the protein content of tissue increased in trees during cold hardening in the autumn. Leucine incorporation into protein occurred more rapidly at this time, so that the increased size of the protein pool was associated with increased protein synthesis. Although it is important to distinguish between hardening against chilling temperatures (Guinn, 1971) and hardening against freezing temperatures (Siminovitch, 1967), these studies illustrate once again that temperature has a variable effect on the protein content of plant tissue.

THE EFFECTS OF WATER DEFICITS

In contrast to the variable effects of temperature, water deficits almost always lead to a reduction in the rate of protein synthesis. Barnett and Naylor (1966), using a pulse-chase technique, showed that the rate of protein synthesis was less in desiccated leaves than in the well-watered controls of Bermuda grass. In addition, there was a decrease in the protein content of the tissue (Barnett and Naylor, 1966), as has also been shown by others (Petrie and Wood, 1938; Shah and Loomis, 1965). The decreased rates of synthesis are accompanied by increased levels of certain amino acids (Petrie and Wood, 1938; Kemble and MacPherson, 1954; Chen *et al.*, 1964; Younis, *et al.*, 1965; Barnett and Naylor, 1966; Stewart, 1973). Barnett and Naylor (1966) showed that amino acid accumulation was not associated with an increased rate of amino acid synthesis but was caused instead by a decreased rate of amino acid utilization for protein synthesis. This was confirmed by a later study that showed decreased rates of leucine incorporation into protein following recovery from desiccation, in comparison with controls that had not been desiccated (Ben-Zioni *et al.*, 1967). In addition, the polyribosome content of desiccated tissue decreased during desiccation (Hsiao, 1970; Morrilla *et al.*, 1973).

Another line of evidence also suggests that protein synthesis is reduced in plant tissue that has been desiccated. NR activity declines soon after the desiccation process begins (Mattas and Pauli, 1965; Huffaker *et al.*, 1970; Bardzik *et al.*, 1971; Morrilla *et al.*, 1973). No evidence could be found for increased rates of degradation or reversible activation–deactivation of the enzyme (Morrilla *et al.*, 1973).

The decline in the rate of protein synthesis that leads to decreased NR activity may not be universal, however. Morilla *et al.* (1973) and Dove (1967) found that RNase activity increased during desiccation. Tvorus (1970) and DeLeo and Sacher (1970) have provided evidence that the increase in RNase is cycloheximide sensitive. Thus, it may be that there is an overall decline in protein synthesis associated with certain proteins but that there are a few proteins that actually appear to be synthesized at an increased rate.

The idea of an overall decrease in protein synthesis but an increase in rates of RNase synthesis may be explained by differential effects of desiccation on protein synthesis. Since NR is so unstable and short-lived, a significant fraction of protein synthesis by the cell must be devoted to this enzyme. The correlation between polyribosome content and NR synthesis by maize (Travis *et al.*, 1970; Travis and Key, 1971) supports this concept. The reduction in the overall rate of protein synthesis might then reflect a reduction in synthesis of unstable enzymes like NR (Bardzik *et al.*, 1971) but might still permit a rise in the rate of synthesis of other enzymes such as RNase.

These enzymatic changes take place at leaf water potentials which are close to those which affect leaf growth. At leaf water potentials of −4 bars, the growth of leaves is 25 percent that of the well-watered controls in maize, soybean, and sunflower (Boyer, 1968, 1970). In maize, NR activity is about 50 percent of the control value at −4 bars (Morrilla *et al.*, 1973). Photosynthesis and dark respiration are virtually unaffected at these potentials (Boyer, 1970). This correspondence between growth and protein synthesis during desiccation in leaves is reminiscent of the correspondence between the two processes at unfavorable temperatures.

THE OUTLOOK

There are certain similarities between the responses of protein production to desiccation and unfavorable temperatures in plants. Both show a relationship between growth and protein synthesis. Both indicate that protein synthesis probably exerts a larger degree of regulation of plant protein content than does protein degradation. However, in higher plants, the control mechanisms for these responses are not known.

The foregoing is not intended to imply that the correlation between growth and protein synthesis in unfavorable environments shows that there is a causal relationship between the two processes. Indeed, Onwueme *et al.* (1971) have shown that NR levels in barley made relatively little difference to the subsequent response of growth to supraoptimal temperatures. However, increased yield generally must be accompanied by increased protein production by the plant, and consequently, at this level, the relationship is important. It suggests that one may continue using growth as a means of selecting for superior plant performance and, in the process, automatically select for improved protein production.

Nevertheless, growth permits only an indirect selection criterion. Probably, higher resolution could be obtained by assaying for the protein content of lines that give high total yields in unfavorable environments. This is time-consuming, but it is a crop-improvement procedure that is being used at this moment.

Time could be saved in selecting lines with improved protein production if a screening process could be applied to young seedlings. There have recently been two interesting developments in this area. Alofe *et al.* (1973) recognized the potential of using NR activity as an indicator of the capability of a plant for protein synthesis. They have identified maize lines with high NR activities at unfavorable night temperatures. When high-NR lines were crossed with inferior lines, the hybrids showed intermediate levels of NR. In some cases, NR activity in the hybrid was higher than could be explained by heterosis. Unfortunately, yield data are not available for the crosses, so it is not possible to determine whether the improved NR levels had beneficial effects on protein levels.

Another screening approach has been suggested by Singh *et al.* (1972). These workers were concerned with the effects of desiccation and showed that varieties of barley that accumulate large quantities of proline in the seedling during water deficiency had a higher stability of yield in a range of desiccating conditions. Proline is known to accumulate as one of the free amino acids in cells during desiccation (Barnett and Naylor, 1966). Once again, however, figures for absolute yield and protein contents are not available for the work with barley (Singh *et al.*, 1972), so it is not possible to completely evaluate the usefulness of proline as a screening tool.

These two studies, while suggestive, nevertheless illustrate a weakness in much scientific data—yields were not measured. No doubt, this was because of the long times needed for these types of experiments. An important problem, of course, is the difficulty in obtaining experimental conditions that are stable enough for long enough to make the data interpretable, since, for crops, the crucial events occur at the end of a long period of development. Field experiments are one way of approaching this problem and are a necessary step in the development of new agricultural procedures, but preliminary work under controlled conditions generally is required before the relevant physiological factors can be tested in the field. This preliminary work should include measurements of grain yield.

What then should be the direction of future research in the area of environment and protein production? The most glaring gap in our knowledge is the question of how environment affects total yield, protein yield, and protein quality in grain. As was pointed out earlier, only a few data are available to indicate the effects of temperature or desiccation on total yield. I know of virtually no data that relate the quality or quantity of grain protein to the environment under which the grain was grown. As a consequence, of course, there are none that deal with the physiological mechanisms that control these parameters. Such data are vital before we can know the validity of such screening procedures as selection for seedlings with high NR or high proline.

I think the most promising means of investigating these questions still lies with the concept of the controlled environment, despite the criticism that this kind of experimentation has received from some quarters. No one disputes the need for field experiments as part of the procedure for the development of agronomic practices. However, the field does not allow environmental parameters to be varied independently and at will. Without the controlled-environment approach, it will take much longer to obtain answers to the above questions and some may not be answerable at all.

Controlled-environment facilities have been justifiably faulted because the environments have generally been different from those in the field and because, in a few cases, the best builders have not necessarily been the best experimenters. Recent developments have shown, however, that controlled environments can now duplicate the light and evaporation conditions present in the field (traditionally the two weak points in older systems) and they can produce crops with yields that equal field performance in every way. A few well-executed experiments in such facilities could contribute a great deal toward answering the questions of how environment interacts with grain production.

Research that deals with the physiological mechanisms that control yield is also in need of further support. From the foregoing, it is apparent that physiological experiments with whole plants are rare, although much is known about the molecular nature of protein synthesis. Nitrogen metabolism, photosynthesis, translocation, and water relations are physiological processes that have major effects on crop production but are influenced in ways that generally cannot be predicted from *in vitro* experiments. For example, still unknown are the degree to which concurrent photosynthesis limits grain production and the importance of translocation for the protein content of the grain. If such knowledge were available for a single crop, it could provide a guide for research with other crops that could reduce the effort needed to rapidly extend the results. Until this kind of research provides direction, the plant breeder will be forced to follow traditional and generally slower ways of improving protein production.

REFERENCES

Alofe, C. O., L. E. Schrader, and R. R. Smith. 1973. Crop Sci. 13:625.

Bardzik, J. M., H. V. Marsh, Jr., and J. R. Havis. 1971. Plant Physiol. 47:828.

Barnett, N. M., and A. W. Naylor. 1966. Plant Physiol. 41:1222.

Beevers, L., and R. H. Hageman. 1969. Annu. Rev. Plant Physiol. 20:495.

Beevers, L., L. E. Schrader, D. Flesher, and R. H. Hageman. 1965. Plant Physiol. 40:691.

Ben-Zioni, A., C. Itai, and Y. Vaadia. 1967. Plant Physiol. 42:361.

Boyer, J. S. 1968. Plant Physiol. 43:1056.
Boyer, J. S. 1970. Plant Physiol. 46:233.
Chen, D., B. Kessler, and S. P. Monselise. 1964. Plant Physiol. 39:379.
Chowdhury, I. R., and J. C. Zubriski. 1973. Agron. J. 65:529.
Croy, L. I., and R. H. Hageman. 1970. Crop Sci. 10:280.
Deckard, E. L., R. J. Lambert, and R. H. Hageman. 1973. Crop Sci. 13:343.
DeLeo, P., and J. A. Sacher. 1970. Plant Physiol. 46:806.
Dove, L. D. 1967. Plant Physiol. 42:1176.
Friedman, H., P. Lu, and A. Rich. 1969. Nature 223:909.
Guinn, G. 1971. Crop Sci. 11:262.
Hsiao, T. C. 1970. Plant Physiol. 46:281.
Hsiao, T. C. 1973. Annu. Rev. Plant Physiol. 24:519.
Huffaker, R. C., T. Radin, G. E. Kleinkopf, and E. L. Cox. 1970. Crop Sci. 10:471.
Kemble, A. R., and H. T. McPherson. 1954. Biochem. J. 58:46.
Kramer, P. J. 1969. Plant and Soil Water Relationships. McGraw-Hill Book Company, New York.
Levitt, J. 1972. Responses of Plants to Environmental Stresses. Academic Press, New York.
Malcolm, M. L. 1969. Nature 221:1031.
Mattas, R. E., and A. W. Pauli. 1965. Crop Sci. 5:181.
Morrilla, C. A., J. S. Boyer, and R. H. Hageman. 1973. Plant Physiol. 51:817.
Nash, C. H., D. N. Grant, and N. A. Sinclair. 1969. Can. J. Microbiol. 15:339.
Onwueme, I. C., H. M. Lande, and R. C. Huffaker. 1971. Crop Sci. 11:195.
Petrie, A. H. K., and J. G. Wood. 1938. Ann. Bot. 2:887.
Rose, A. H. (ed.). 1967. Thermobiology. Academic Press, New York.
Shah, C. B., and R. S. Loomis. 1965. Physiol. Plant. 18:240.
Siminovitch, D., B. Rheaume, and R. Sachar. 1967. *In* Molecular Mechanisms of Temperature Adaptation (C. L. Prosser, ed.). American Association for the Advancement of Science, Washington, D.C., p. 3.
Singh, T. N., D. Aspinall, and L. G. Paleg. 1972. Nat. New Biol. 236:188.
Slatyer, R. O. 1967. Plant–Water Relationships. Academic Press, New York.
Stewart, C. R. 1973. Plant Physiol. 51:508.
Travis, R. L., and J. L. Key. 1971. Plant Physiol. 48:617.
Travis, R. L., R. C. Huffaker, and J. L. Key. 1970. Plant Physiol. 46:800.
Tvorus, E. K. 1970. Fiziol. Rast. 17:787.
Watts, W. R. 1972. II. Journ. Exp. Bot. 23:713.
Wilson, G. L., and M. M. Ludlow. 1968. Journ. Exp. Bot. 19:309.
Yoshida, S. 1972. Annu. Rev. Plant Physiol. 23:437.
Younis, M. A., A. W. Pauli, H. L. Mitchell, and F. C. Stickler. 1965. Crop Sci. 5:321.

DISCUSSION

DR. SCHRADER: I would like to make a few comments with regard to some of the gaps Dr. Boyer mentioned. This research was done on 16 genotypes (a four-parent diallel cross) of corn in the Biotrol at the University of Wisconsin at Madison. We used five different temperature regimes, but in the published report (C. O. Alofe, L. E. Schrader, and R. R. Smith. 1973. Crop Sci.

13:625–629), we discussed only the three highest temperatures. These plants were grown in the Biotrol for about 30 days. We were not able to grow these plants to maturity because the ceilings are only 7 feet high in these environmental rooms. Fresh and dry weights were recorded at each sampling. These data will be in Mr. Alofe's doctoral dissertation and, it is hoped, will be published soon.

Combining-ability analyses showed that general combining ability for nitrate reductase activity was extremely high in these plants. Thus, tolerance of heat stress in certain inbreds was reflected in the performance of F_1 progenies from those inbreds. This suggests that one could develop F_1 hybrids of corn that are more tolerant of the high-temperature stress.

We observed that growth rates decreased appreciably during the high-temperature stress. The stress was imposed for 15 hours per day. With this stress for a 30-day period, it really began to show on the plants at the highest temperature.

DR. BOYER: Larry, I thought that this was a very valuable study, and I am pleased to hear that you are carrying it on. I would think that this sort of approach could be very important, particularly if we can carry it a lot further and look at what happens at the end of the growing season.

DR. DIECKERT: We have a fortuitous experiment going on in the Virgin Islands. I would like to talk to you about it. It takes about 13 months to grow coconuts, and these fruit are growing on trees along the beach on Saint John Island. This island is a desert the first 6 months of the year, and then it gets barely adequate rainfall the last 6 months of the year. I have been following the effects on the growth of the fruit.

The ovule grows first and then the endosperm fills the cavity or partially fills the cavity. There is one stalk of flowers put on every month. So you have a monthly arrangement to study. You can tell approximately when the fruit is growing and when the ovular cavity is growing. You can tell when the endosperm is growing just by counting down on the fruit of the tree.

Coconut trees on the beach get all the seawater they can use. They cannot use seawater too well. When it rains, there is a little pulse of fresh water that dilutes out the seawater for a week or 3 or 4 days, and then the seawater comes in and washes it back out.

During parts of the year when rains are sporadic, there is inhibition of growth. When growth of the central cavity is occurring, that is when the ovule is growing. Drought checks growth. No matter what happens after that, that is the limiting factor for the endosperm. That loss of growth during the time the ovular cavity develops determines in a large way how much endosperm you are going to get at the end of the growth period, and it apparently also is having an effect on how much protein is laid down in the endosperm during this period.

I think that seed crop total protein production in the field will depend on what sort of water stress is put on the plant during the time the ovule is reaching its maximum growth. It remains to be seen whether that hypothesis is correct, but it sure looks good for coconuts.

DR. BOYER: In our experience, leaf growth is the most sensitive indicator of desiccation that we have been able to find. Inhibition of growth can occur in the light or at low relative humidity, even in an otherwise well-watered plant.

In other words, the conditions of the atmosphere that crops normally meet out in the field even when well watered are enough to radically affect the rate of leaf growth. Ted Hsiao (E. Acevedo, T. C. Hsiao, and D. W. Henderson. 1971. Plant Physiol. 48:631) has shown this and has also shown that after short periods of slow rates of growth, leaf growth will often recover completely. If desiccation occurs over a long period, the leaves often do not completely recover. Thus, desiccation can also predispose leaves to develop to a smaller final size than those that have not been subjected to desiccation.

DR. MUNRO: In mammalian tissues and probably in prokaryotes, too, the rate of protein synthesis often correlates closely with the amount of RNA per unit of tissue or per DNA or per cell. Could you use RNA measurements to help you select plants? RNA measurements are relatively simple though not so simple in plant materials as they are in animal tissues.

DR. BOYER: There are differences in total RNA content of tissue that has been desiccated. In general, total RNA content goes down, but as I mentioned earlier, there are differences in the subfractions that may be significant. Some fractions change, even increase; others decline in size. Clearly, there are differences in RNA that are brought about by desiccation, and it may be that these could be utilized in breeding, but thus far there has been very little work done on this. I think we need to know more about what the changes are and whether there are differences between varieties with regard to them.

DR. DOY: I would like you to comment on the difference between C-4 photosynthetic plants and C-3. Your talk has been concerned with C-4. What happens with C-3 plants?

DR. BOYER: In Illinois C-3 soybean and C-4 maize are always competing for top billing as far as acreage is concerned, and so we did do a little study to compare soybean response and maize response to drought. In comparably aged tissue, soybean photosynthesis is less sensitive than maize photosynthesis. Furthermore, maize seems to have a senescence system that causes the leaves to senesce during drought. This does not happen in soybeans until conditions are very extreme. I should point out, however, that although maize shows a high sensitivity of photosynthesis to desiccation, nevertheless the total rate of photosynthesis is so superior to soybean that, even with this sacrifice in rate, it is accumulating dry matter more rapidly than soybean is.

DR. CANVIN: I would like to expand on the topic Dr. Dieckert introduced because it is a fairly important one, not limited to the coconut. It bears on having an understanding of how seeds develop, and two other seeds that fit into the same category as the coconut that immediately come to my mind are the castor oil bean plant and sunflowers, where you have hard integument surrounding the seed. The integument develops before you get any storage material laid down at all. It is not only water stress that will limit the size of this integument but also a lack of carbohydrate.

If you put them under low light you limit the size of the seed. Once the seed coat is laid down and hardens, and it hardens very quickly when only about 5 percent of the storage product is laid down, there is nothing you can do to increase the size of that seed. All you can do is fill it up.

On the other hand, if you have extremely good conditions when this integument is laid down, you get very large seeds produced, and then you expose it to either water stress or carbohydrate stress, a large number of these seeds abort. They do not fill up, and all you have left is the seed coat, with nothing inside. This is very common in castor bean seeds and in sunflowers. This type of seed development contrasts very strongly with the type of seed development such as the pea where the seed starts small and just keeps growing and growing. If you impose temporary stress on such plants, what happens is that the seed stops growing, and if the stress is not too long, growth will start again.

The seed does expand before all the storage product is in; there is no question of this. It is just like wheat: You have a large seed, but there is not very much storage product there; the plant does have to fill this up. If you impose the stress later, then what results is a shrunken seed. But in sunflower or castor oil plants you will never see a shrunken seed because the integument is solid.

DR. MIFLIN: In Britain we have a problem in terms of importing wheat. In terms of bread-making quality, the wheat that we grow in Britain is inferior and we have to import hard Canadian wheats. Now, this difference is apparently due to the superior content and quality of the protein in the Canadian wheats, and this is not a genetic difference; it is an environmental difference. It would appear that in Britain, where one thing we are not short of during the summer is rain, we get wheat that is deficient both in protein content and quality for bread making, which goes against the general thesis. I wonder if any bread technologists have any comment on this?

DR. HEYNE: You stated that 35°C to 38°C was the optimum temperature for protein synthesis in maize in laboratory studies. This would be considered a high temperature in the field, especially in Illinois. Cultivated crops do poorly in Kansas at such temperatures even if adequate moisture is available. We found that protein degradation occurs in wheat, as measured by its physical properties, when the air temperature was over 32°C and the relative humidity 20 percent or less. The temperature within the leaf would be expected to be less than 32°C due to transpiration. We assumed that optimum conditions for protein synthesis in the field for wheat would be less than 32°C. Protein synthesis of hard winter wheat is associated with temperature, relative humidity, and moisture availability, and *in vitro* studies in the laboratory are not always realized under field conditions.

DR. BOYER: There are a lot of physiological processes in maize that show the same temperature optimum, however. Photosynthesis, for example, will peak at about 40°C.

DR. HEYNE: What about relative humidity?

DR. BOYER: Relative humidity would have been about 50 to 60 percent in my experiments. However, the low humidities of which you speak would cause a large evaporative demand and would tend to cause drought effects. I should point out, however, that, of course, there are other species, such as radish, bean, and probably wheat, that tend to have lower temperature optima than the one that I quoted for maize. C_4 plants tend to have higher temperature optima than C_3, and in the C_3 radish, for example, the optimum for protein synthesis is 31°C. The optimum for leaf expansion in C_3 species is more on the order of 25 to 30°C. So, I think this depends on the species.

DR. EVERSON: I would just like to answer Dr. Miflin. What you are talking about is something other than protein. It is a lack of dormancy in wheat that is very common in England and in Sweden, and if you have a wet season as wheat is maturing, you get a high alpha amylase content, and flour from this wheat makes very poor bread. It is a common problem in the eastern United States. It has nothing to do, really, with the protein content. You may have, in those conditions, lower protein content, which might accentuate the poor quality, but I think this other is the answer.

DAVID T. CANVIN

Interrelationships between Carbohydrate and Nitrogen Metabolism

The topic that I was asked to speak on, in its broadest sense, encompasses important practical objectives of all breeding programs that have a stated interest in the nutritional quality of the edible seeds produced. We might easily show this, as yield (or weight of seeds produced per unit ground area), which must have high priority for economic reasons, is nothing more than the sum total of carbon and nitrogen that finally gets deposited in the seed. Of course, there are oxygen and hydrogen and a few mineral salts, but these are seldom determining factors. Since one cannot be expected to consume excessively large amounts of carbon in the form of carbohydrates to obtain the amount of nitrogen that one requires, we must also consider the nitrogen/carbon ratio in the seeds, or the percent protein. The total protein yield is, of course, just a product of these two quantities. Finally, because of the requirements in human metabolism, we must add to sufficient protein supplies the requirement for certain amino acids, and so the form of nitrogen or the quality of the protein is also of extreme importance.

In any extensive treatment of the subject, it would be proper to consider the interrelations between carbohydrate and nitrogen metabolism during the total life of the plant until the seeds are harvested. This is clearly an impossibility in the time allowed for this paper, so only certain aspects of the problem have been chosen. These have been chosen not only because they are of academic or theoretical import but also because they illustrate the immensity and complexity of the problem and because they illustrate areas that seem to merit research or that seem amenable to manipulation to obtain the practical objectives of a program designed to improve and increase the protein food supplies of the world.

Yield is a direct function of the total number of seeds produced per unit ground area, and this must be a result of the total carbon and nitrogen that is put into these developing sinks. I have chosen, however, not to deal at length with yield but rather to introduce aspects of it, where necessary, for the interpretation of the amount of protein per seed. This characteristic is a direct function of seed composition, and in a general sense we should examine what has occurred in the plant world and ask what is theoretically possible. All seeds contain three major components—protein, oil, and carbohydrate—but the relative amount of these components in different seeds varies greatly (Figure 1). Ignoring thick seed coats because they tend to distort the picture, we see that oil can vary from about 70 percent to about 2 percent of the total seed, carbohydrate from about 5 percent to over 80 percent, and protein from about 3 percent to 50 percent.

I show you this figure not as a basis to propose that we can in the near future produce corn with 60 percent oil or 40 percent protein but

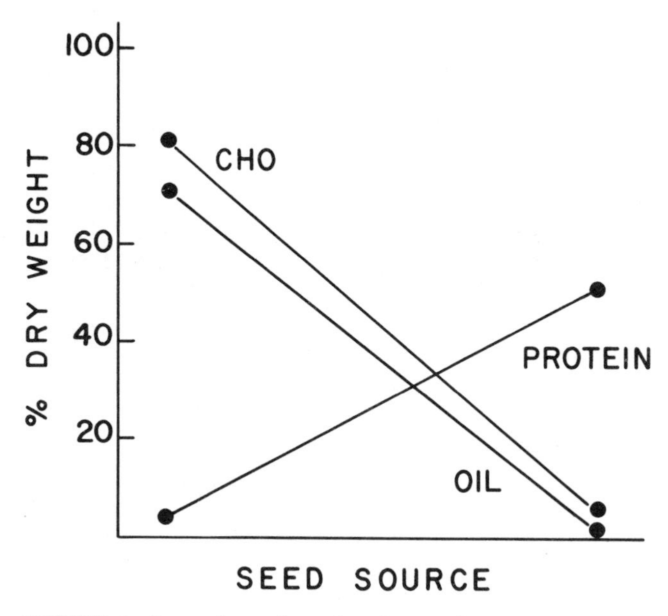

FIGURE 1 Proportions of protein, oil, or carbohydrate observed in plant seeds.

rather to emphasize that there does not seem to be any particular combination or composition that is particularly advantageous to the plant, and we must conclude that our present seed compositions reflect random walks down evolutionary roads, with each walk, when reinforced with suitable anabolic and catabolic machinery, as successful as any other from the plant's standpoint. There would thus appear to be no major biological or viability barrier that would prevent us from bringing about over a period of time even greater changes in seed composition than we have achieved to date.

Let us look at a few examples of what we have achieved to date in our efforts to manipulate the seed composition for our purposes (Figure 2). In wheat we have protein ranging from about 8 percent to 24 percent, oil from 1.25 percent to 2 percent, and carbohydrate essentially making up the remainder of the seed (Pomeranz *et al.*, 1966; Sosulski *et al.*, 1963). In soybean we have protein from about 33 percent to 50 percent, oil from 14 percent to 23 percent, and carbohydrate again making up the remainder of the seed (Hymowits *et al.*, 1972). In sunflower we have protein from 19 percent to 30 percent, oil from 17 percent to 46 percent (Cummins *et al.*, 1967; Kinman and Earle, 1964), with carbohydrate again comprising the remainder.

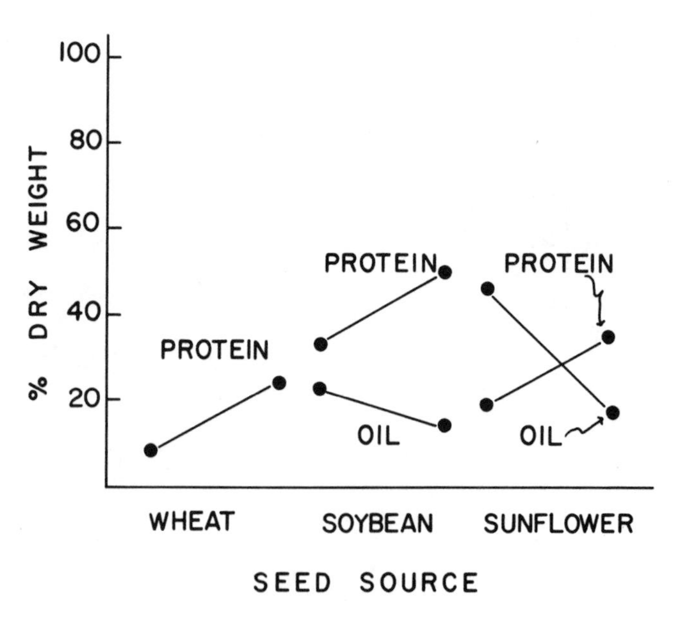

FIGURE 2 Proportions of oil or protein observed in seeds of wheat, soybean, and sunflower.

The presentation shows that there is, within any species, considerable variation in the relative amounts of the three components. It further shows that through systematic and determined effort, scientists have been able to select plants with higher protein contents in the seed. This increase in protein content is at the expense of a carbon store, either oil or carbohydrate, but in all cases carbohydrate in one form or another continues to make up a sizable portion of the seed. Some of this carbohydrate is still storage carbohydrate, and it would seem that it might be possible to further reduce the proportion of this component.

There seems to be little doubt that most of our attention will continue to be directed toward the major crops of the world such as soybeans and wheat and other cereals, but I have included sunflower in this example for a number of reasons. I believe we must begin to direct more attention to minor crops. Sunflowers are cross-pollinated and have a great deal of heterogeneity, so that even with only a relatively small effort in manpower, very large alterations in the seed composition have been achieved. Similar results might be expected with other cross-pollinated minor crops, and thus, with a relatively modest effort, major gains may be made in protein production in some regions of the world.

While these changes in carbon/nitrogen ratio in the seed have certainly been noticed, perhaps the most attention has been directed toward the discovery of genetic factors in corn (Nelson, 1969), barley (Ingversen *et al.*, 1973), and sorghum (Singh and Axtell, 1973) that result in changes in the proportion of the protein fractions in these cereal seeds with a resultant marked increase in nutritional value. The regulation of relative amounts of different proteins in cereals or, for that matter, of the carbon/nitrogen ratio in seeds is unquestionably under genetic control, and this control is expressed only in the seed and not in the vegetative plant. Apart from being able to change the carbon/nitrogen ratio by altering the amount of nitrogen supplied to the plant, we have no biochemical understanding of how this composition is achieved and regulated—we only know from the final compositions that it is regulated—and it does not appear that we will know much about it in the near future, as extremely little physiological and biochemical work is being done on developing seeds.

In spite of this lack of knowledge, though, further progress in altering the quantity and quality of seed protein will be made. We may not and probably will not achieve it in one step, but rather through many small steps reflecting the efforts of many men. I suggest that the screening techniques are even now sufficient for this objective, and it could be achieved through greater support and effort toward, first, increasing

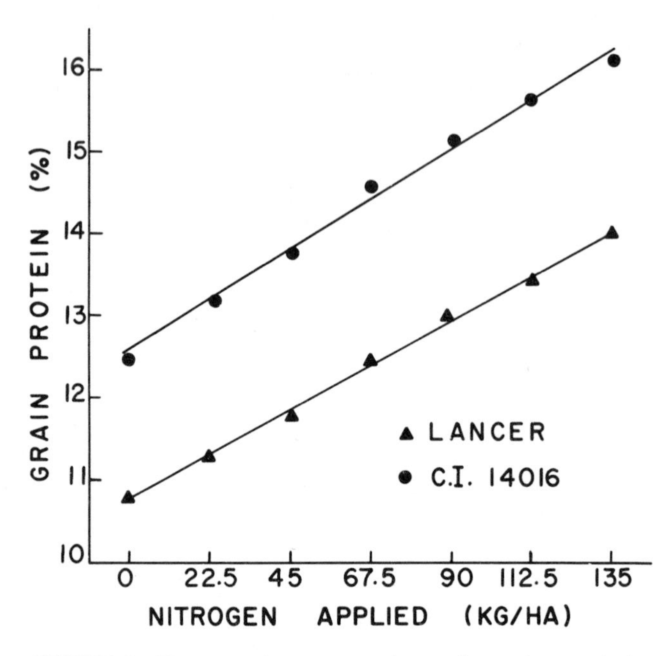

FIGURE 3 Mean protein response of two winter wheat varieties to nitrogen fertilizer in six experiments in Nebraska 1969–1970. (Redrawn from Johnson *et al.*, 1973.)

the variability in the plant populations and, second, methodically screening and selecting for those seeds possessing the desired characteristics.

Efforts towards these objectives will undoubtedly be hampered by the fact that the regulation of the carbon/nitrogen ratio in seeds is not precise but is rather sloppy. In Figure 3, the genetic regulation of protein/carbohydrate ratio is obvious as the difference between the two varieties, but it can also be seen that the protein content of either variety can be increased rather markedly as a result of increasing the nitrogen supply that can come from the vegetative plant (Anderson and Peterson, 1973; Johnson *et al.*, 1973). In part, then, the composition reflects the supply from the vegetative plant, and it would seem appropriate to turn to the carbon/nitrogen relationships of the vegetative plant and try to assess the influence of these relationships on yield of seed and yield of protein.

I do not have a definitive answer to this question. A large part of the difficulty is due to a feature called compensation. To illustrate this we can go back to yield for a moment, as it has frequently been subjected to what is called component analysis, or path coefficient analysis (Duarte and Adams, 1972; Grafius, 1956; Pandey and Torrie, 1973). It is perfectly logical to propose that yield in wheat is a product of number of plants per unit ground area × number of spikes × number of seeds × size of seeds, and so on. To get the highest yields, we should combine the best of each of these characteristics, but we know also that the combination may not yield better because of compensation. Thus, if we increase the number of plants, we decrease the number of spikes per plant. This phenomenon of compensation and the relative magnitude of it is due in large part to the relative supply of carbohydrate and nitrogen in the vegetative plant. It operates at all stages in the life cycle of the plant; for example, it is well known that there is a negative correlation between seed protein content and yield, i.e., there is some compensation, to the extent that, if nitrogen is limiting and more seeds are obtained, less nitrogen is available for each seed. Fortunately, it is equally well known that this negative correlation can be broken rather easily either by supplying extra nitrogen, which seems to be the factor in least supply, or by finding appropriate germ plasm that dictates a higher nitrogen/carbon ratio in the seed. Nevertheless, compensation remains a complicating factor in understanding carbon/nitrogen relationships and in selection of high-protein seeds.

In the 1920's and 1930's a great deal of effort was devoted to trying to determine carbon/nitrogen ratios in the vegetative plant and to attach precise significance to them (Curtis and Clark, 1950; Miller, 1938). It was known then and it is known now that without nitrogen there is little yield and little seed protein. It was also eventually realized that above the situation of nitrogen deficiency, i.e., when nitrogen is more or less adequate, a general rather than precise balance had to be maintained for good growth. At luxurious nitrogen levels it is extremely important that carbon assimilation not be curtailed, and this finally was at least partially achieved by altering the architecture of the plant. Thus, in cereals short, stiff, erect leaves and short strong straw (Athwal, 1971; Cummins *et al.*, 1967; Wallace *et al.*, 1972) allowed the carbohydrate-accumulating power of the plant to remain high and offset the detrimental effects (i.e., mutual shading of leaves and lodging) that would normally follow from luxurious supplies of nitrogen. Without these architectural changes to assure carbohydrate supply, a luxurious supply of nitrogen is useless; thus, the first thing that may be required in, for instance, a legume program, is the modification of plant architecture to reduce mutual self-shading of leaves and assure maximum insolation of the vegetative plant.

We know from other types of experiments that to obtain maximum beneficial use of supplied nitrogen, we must retain maximum carbohydrate-accumulating power. Light supplementation experiments (Johnson *et al.*, 1969; Waggoner, 1969) and CO_2 supplementation experiments (Hardman and Brun, 1971; Hardy and Havelka, 1973) result in both greater yield and greater protein content, but they are hardly economical. Thus, even though one may have a maximum amount of metabolic machinery to assimilate nitrogen (e.g., nitrate reductase), it can only be expressed in yield if carbohydrate assimilation is also allowed to continue maximally. In the past, these two features have probably been selected together, but if we now emphasize one, we must not lose sight of the other.

It should be obvious by now that over a very wide range some sort of balance must be maintained in the plant between carbon assimilation and nitrogen assimilation for good growth and high yield. How is the balance maintained? It has already been pointed out that the balance can be maintained and maximum yield attained only if either carbon assimilation or nitrogen assimilation is not curtailed or hampered. There exists a large literature (Murata, 1969) dealing with the importance of nitrogen supply on various developmental stages of the plant such as spikelet number and pod number, but these analyses, although enlightening, are hardly of much consequence to the problem of obtaining maximum protein yields since the results can be obtained only because nitrogen was limiting. In order to select for maximum protein-producing capability, it is self-evident that nitrogen should not be limiting.

Given, then, that nitrogen is not limiting but rather that carbon assimilation is limiting, a balance is maintained because most nitrogen assimilation from nitrate occurs in the leaves (Beevers and Hageman, 1969). In the leaves, the reduction of nitrate is as closely dependent on light as is CO_2 assimilation (Figure 4). Details of the methods are available (Canvin and Atkins, 1974). The close connection of NO_3 assimilation to light is further verified by the use of the inhibitor DCMU (Figure 5) where it can be seen that similar inhibition of NO_3^- or NO_2^- assimilation and CO_2 assimilation is obtained in response to a given concentration of inhibitor. Much earlier studies (Stoy, 1955) also showed that nitrate assimilation and CO_2 assimilation were similarly affected by changes in light intensity. While the level of nitrate reductase provides coarse control and indeed may be the major control of nitrate assimilation (Beevers and Hageman, 1969; Eilrich and Hageman, 1973), it is also probable that the fine control via light also plays a role, especially in situations where light may be limiting.

The reduction of nitrate in leaves produces ammonia, which is

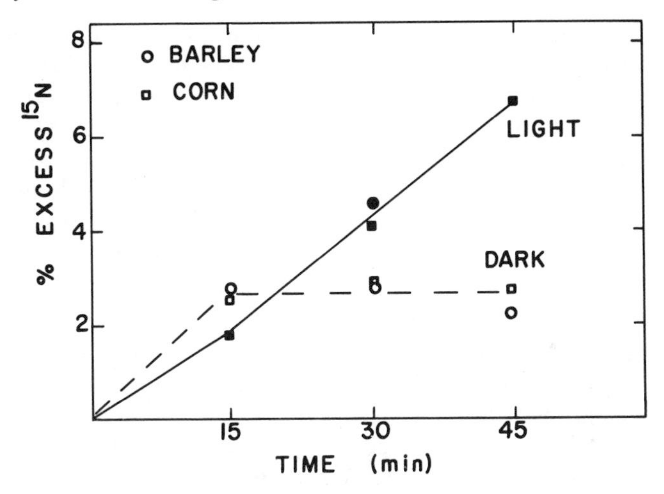

FIGURE 4 ^{15}N incorporation from $^{15}NO_3$ into soluble amino acids of barley or corn leaves in light or darkness. Dark leaves placed in the dark for 15 min. $^{15}NO_3^-$ uptake through base of leaf segments.

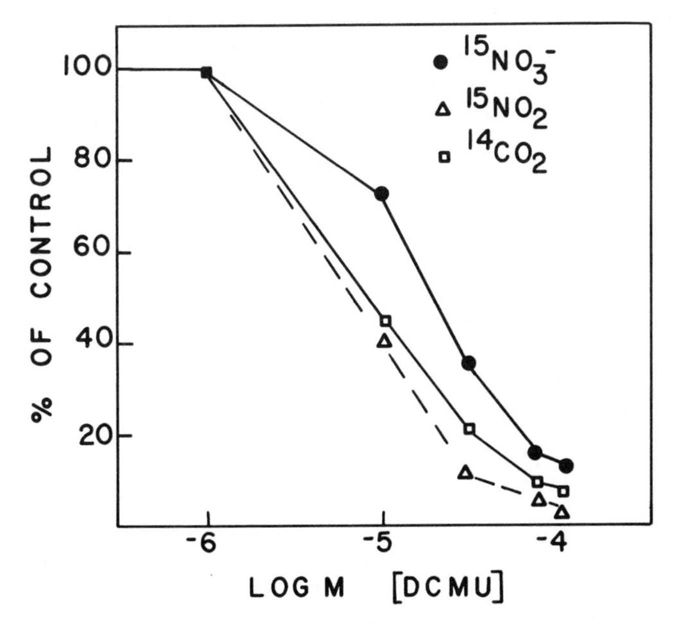

FIGURE 5 ^{15}N incorporation from $^{15}NO_3$ or $^{15}NO_2$ into soluble amino acids of barley leaves and $^{14}CO_2$ fixation as affected by DCMU. Leaf samples infiltrated and placed in light.

subsequently incorporated into amino acids. The subject of amino acid synthesis has already been covered, but I would like to say a few words on the subject of amino acid synthesis in the light in relation to photosynthesis and photorespiration, two other processes that also occur in the light. Glutamate or glutamine appears to be a major site of initial incorporation of ammonia in leaves, but the carbon skeletons for this synthesis do not come from recent products of photosynthesis, since even after 15 minutes' supply of $^{14}CO_2$, the specific activity remains low (Figure 6). Dark respiration of unlabeled compounds probably supplies the carbon skeletons here as, inasmuch as we can assume that dark respiration supplies essential intermediates, the process must continue in the light because plants can be grown in continuous light with no detrimental effects.

I am not saying, of course, that dark respiration continues in the light at the same rate as in darkness, but it, nevertheless, must continue. Thus much of amino acid synthesis in leaves may only involve dark metabolic reactions. Certain amino acids, however, always occur as

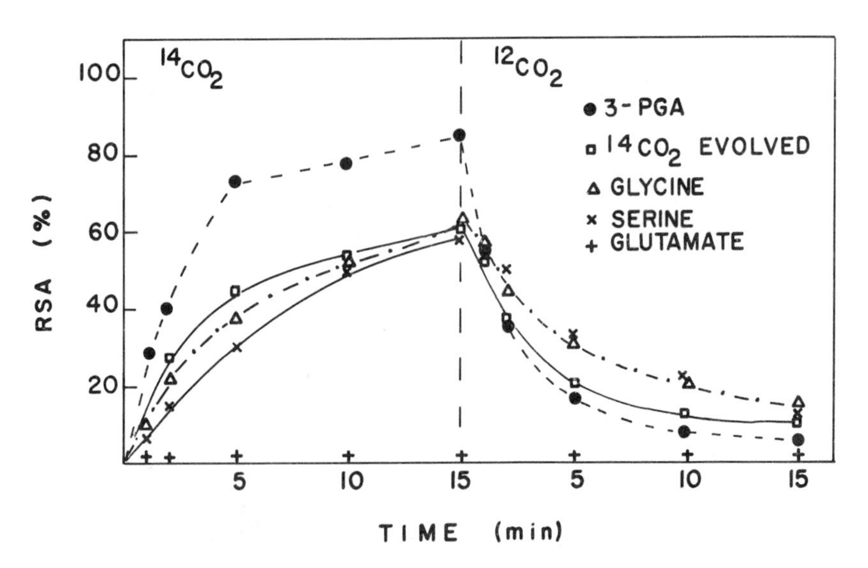

FIGURE 6 The specific activity of several compounds isolated from sunflower leaves exposed to $^{14}CO_2$ and 400 ppm CO_2 for 15 min and then exposed for a further 15 min to 400 ppm CO_2. Oxygen concentration, 21 percent; high intensity, 3,500 ft-c. RSA stands for relative specific activity.

$$\text{RSA} = \frac{\text{specific activity of compound}}{\text{specific activity supplied } ^{14}CO_2} \times 100$$

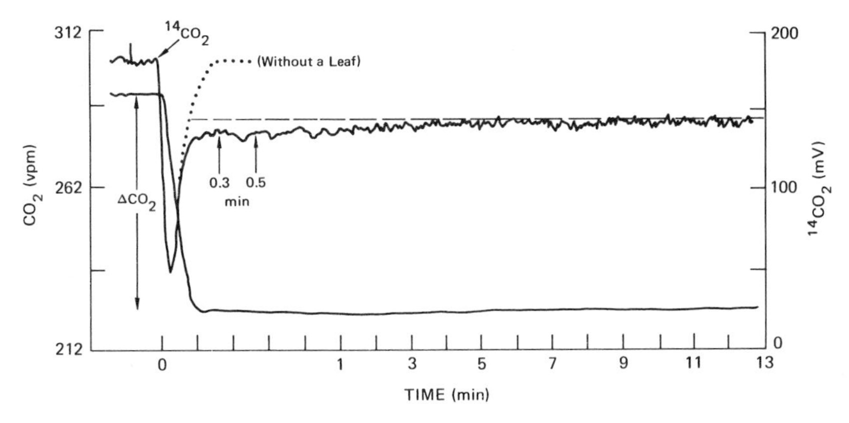

FIGURE 7 Recorder tracing of CO_2 and $^{14}CO_2$ uptake by a sunflower leaf. Conditions were 290 vpm CO_2, 21 percent O_2, 25°C, and 3,500 ft-c. (From D'Aoust and Canvin, 1972).

labeled products of photosynthesis in $^{14}CO_2$. Inasmuch as there are no nitrogen-containing compounds in the photosynthetic carbon cycle of C_3 plants, the amino acids cannot be produced here. Rather, they are produced in concurrent metabolic reactions. Two of them, glycine and serine, are produced as intermediates in the process of photorespiration.

Photorespiration occurs in all C_3 higher plants and has as its external manifestations or end results the uptake of O_2 and evolution of CO_2 in the light. The process can be measured by a variety of methods (Canvin and Fock, 1972), but the method I have chosen involves the simultaneous measurement of CO_2 and $^{14}CO_2$ fluxes from a leaf in the light (Ludwig and Canvin, 1971). The leaf is placed in a leaf chamber and preconditioned under the temperature, light, and gas composition that is to be used in the experiment. At time zero, or the start of the experiment, the gas stream is quickly changed to a gas stream of identical composition but containing a trace of $^{14}CO_2$. A recorder trace of the results with a sunflower leaf at 290 ppm CO_2, 21 percent O_2, 25°C, and 3,500 ft-c is shown in Figure 7. The first accurate measurements can be obtained 15 seconds after the introduction of the $^{14}CO_2$ gas stream and then the net uptake of both $^{14}CO_2$ and CO_2 can be continuously measured. It should be noted that the uptake of CO_2 is constant during the experiment but that the uptake of $^{14}CO_2$ is greatest at the beginning.

What these changes mean can be better seen in a diagram (Figure 8). The uptake of CO_2 as measured by infrared gas analyzer is 28 mg CO_2

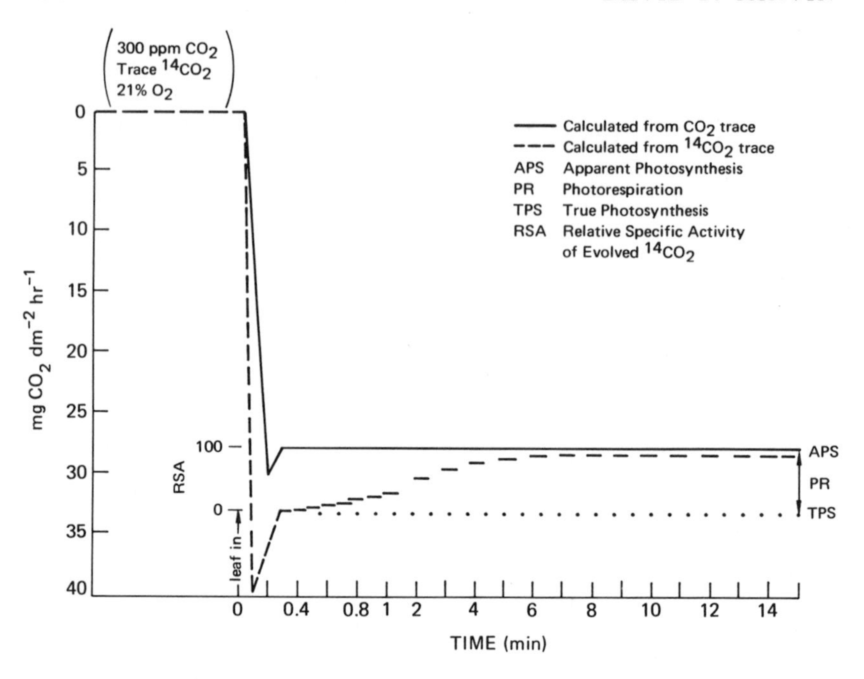

FIGURE 8 Normalized CO_2 and $^{14}CO_2$ gas-exchange from 30 experiments with sunflower leaves at 3,500 ft-c. (From D'Aoust, 1970).

dm^{-2} hr^{-1} and is apparent photosynthesis. The initial uptake of $^{14}CO_2$ is equal to 34 mg CO_2 dm^{-2} hr^{-1} and is a minimum measurement of true photosynthesis. The difference between these two measurements (6 mg CO_2 dm^{-2} hr^{-1}) is CO_2 evolution from the leaf or a minimum estimate of the rate of photorespiration. As time goes on, $^{14}CO_2$ uptake by the leaf diminishes until there is little difference between the CO_2 and $^{14}CO_2$ measurements. The decrease in $^{14}CO_2$ uptake is due to $^{14}CO_2$ evolution from the leaf, and by making certain assumptions we can calculate the specific activity of this $^{14}CO_2$. When this is expressed as a percentage of the specific activity of the $^{14}CO_2$ supplied, we have the *relative specific activity* (RSA).

Two major points should be made: (a) The minimum rate of photorespiration in this case is large, amounting to 17.5 percent of true photosynthesis or 21 percent of apparent photosynthesis, and (b) within 7 minutes of exposure of the leaf to $^{14}CO_2$, the $^{14}CO_2$ is evolved at 90 percent RSA. The magnitude of photorespiration in relation to dark respiration can be clearly seen in Figure 9. The maximum rate of CO_2 evolution into CO_2-free air in the first 5 minutes, and only at that time

(Canvin and Fock, 1972), corresponds very well with the rate of CO_2 evolution measured during photosynthesis. It is more than double the rate of dark respiration and is considerably greater than dark respiration even after the leaf is flushed with CO_2-free air for 50 minutes.

The CO_2 is thought to be evolved from the glycolate pathway (Figure 10), and Tolbert (1971) has suggested that it originates from the specific

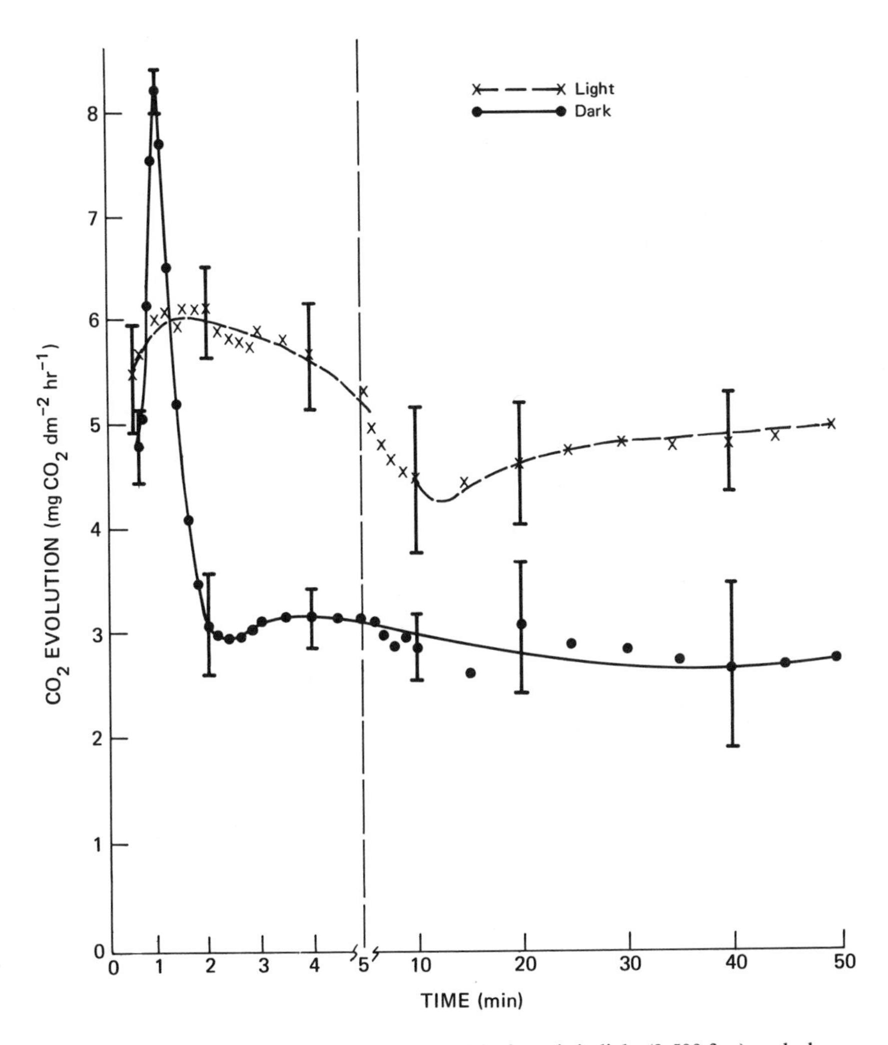

FIGURE 9 The rate of CO_2 evolution into CO_2-free air in light (3,500 ft-c) or darkness by sunflower leaves at 25°C. Vertical lines represent 95 percent confidence limits for n = 6 or more. (From D'Aoust and Canvin, 1972).

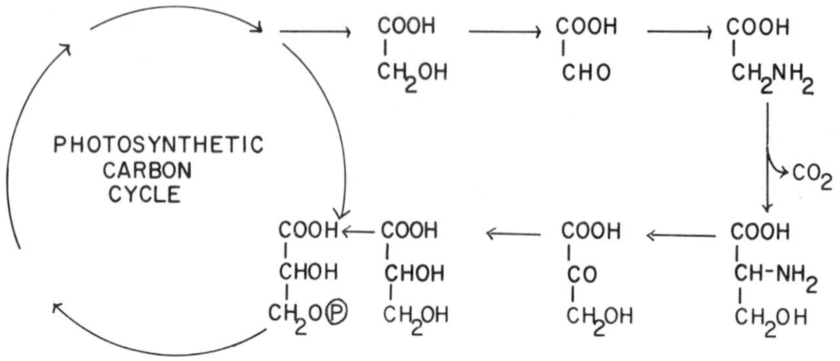

FIGURE 10 Outline of the glycolate pathway.

step of converting glycine to serine. This means in terms of the previous experiments with sunflower (Figure 8) that carbon flow through glycine must be equivalent to 24 mg CO_2 dm^{-2} hr^{-1}, and this rate is almost equal to apparent photosynthesis and is 70 percent of true photosynthesis. From the outline (Figure 10) it should also be apparent that the specific activity of the $^{14}CO_2$ evolved would be equal to that of glycine and serine and further that the specific activity of glycine and serine should be similar to the specific activity of 3-phosphoglyceric acid.

With full awareness of the possibility of multiple pools and the problems of variability of samples, the specific activity of these compounds was measured during $^{14}CO_2$ feeding and a subsequent CO_2 flush (Mahon *et al.*, 1974). The specific activities of glycine and serine were always very similar (Figure 6), even though the amount of glycine (34.0 ± 1.0 μg C) and serine (48.5 ± 2.2 μg C) per sample of leaf disks (0.554 dm^{-2}) was variable and different. Further, the specific activity of the $^{14}CO_2$ released was very similar to the specific activity of glycine and serine. But the specific activity of glycine and serine was much lower than the specific activity of 3-phosphoglyceric acid, a point not entirely consistent with a common origin in the photosynthetic carbon cycle and especially not consistent with a common origin from ribulose diphosphate (Andrews *et al.*, 1973).

The data, however, were consistent with a close relationship between glycine, serine, and the evolved CO_2, and, from the changes in radioactivity and the specific activity, it was possible to make some calculations regarding the flux of carbon from glycine to serine. The rate of carbon flux from glycine to serine was calculated as 40 μg min^{-1} 0.55 dm^{-2} of leaf. The rate of CO_2 evolution from the leaf was 13 μg C

min^{-1} 0.55 dm^{-2} and the rate of apparent photosynthesis was 53 μg C min^{-1} 0.55 dm^{-2} leaf. Thus the flow of carbon through the glycolate pathway was indeed very substantial.

While it has been widely thought that the amount of ^{14}C found in glycine and serine represented net synthesis, this view is probably incorrect, since the label was rapidly flushed from these compounds when CO_2 was supplied after the $^{14}CO_2$ (Figure 6). Glycine and serine are then merely transit points for a very substantial flow of carbon, and while there is no question that glycine and serine can be synthesized in this manner, this would not seem to be the principal function of photorespiration.

Indeed, the principal function of the pathway is not firmly established at the moment, but two views seem to predominate: (a) that it is a means of salvaging carbon compounds produced as an inevitable result of the fact that the plants are growing in oxygen (Lorimer and Andrews, 1973), and (b) that it is completely without purpose and could be eliminated (Zelitch, 1973). Attempts to eliminate it, or to find plants without it, have been singularly unsuccessful (Moss and Musgrave, 1971; Widholm and Ogren, 1969; Zelitch, 1973). This might well have been predicted, as any plant without it could possibly produce dry matter at a rate 20 percent greater than those with it and such plants would unquestionably have been noticed earlier.

In view of the magnitude of the process, I am reluctant, as yet, to accept the theory that the pathway is solely a means of salvaging carbon compounds in normal oxygen atmospheres, but the increased growth of plants in low-oxygen atmospheres would seem to support this view (Bjorkman *et al.*, 1966; Quebedeaux and Hardy, 1973). If the salvage operation is the function of photorespiration, it means, of course, that it will be impossible to select for low-photorespiration plants and also that involvement of nitrogen in the cycle is merely incidental. I do not think that enough is yet known about photorespiration to allow an unequivocal conclusion regarding its function, and I suggest that the relationship of photorespiration to amino acid synthesis or, for that matter, to nitrogen assimilation should be further explored.

I know that I have omitted changes in the levels of enzymes as a means of achieving a balance between carbon and nitrogen assimilation. This is an important mechanism, but it is straightforward in concept and has already been dealt with to some extent at this workshop.

Assuming now that we have adequate supplies of both nitrogen and carbon in the vegetative plant, a very important next step is the transfer of this material to the portion of plant that is harvested—the seed. In a general sense, we would like to have a minimum amount of non-reproductive material present at harvest, and thus a determinate type of plant growth would seem preferable.

In a more specific sense, the total dry matter in the seeds versus that in the above-ground portion of the plant is known as the Harvest Index, and a high value of Harvest Index would seem desirable. There is considerable variation in the Harvest Index of a number of cereals (Singh and Stoskopf, 1971), and there is no doubt that the Harvest Index of the semidwarf wheats is much greater than those of the normal tall varieties (Table 1). This increase in Harvest Index would appear to be a direct result of shortening the straw (Singh and Stoskopf, 1971), but one should notice that the actual yield in these cases is lower than that in the tall varieties. The Harvest Index, then, should be interpreted with caution, since selection on this basis alone will not necessarily result in higher yields; the total dry matter production per unit ground area must also be considered.

We could deal with nitrogen distribution in the mature plant in the same way and call this the Harvest Nitrogen Index. It has been suggested that if we could get a better transfer of the nitrogen from the vegetative portion of the plant to the seed, there would be an increase in protein content and protein yield.

In an experiment in which lodging resulted from application of

TABLE 1 Harvest Index and Relative Yield of Some Wheat Varieties[a]

Variety	Relative Yield	Harvest Index (%)
Tall		
Jufy I	100[b]	43
Kloka	96	42
Short		
Penjamo 62	92	46
Lerma Roja 64A	86	49
Sonora 64	81	52
Mexico 120	83	49

[a] Calculated from data of Thorne *et al.* (1969).
[b] 100 equals 582 g/m².

nitrogen fertilizer, the data of Eilrich and Hageman (1973) show that carbon assimilation was curtailed (Table 2). This curtailment in carbon assimilation also resulted in a marked decrease in the amount of nitrogen that was translocated to the grain, a fact that re-emphasizes that a proper balance must be maintained between carbohydrate and nitrogen assimilation and that the curtailment of one will affect the other at any stage in the development of the plant. The transfer of nitrogen to the grain is then as much a result of carbon metabolism as it is of nitrogen metabolism. In the same paper the authors also mention that changes in the proportion of nitrogen left in the straw with fertilizer treatments were primarily due to increased straw yield or to the development of numerous late tillers. Such development is a disadvantage of indeterminate growth.

From the data (Table 3) of Hucklesby *et al*. (1971) it is also apparent that there are differences among varieties in relation to the percentage of nitrogen transferred to the grain. These differences are small, but even 1 or 2 percent will make large differences in the nitrogen yield. The Harvest Nitrogen Index, though, is not a constant feature of a variety, and there is as much variation within a variety as there is between varieties. There would seem to be two patterns, one shown by Arthur, in which the Harvest Nitrogen Index does not decline with increasing amounts of fertilizer, and the second shown by Blueboy, in which the Harvest Nitrogen Index does decline. Not being familiar with these

TABLE 2 Relationship between Yield, Nitrogen Yield, and Percent Nitrogen Translocation from Vegetative Plant and Lodging in Gage Wheat[a]

Lodging (%)	Relative Yield	Relative Grain N Yield	Harvest Nitrogen Index (%)
0	100[b]	100[c]	69
6	86	100	64
13	82	95	61
15	94	111	65
30	73	87	57
40	71	90	53

[a] Compiled from data of Eilrich and Hageman (1973).
[b] 100 equals 4,430 kg/ha.
[c] 100 equals 102 kg N/ha.

TABLE 3 Relative Nitrogen Yield and Percent Translocation of Nitrogen to the Grain in Fertilized Wheat Varieties[a]

Variety	Rate of N (kg/ha)	Relative Nitrogen Yield	Harvest Nitrogen Index (%)
Blueboy	0	100[b]	81
	56	142	77
	112	174	73
	224	174	67
Arthur	0	84	79
	56	120	78
	112	151	76
	224	181	77
Parker	0	90	74
	56	125	76
	112	157	75
	224	171	71

[a] Calculated from data of Hucklesby *et al.* (1971).
[b] 100 is equal to 88 kg grain N/ha.

varieties, I am unable to say whether this is due to differences in the habit of growth of the two varieties (e.g., nitrogen fertilization of Blueboy may result in production of more late tillers that do not mature) or to differences in the capabilities of nitrogen translocation.

No change in the Harvest Nitrogen Index of a spring wheat in response to fertilizer was also obtained by Spratt and Gasser (1970) under conditions where the nitrogen yield was doubled (Table 4). The proportion of the seed nitrogen that was absorbed after anthesis, though, was quite different in the various treatments, ranging from 50 percent to 0 percent.

Drought at any time resulted in a marked suppression of the nitrogen yield (Table 5). Since the Harvest Nitrogen Index for drought at tillering or heading is actually greater than the control, the decrease in nitrogen yield is primarily due to interference with the absorption of nitrogen from the soil. When the drought was at stem extension, absorption from the soil was decreased and the Harvest Nitrogen Index was also markedly reduced.

I must, of course, mention Johnson's work on high-protein wheats, since he suggested that Atlas-66-derived high-protein lines were more

TABLE 4 Origin and Distribution of Nitrogen in Kloka Spring Wheat at Harvest[a]

Fertilizer Treatment	Relative Nitrogen Yield	Source of Seed Nitrogen[b]		Harvest Nitrogen Index (%)
		Plant (%)	Soil (%)	
None	100[c]	50	50	66
$Ca(NO_3)_2$:				
112 kg N/ha	200	78	22	67
$(NH_4)_2SO_4$:				
112 kg N/ha	200	100	0	67

[a] Compiled from data of Spratt and Gasser (1970).
[b] Percent of nitrogen in the seed transferred from the plant from nitrogen absorbed prior to anthesis (plant) or absorbed after anthesis (soil).
[c] 100 equals 28 kg seed N/ha.

efficient in translocating nitrogen from the vegetative plant to the seeds than were low-protein varieties such as Warrior (Johnson *et al.*, 1967; Murata, 1969). The percentage of nitrogen in the straw (Johnson *et al.*, 1967) and the distribution of nitrogen in the mature plant (Johnson *et al.*, 1969) support this interpretation, but I cannot understand why,

TABLE 5 Effect of Time of Drought on Origin of Seed Nitrogen and Distribution of Nitrogen in Kloka Wheat Fertilized with 112 kg N/ha as $Ca(NO_3)_2$[a]

Time of Drought	Relative Nitrogen Yield	Source of Seed Nitrogen[b]		Harvest Nitrogen Index (%)
		Plant (%)	Soil (%)	
None	100[c]	76	24	60
Tillering	67	69	31	66
Stem extension	42	60	40	50
Heading	75	80	20	64

[a] Compiled from data of Spratt and Gasser (1970).
[b] See Table 4.
[c] 100 equals 60 kg seed N/ha.

even in the high-protein lines, he obtained only 60 percent of the nitrogen in the grain (Johnson *et al.*, 1969).

From the data presented and other data it is clear that the Harvest Nitrogen Index is subject to considerable variation and its use as a selection tool may not be too profitable. It would seem, though, that the relative importance of the type of response of the Harvest Nitrogen Index to fertilizer application should be determined. In spite of the fact that the index will be affected by such factors as water supply and carbohydrate supply, there may be a basic difference in plant capability.

It is also clear from the literature that the capability of effective transfer of nitrogen to the seed already exists in many wheat varieties that show values for the Harvest Nitrogen Index of 80 to 84 percent (McNeal *et al.*, 1971, 1972). In these varieties it is doubtful if a further portion of the nitrogen remaining in the straw could be transferred to the seed. Since this is not the case, though, in many varieties, it would seem that Harvest Nitrogen Indices should be determined and improved if necessary, to perhaps result in some gain in grain protein.

Now that we have returned to the seed, let me point out that we know much more about carbon transfer from the various parts of the plant and carbon interconversions in the seed than we do about nitrogen transfer or interconversion. This latter area would again seem to be one that should merit serious study, especially amino acid metabolism of developing seeds and the contribution of glumes, awns, and other parts in supplying reduced nitrogen to the seed.

REFERENCES

Anderson, F. N., and G. A. Peterson. 1973. Agron. J. 65:697.

Andrews, T. J., G. H. Lorimer, and N. E. Tolbert. 1973. Biochemistry 12:11.

Athwal, D. S. 1971. Quart. Rev. Biol. 46:1.

Beevers, L., and R. H. Hageman. 1969. Nitrate reduction in higher plants. Annu. Rev. Plant Physiol. 20:495.

Bjorkman, O., W. M. Hiesey, M. Noks, F. Nicholson, and R. W. Hart. 1966. Carnegie Institution Annual Report 1966–67, p. 228.

Canvin, D. T., and C. A. Atkins. 1974. Planta 116:207.

Canvin, D. T., and H. Fock. 1972. *In* Methods in Enzymology (A. San Pietro, ed.). Academic Press, New York, p. 246.

Chandler, R. F. 1969. *In* Physiological Aspects of Crop Yield (R. C. Dinauer, ed.). American Society of Agronomy, Madison, Wisconsin, p. 265.

Cummins, D. G., J. E. Marion, J. P. Craigmiles, and R. E. Burns. 1967. J. Am. Oil Chem. Soc. 44:581.

Curtis, O. F., and D. G. Clark. 1950. Plant Physiology. McGraw-Hill Book Co., New York.

D'Aoust, A. L. 1970. Ph.D. thesis, Queen's University, Kingston, Ontario.

D'Aoust, A. L., and D. T. Canvin. 1972. Photosynthetica 6:150.

Duarte, R. A., and M. W. Adams. 1972. Crop Sci. 12:79.
Eilrich, G. L., and R. H. Hageman. 1973. Crop Sci. 13:59.
Grafius, J. E. 1956. Agron. J. 48:419.
Hardman, L. L., and W. A. Brun. 1971. Crop Sci. 11:886.
Hardy, R. W. F., and V. D. Havelka. 1973. Plant Physiol. Suppl. 51:35.
Hucklesby, D. P., C. M. Brown, S. E. Howell, and R. H. Hageman. 1971. Agron. J. 63:274.
Hymowitz, T., F. E. Collins, J. Panczner, and W. M. Walker. 1972. Agron. J. 64:613.
Ingversen, J., H. Anderson, and B. Køie Doll. 1973. *In* Nuclear Techniques for Seed Protein Improvement. IAEA, Vienna, p. 193.
Johnson, V. A., A. F. Drier, and P. H. Gabouski. 1973. Agron. J. 65:259.
Johnson, V. A., P. J. Mattern, and J. W. Schmidt. 1967. Crop Sci. 7:664.
Johnson, V. A., P. J. Mattern, D. A. Whited, and J. W. Schmidt. 1969. *In* New Approaches to Breeding for Improved Plant Protein. IAEA, Vienna, p. 29.
Johnston, T. J., J. W. Pendleton, D. B. Peters, and D. R. Hicks. 1969. Crop Sci. 4:577.
Kinman, M. L., and F. R. Earle. 1964. Crop Sci. 4:417.
Lorimer, G. H., and T. J. Andrews. 1973. Nature 243:359.
Ludwig, L. J., and D. T. Canvin. 1971. Can. J. Bot. 49:1299.
Mahon, J. D., H. Fock, and D. T. Canvin. 1974. Planta 120:245.
McNeal, F. H., M. A. Berg, P. L. Brown, and C. F. McGuire. 1971. Agron. J. 63:908.
McNeal, F. H., M. A. Berg, C. F. McGuire, V. R. Stewart, and D. R. Baldridge. 1972. Crop Sci. 12:599.
Miller, E. C. 1938. Plant Physiology. McGraw-Hill Book Co., New York.
Moss, D. N., and R. B. Musgrave. 1971. Adv. Agron. 23:317.
Murata, Y. 1969. *In* Physiological Aspects of Crop Yield. American Society of Agronomy, Madison, Wisc. p. 235.
Nelson, O. E. 1969. Adv. Agron. 21:171.
Pandey, J. P., and J. H. Torrie. 1973. Crop Sci. 13:505.
Pomeranz, Y., O. Chung, and R. J. Robinson. 1966. J. Am. Oil. Chem. Soc. 43:511.
Quebedeaux, B., and R. W. F. Hardy. 1973. Plant Physiol. Suppl. 51:51.
Singh, I. D., and N. C. Stoskopf. 1971. Agron. J. 63:224.
Singh, R., and J. E. Axtell. 1973. Crop Sci. 13:535.
Sosulski, F. W., E. A. Paul, and W. L. Hutcheon. 1963. Can. J. Plant Sci. 43:219.
Spratt, E. D., and J. K. R. Gasser. 1970. Can. J. Plant Sci. 50:613.
Stoy, V. 1955. Physiol. Plant. 8:963.
Thorne, G. N., P. J. Welbank, and G. C. Blackwood. 1969. Ann. Appl. Bio. 63:241.
Tolbert, N. E. 1971. *In* Photosynthesis and Photorespiration (M.D. Hatch, C. B. Osmond, and R. O. Slatyer, ed.). Wiley Interscience, New York, p. 458.
Waggoner, P. E. 1969. *In* Physiological Aspects of Crop Yield. American Society of Agronomy, Madison, Wisc., p. 343.
Wallace, D. H., J. L. Osbun, and H. M. Munger. 1972. Adv. Agron. 24:97.
Widholm, J. M., and W. L. Ogren. 1969. Proc. Nat. Acad. Sci. U.S.A. 63:668–675.
Zelitch, I. 1973. Proc. Nat. Acad. Sci. U.S.A. 70:579.

DISCUSSION

DR. MUNRO: What are the reasons for the considerable differences within the same species in protein content? Is it the difference in the proportion of certain cell types, of certain organelles like storage granules, or is it due to the accumulation of protein in the cytoplasm to different extents?

DR. CANVIN: I cannot answer that. There are experts on seeds here that can answer that. Maybe Dr. Johnson can answer that, or Dr. Dieckert.

DR. DIECKERT: We cannot answer that question yet.

DR. RABSON: Dave Canvin mentioned a method that we are working on that I did not intend to discuss, but since he provoked me, I will respond. At the International Atomic Energy Agency Seibersdorf Lab, we are interested in induced mutations controlling protein quantity and quality. One of the things we have found to be a limitation has been the question of rapid and accurate analysis for screening. Earlier, I mentioned that we were confronted with detecting a rather weak genetic signal among a lot of environmental noise. We thought about this problem for a number of years and finally decided that we would try a new tack in approaching the whole question of analysis.

The method is still in the early stages of development. It involves a procedure of hydrolyzing seeds or grains with hydrochloric acid without grinding, and analyzing the hydrolysate with two different ninhydrin reactions. One is the typical ninhydrin reaction. The other is an acidic ninhydrin reaction that is more or less specific for proline. In this way, using the same sample, we are able to get a ratio. We feel that, judging by the information available on opaque-2 and some of the high- and low-lysine genotypes of barley, changes in the prolamins, which are rich in proline, can be detected in this way.

We have tried the method with opaque-2 and floury-2 maize against a wild-type background on some material from Ollie Nelson. We also tried the technique on some of the barley genotypes with altered lysine and indeed the different genotypes can be distinguished in most, but not all, cases. We do not know why the method does not identify differences in every case, but this may indicate that there is not a significant change in the prolamins in some of the mutants.

Part of the rationale of this technique is that the ratio of these reactions is less affected by environmental conditions than are the absolute levels of constituents measured by dye-binding capacity or other procedures.

We are working on the method. We hope in the near future to determine whether the method is indeed a useful technique for screening rather large numbers of seeds. By large numbers, we are talking about a thousand analyses a day. This method is designed for screening a mutagenized population, not a world collection. We are starting with the same genetic material and inducing mutations in that. We look for differences within that population. I do not want to give the impression that this is a method that can be applied to each and every analytical situation.

DR. INGVERSEN: Dr. Hageman, you asked about how many entries you should screen to have a chance to find a changed protein composition. We have some data from our screening. Henstart at Risø has screened 15,000 mutants of barley, and he found six mutants with very clear-cut changes in protein composition. So that is approximately one per 2,500.

You said what we really need is a higher protein production and you did not really mind how we reached that higher protein production. I agree with you, but I think you have to go a little further. You told us about the failure of the composition.

Sprague said earlier that we should not expect plant breeders to come to biochemists and molecular biologists and tell us that they want this or that screening technique. They want us to find out for ourselves. But I find that very hard because we have to rely on nutritionists, and I do not think they have specified protein requirements very well. I think we are concentrating too much on lysine and high protein production without taking account of digestibility and very many other nutritional factors, and I think that the nutritionists should specify their demands much more precisely than they have. We did not get yesterday a specific statement from Dr. Altschul about what the nutritionist really wanted of us.

DR. CANVIN: I agree with you entirely. The nutritional aspect is something that I did not get into. We should be assembling as many different mutants or strains as possible with different protein contents and different changes in composition, regardless of whether it is a desirable one or not a desirable one. You will never know what will happen when you start combining these things. All you really want now is to get a bunch of mutants so that you can start working with them. That is my point, that you pick out everything that has an altered protein composition or an altered protein amount and go from there.

DR. MUNCK: I think that adequate funding of work with plants that could be practically used is not the only problem. I also think it is a matter of education. Most of us here are analysts for whom it is more rewarding to devise new methods. We are only searching for new methods, and there are very few of us who want to use these methods in a large-scale screening work to explore the genetic variation.

Let me take an example. In 1962, Roger Mossberg and I were convinced that the dye-binding (DBC) method (R. Mossberg. 1969. pp. 151–161 *in* New Approaches to Breeding for Improved Plant Protein. IAEA/FAO, STI/PUB 212. Vienna) or the Udy analysis for protein was in fact measuring the basic amino groups and thus also lysine. We started to use the ratio between dye binding and Kjeldahl as screening for lysine. At that moment nobody believed in us.

Applying the method to the world barley collection we found the first high-lysine gene in barley Hiproly at Svalöf, Sweden, in 1967 (see Munck and Wettstein, pp. 71–82, this volume). Later the research group at Risø, Denmark, took up this dye-binding procedure and obtained 12 high-lysine mutants out of 10,000 plants.

The dye-binding procedure is now used in high-lysine maize breeding by Professor Oliver Nelson at Madison and by the Centro Internacional de Mejoramiento de Maiz y Trigo (CIMMYT), Mexico, but it is not used to the extent it should be, simply because biochemists think it is too nonspecific. That is not the question. When you find that all high-lysine mutants hitherto found (with other often more tedious methods in maize, sorghum, and barley) could be selected by the DBC method, then there must be something in it. As a proof, this must be sufficient. We cannot as chemists go round with the second decimal in all our lives, only *hunting* for new methods. We must get them used, too, in order to explore the genetic variation. Where on earth should we stand today if, for instance, Tatum had said to himself in the thirties, "Well, biochemistry is so complicated that I must study this specific strain of *Escherichia coli* during my whole life in order to grasp its metabolism." This was not the way in which biochemistry evolved. Biochemistry evolved through selection of radically different mutants—with the help of potent screening methods.

One of my colleagues here said to me that he would not go into this problem of selection because he was more keen on methods, but I do not agree with him. If he is interested in a certain metabolic pathway, for instance, in a higher plant, then he should devise screening methods to get these mutants that could answer the question of how this pathway is constructed. What I mean is that selection is just as important to get sophisticated answers on molecular biology as it is to get good results in plant breeding. We have very good screening methods today. Let us use them.

DR. DOY: I support you absolutely.

DR. MARCUS: The thesis that the manipulation at the genetic level is going to provide you with a higher-protein product, I think, is the theme that we are pushing. I think that there is an underlying assumption that compensation and all those ancillary things are going to be taken care of on the side. Starting from that thesis, I want to make one further suggestion.

There has been a lot of belaboring of the point that one is going to be able to mutate seeds, grow these seeds, and then look at the seed product and select something out with either high protein or high-quality protein, perhaps by a proline test like the one Dr. Rabson talked about. I want to focus in on another possibility, which requires growing up all of these mutants and then looking at the seed products of these mutants. The rate of mutation, of course, is very low. So, obviously, a major advantage would be gained if one could get a good selection before growing these things up. The culture method approach seems to be worthwhile. One takes somatic cells, and, whether one does mutation at the level of somatic cells or uses cell fusion, the mutants are developed at the level of somatic cells.

The problem is that the end product that you are looking for—namely, the high protein or the high-quality protein—is the seed. The metabolism of the developing seed is clearly different from that of the somatic cells of the vegetative plant; therefore, the argument is that you have to study things at the level of the developing seed. My thesis is that the selection has to be made at the level of the vegetative cell if you are going to do it in culture.

The point that I am making, then, is that one has to correlate. But the metabolism of a soybean, let us say, that ends up with 20 percent protein in its seed and another soybean that makes 40 percent in its seed has got to be manifested in a much more subtle way at the level of the vegetative cell because something is different about that plant to begin with. If it is all in the genetics, then the genetics is going to manifest itself in a much more subtle way at the level of the vegetative cell, and you have to develop your selection at that level.

What I am appealing for is another approach: Try to use the lines that give a higher protein, for example, in the seed product; take those at the level of the somatic cells, and if you can find differences at that point, then you can develop your screening to select out something that correlates, or appears to correlate, with a cell that will eventually give you the desired seed protein. You can do a lot more screening at that point because you can select out and you only have to test 10 or 20 lines and not 15,000 or 20,000 at that level. I suggest this as an alternative approach to high efficiency in terms of getting what you are looking for.

DR. CANVIN: I was hoping some of the experts could respond to that.

DR. BOULTER: This is an idea that I think crosses quite a few peoples' minds, but we are going to have to understand the fundamental biochemistry of the whole of development in order to correlate what is going to happen in the seed with what is happening in the seedling. Unless Dr. Marcus has some specific points where the correlation can be made, until we really understand the whole of the biochemistry of the development from the word "go" and can correlate differences with the seed, I think we have a long way to go.

DR. MARCUS: I am suggesting shortcutting. Let us do what the drug companies do. They miss most of the time, but the idea is simply to try to look for correlation with a little sense, rather than waiting until the whole textbook of biochemistry is written on the developing vegetating cells and the developing seeds, which in the end are going to be 90 percent a copy of the bacterial cells. Rather than waiting until all that is done, I think one might start to look—it is a very inexpensive kind of thing to do at the level of vegetative cells. I could be all wrong, you know, if the cell fusion business does not really turn out to be as good as it looks. But I think it is worth investing a fair amount, just simply taking educated guesses and looking for correlations. You have good end products already; try to work back from them.

DR. DOY: I acknowledge the implications of most of what all three previous people have said, and tomorrow my task is to try to talk about some of these things. I want to restate this more clearly. We have got to try to look at one character at a time. In addition to what has been said, haploid cells in culture may be an even greater tool in getting to this, because there is just a chance there that you can mutate one gene and demonstrate mutations that otherwise would not show up in physical situations. But it is all easier said than done. You can see how the idea of doing these things turns people on. We have to think simply but not too simply; we must not ignore known facts and difficulties.

R. W. F. HARDY, U. D. HAVELKA,
and B. QUEBEDEAUX

Opportunities for Improved Seed Yield and Protein Production: N_2 Fixation, CO_2 Fixation, and O_2 Control of Reproductive Growth

The characteristics of an ideal crop may include high total growth rate, Harvest Index and nutritional quality produced at the most favorable economics, and environmental impact. We will consider the relationship of three processes, N_2 fixation, CO_2 fixation, and O_2 control of reproductive growth, to the first two above characteristics. Our objective is to identify chemical, biochemical, or physiological opportunities for the improvement of seed yield and protein quantity by either genetic or synthetic chemical growth-regulating means.

Specifically, we will outline novel biological and abiological approaches that are being explored to improve utilization of N_2 by crops. Unfortunately, many investigations concerned with improvement in food protein production ignore the primary source of nitrogen, N_2, and the process that makes this nitrogen available to the food chain (Figure 1). Experiments under field conditions will be described in which the total nitrogen input into a grain legume was almost doubled by a multifold increase in symbiotic N_2 fixation. This first major breakthrough in increasing nitrogen input into soybeans was achieved by CO_2-enrichment, and the effect is attributed to increased production of photosynthate through a decrease in photorespiration. Thus, N_2 fixation appears to be a source-limited rather than a sink-limited process, and improved carbon nutrition is a key to increased nitrogen input. It is proposed that improved nitrogen nutrition may be sought by practical approaches to improve carbon nutrition through genetic or chemical manipulation of photorespiration. Finally, effects of O_2

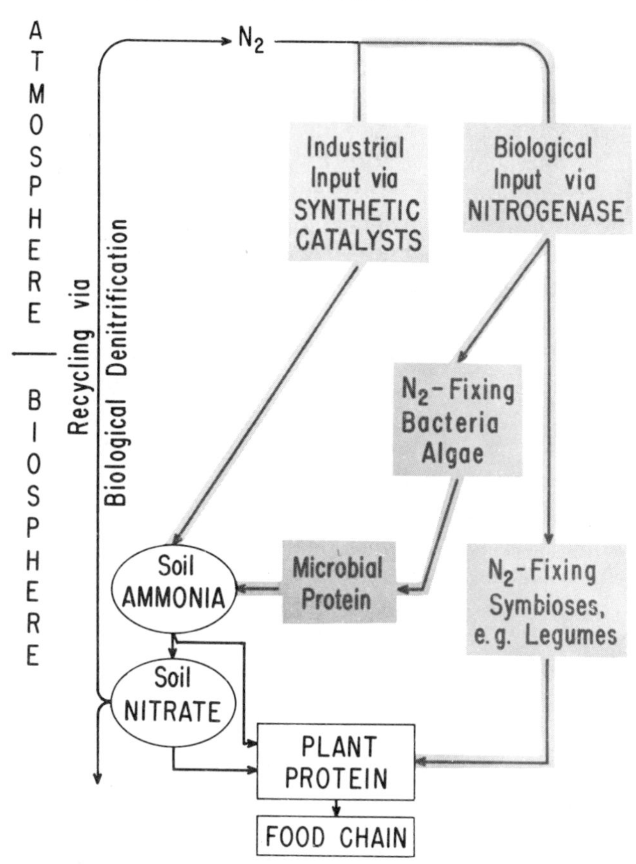

FIGURE 1 A simplified nitrogen cycle indicating coupling of atmospheric N_2 to the plant protein food chain by industrial N_2 fixation or biological N_2 fixation (Burns and Hardy, 1974).

concentration on the distribution of assimilates between reproductive and vegetative growth will be described, and this heretofore unrecognized oxygen system may control Harvest Index and thereby seed yield and protein productivity.

CONTRIBUTIONS AND NOVEL APPROACHES TO UTILIZING N_2 FOR CROPS

There have been several practical contributions to utilizing N_2 for improved nitrogen nutrition. These may be contrasted with carbon nutrition where use of CO_2 enrichment for production of greenhouse crops may be the only major practical contribution (Wittwer and Robb, 1964). For centuries N_2-fixing legumes have been used as "green manure" crops to add nitrogen to the soil; however, the unacceptable economics of spending an entire crop for nitrogen fertilization has eliminated this cultural practice except in developing countries, where nitrogen fertilizer is less available or too expensive, and, in fact, legume acreage even in developing countries is decreasing, presumably because of expanded acreages of the higher-yielding Green Revolution crops (Nutman, personal communication).

During this century improved strains of *Rhizobia* have been selected and used to inoculate legume seed, and an inoculum industry has developed to serve this need; however, the inability to demonstrate substantial yield increases from inoculation of soybeans sown on fields on which soybeans have been grown previously documents the need for new approaches in rhizobial research, at least in developed countries. Furthermore, the source rather than the sink with which the *Rhizobia* are associated may be the limiting factor in N_2 fixation as described below, and this site of limitation suggests the need for increased emphasis on the plant rather than the bacterial partner. A comparison of the N_2-fixing characteristics of different soybean varieties in which there were examples of similar and different bacteroidal strains supports this suggestion, since it was concluded that the plant variety is more dictative of N_2 fixation characteristics than the bacterial strain (Hardy *et al.*, 1973a).

The development of the Haber-Bosch abiological N_2-fixing system in 1914 provided the basis for the manufacture of synthetic nitrogen fertilizer. Its use has become an accepted and necessary practice for production of high yields of corn and the nitrogen-responsive varieties of wheat and rice. Synthetic nitrogen fertilizers such as ammonia, nitrate or urea would even be used to supply the total nitrogen needs for N_2-fixing legumes such as soybeans (see below) if a positive yield response occurred. Synthetic nitrogen fertilizers provide the major source of exogenous nitrogen for crop production, and expansion of this source will be required to meet crop needs during this decade. However, there are disadvantages associated with the use of nitrogen fertilizers, and exploratory research is needed to seek alternate approaches. These disadvantages include an inefficient uptake by crops with an average recovery of only 50 percent, a distribution cost that approaches the manufacturing cost (Hardy *et al.*, 1971), and a high energy requirement for manufacture, with a recent estimate assigning one-third of the commercial energy input for corn production to the nitrogen fertilizer component (Pimentel *et al.*, 1973).

The renaissance in research in both biological and abiological N_2 fixation since 1960 has exposed for investigation three possible objectives that may lead to novel routes for utilizing N_2. Enhancing symbiotic N_2 fixation in leguminous crops such as soybeans may be the most readily attained objective, and success in this area has already been predicted: "Within the next 10 years a method will be developed to increase the nitrogen fixing ability of soybeans" (Anon., 1971). The experimental success in the field as described below indicates that this is a real possibility and defines an approach that may lead to practical success.

Extending biological N_2 fixation to crops that must now utilize either soil or fertilizer nitrogen is a second major objective. At least three types of approaches are being explored; that the problem has received much attention is shown in the following selected examples.

N_2-Fixing Chloroplasts. "I would urge botanists to tackle the problem of nitrogen deficiencies that exist in the living world with boldness. We need to devise more efficient ecosystems, put root nodules on cereal crops, and put nitrogen fixing chloroplasts in their leaves" (Van Overbeek, 1969).

Rhizosphere Association. "Also a wide-ranging search should be made for monocots that show evidence of nitrogen fixation capacity, by way of either nodulated structures or rhizosphere relationships. Discoveries in this area could open up new avenues for extending the capacity to fix nitrogen to the major food crops" (Phillips *et al.*, 1971).

N_2-Fixing Agrobacterium. "Also, harmful bacteria, such as *Agrobacterium tumefaciens* which can infect non-legumes and which is apparently related to *Rhizobium,* might someday be genetically deprived of harmful traits and given nitrogen-fixation capabilities from donor *Rhizobium*. In this way, nitrogen fixation might be introduced into non-legume crops" (Anon., 1972).

Rhizobium cerealis. "In my dream I see green, vigorous, high-yielding fields of wheat, rice, maize, sorghum and millet which are obtaining, free of expense, 100 kilograms of nitrogen per hectare from nodule-forming, nitrogen-fixing bacteria. These mutant strains of '*Rhizobium cerealis*' were developed in 1990 by a massive mutation breeding program with strains of '*Rhizobium*' obtained from roots of legumes and other nodule-bearing plants" (Borlaug, 1971).

Protoplast Fusion. "What a boon it would be if we could induce corn, wheat, or rice to become nitrogen fixing in this way! Not only would we save the expense of tons and tons of nitrogenous fertilizer but we would also avoid the pollution caused by fertilizer salts . . ." (Galston, 1972).

One approach seeks naturally occurring associative symbioses in the hope that they can be used with crop plants; a second attempts to extend the rhizobial—legume symbiosis to nonleguminous crops; and a third hopes to transfer the *nif* genes so that N_2-fixing activity is expressed in crop plants other than legumes. Associative N_2-fixing symbioses involve the association of a free-living N_2-fixing bacteria with plants. The various types of N_2-fixing relationships are diagrammed in Figure 2, and various examples such as the *Paspalum–Azotobacter paspalum* association in South America (Dobereiner *et al.*, 1973) have been demonstrated. Several groups are exploring this possibility; for example, one of us (Hardy) in collaboration with Dr. Peter Graham of the Centro Internacional de Agricultura Tropical (CIAT) has measured N_2 fixation of non-nodulated plant roots in the field in Colombia and found evidence for fixation in several cases, with the highest activity observed in sugarcane (Hardy and Graham, personal communication). New techniques, including tissue culture, are being applied to elucidate the nature of the *Rhizobium*–legume symbiosis (Holsten *et al.*, 1971), and the first example of a N_2-fixing symbiosis between a nonlegume and a *Rhizobium* has been found in nature (Trinick, 1973). The successful transfer of *nif* genes to *Escherichia coli* demonstrates that N_2-fixing activity can be given to non-N_2-fixing organisms (Dixon and Postgate, 1972).

Obviously, an unlimited number of possibilities now exist. Nonetheless, we believe that, with the exception of the natural associative symbioses, it may be as long as a quarter-century before practical results, if any, will be demonstrated in this area. As we will show in the case of the soybean, other physiological limitations may need to be eliminated before N_2-fixing wheat or rice could be other than an experimental curiosity.

Novel abiological systems that more effectively couple N_2 to the crop are a third objective. Chemists have developed systems that catalyze the reduction of N_2 to NH_3 under ambient conditions and in aqueous media (Schrauzer *et al.*, 1971, 1974; Shilov *et al.*, 1971; Hardy *et al.*, 1973b; Hardy and Burns, 1973). An irrigation–membrane-fixation system such as that diagrammed in Figure 3 is an example of the manner in which such catalysts may be utilized (Hardy *et al.*, 1975). Fixation units containing membrane-enclosed catalysts might be localized at farm sites, thereby eliminating transportation costs. The actual time of fixation could be synchronized with that of crop need in order to improve efficiency of nitrogen utilization by the crop. This area, like that of extending biological N_2 fixation, has not been transferred from the laboratory to the field, although each step can be performed in the

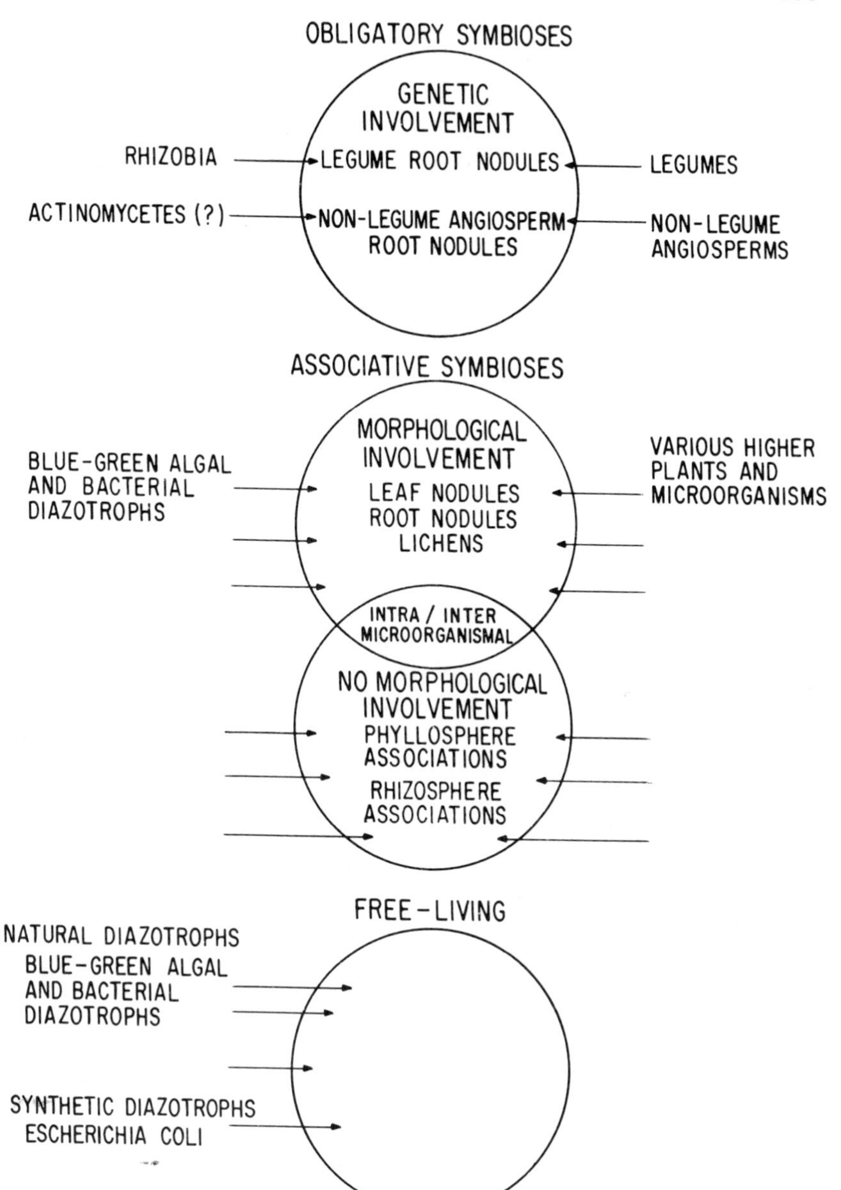

FIGURE 2 Diagrammatic representation of the various biological N_2-fixing relationships extending from free-living forms to obligatory symbioses with intermediate stages referred to as associative symbioses. Some of these associative symbioses may be developed to provide important new N_2 inputs for crop plants (Burns and Hardy, 1974). Diazotrophs are N_2-fixing organisms.

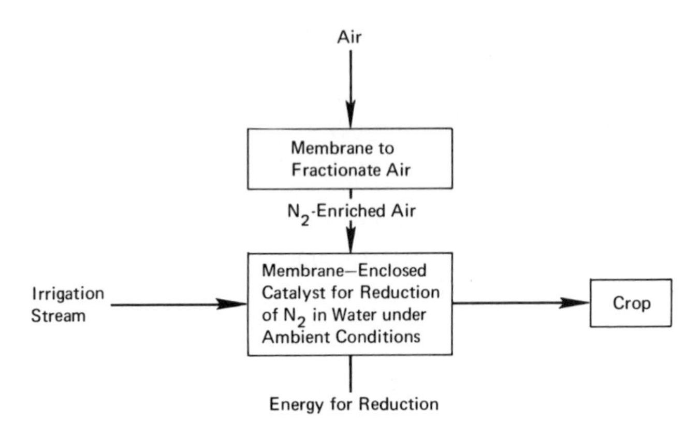

FIGURE 3 A possible system for improved abiological coupling of N_2 to crops by membrane-enclosed catalysts that function in an aqueous medium at ambient conditions (Hardy *et al.*, 1975).

laboratory. It is emphasized that the turnover rates of the catalysts are impractically low and improvements of several orders of magnitude will be required before this area could be of practical significance. Nevertheless, the reduction of N_2 to NH_3 by abiological catalysts at other than high temperatures and pressure was considered impossible before the last decade.

In summary, the plant geneticist seeking improvement in seed protein should be aware that there are a number of new biological and abiological approaches to utilizing N_2 that are being actively investigated and the practical success of one of these could supplement or dramatically change our source of nitrogen to forms other than conventionally produced synthetic nitrogen fertilizers. We will now consider in more detail the opportunity for increasing nitrogen input by enhancing symbiotic N_2 fixation.

INCREASED NITROGEN INPUT INTO LEGUMES (SOYBEANS)

SOYBEAN YIELD PLATEAU AND NITROGEN NEED

It is well known that U.S. soybean production has increased dramatically during the past decade; this increase is mainly a product of increased acreage, with only a modest improvement in yield per acre (Figure 4). The breakthroughs in corn, wheat, and rice yields are attributed in large part to the development of hybrids and selection of varieties that respond to nitrogen fertilizers. Consideration of the

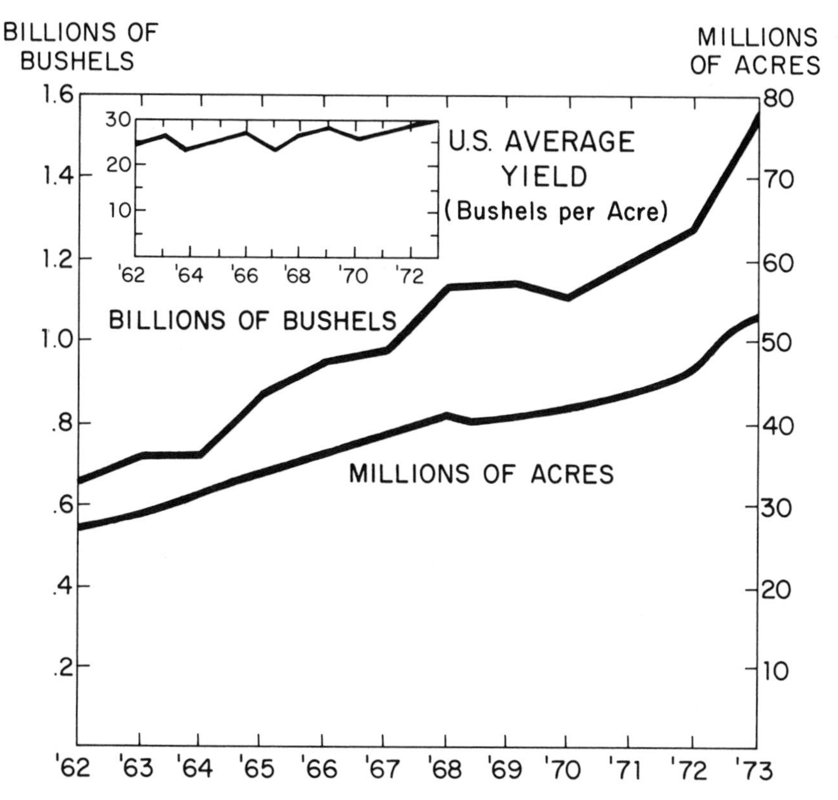

FIGURE 4 Increase in soybean production parallels increase in acres used for soybean production during the last decade with only a modest increase in yield/acre.

nitrogen content of protein-rich legumes suggests that nitrogen should be of even greater importance for these crops than for the protein-low cereals. For example, each bushel of soybean seed contains 4 lb of nitrogen, and the supporting plant structure contains an additional 2 lb of nitrogen. This nitrogen content translates to a need for 300 lb of nitrogen for an above-average yield of 50 bu/acre and 600 lb of nitrogen for the experimental goal of 100 bu/acre.

N_2-FIXATION CHARACTERISTICS

We have determined the source of this nitrogen for field-grown soybeans by measuring their N_2-fixing activity over the complete growth cycle during several seasons (Hardy *et al.*, 1968, 1971, 1973a). A typical age–activity profile is shown in Figure 5 for soybeans and

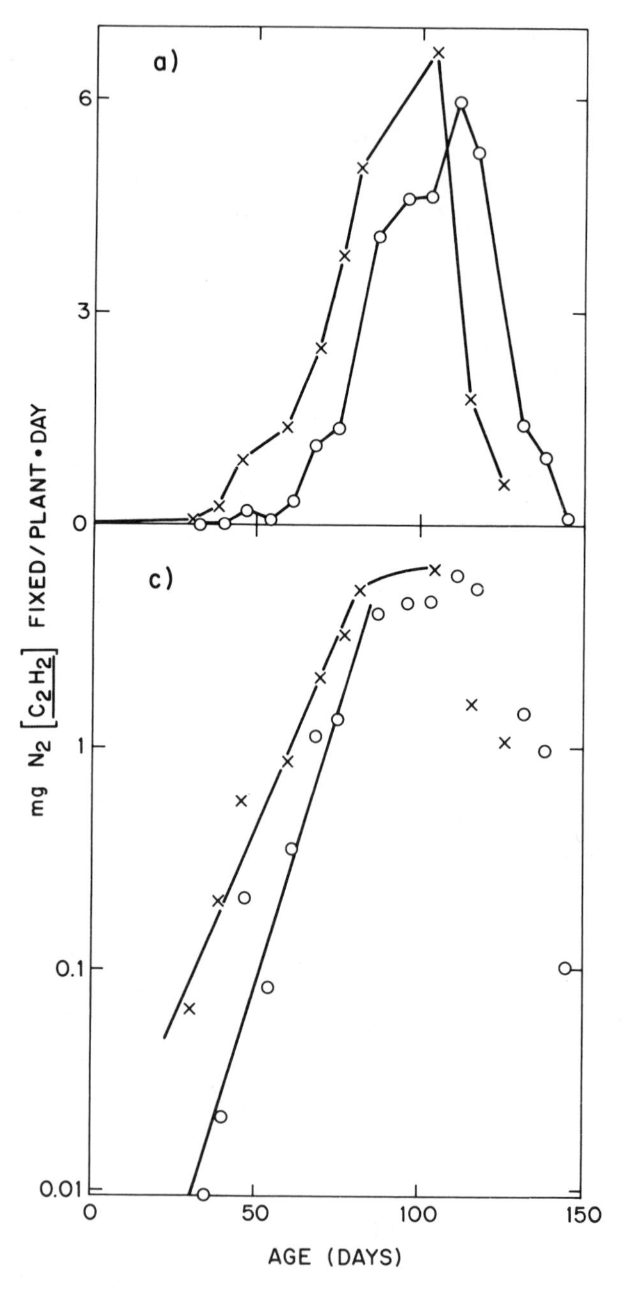

FIGURE 5 Profiles of N_2 fixation during the complete cycle for field-grown soybeans and peanuts (Adapted from Hardy *et al.*, 1971, 1975).

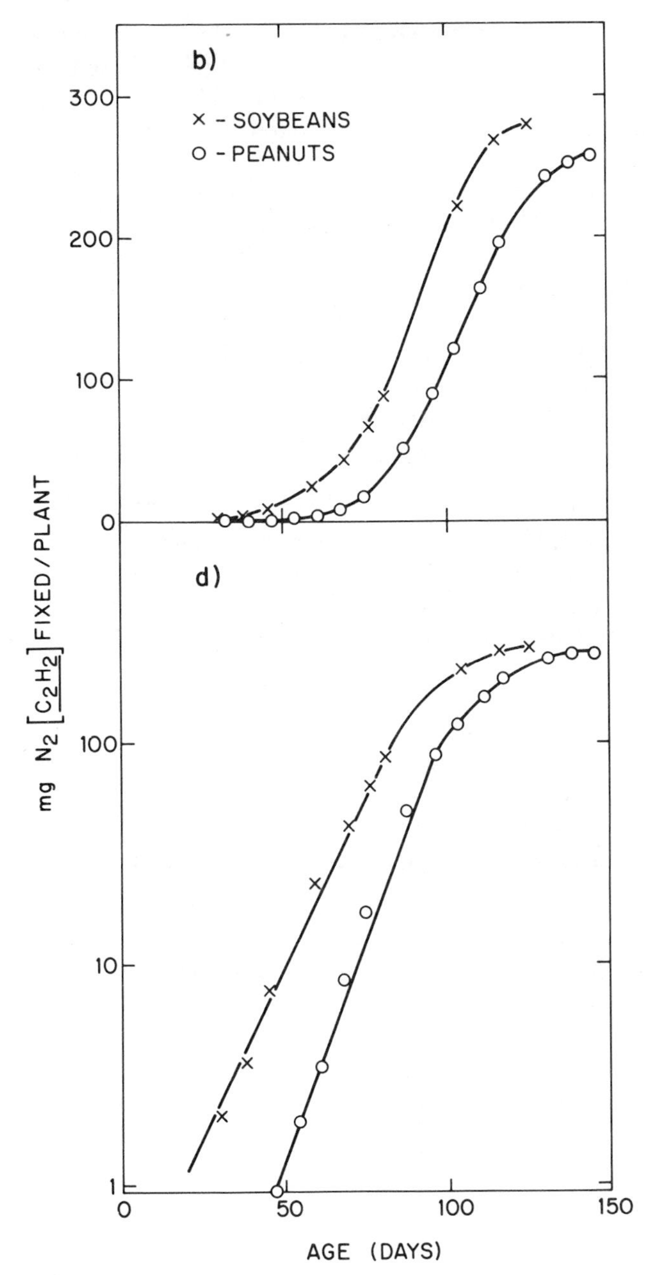
b)
300
200
100
0
× - SOYBEANS
O - PEANUTS
mg N_2 [C_2H_2] FIXED / PLANT
d)
100
10
1
0
50
100
150
AGE (DAYS)

peanuts. In the case of soybeans less than 10 percent of the N_2 is fixed during the first 50 days of vegetative growth, with more than 90 percent during the period of the maximum nitrogen need for reproductive growth. The total amount of N_2 fixed accounts for only about 25 percent of the nitrogen in the mature plant (Table 1). Thus, soil nitrogen is the primary source and atmospheric N_2 the secondary source for soybeans grown on agricultural soil of normal nitrogen fertility. The soybean should be considered much less than self-sufficient with respect to nitrogen nutrition. The results for peanut are similar, and other workers (T. A. G. La Rue, National Research Council, Saskatoon, Saskatchewan, personal communication) have reached the same conclusions for peas.

The development of N_2-fixing activity in the soybean is exponential from around 25 days of age to about 75 days of age; during this period the total N_2 fixed doubles about every 8 days. In theory, increasing nitrogen input from N_2 fixation may be achieved by an increased rate of development, a delay in loss of exponential phase or earlier initiation of the fixation system. The cessation of the exponential phase around 75 days of age suggests that competitive sinks, such as reproductive structures, out-compete the nodules for available photosynthate.

FERTILIZER NITROGEN

Before considering the opportunity for increasing nitrogen input from N_2, let us examine the response of soybeans to fertilizer nitrogen. Table 2 shows the effect on N_2 fixation of various conventional forms of fertilizer nitrogen and the absence of a substantial yield response to

TABLE 1 N_2 Fixation Characteristics of Field-Grown Legumes, including N Input from N_2[a]

	Exponential Phase			$N_2[C_2H_2]$ Fixed		
Legume	Initiation Age (days)	Termination Age (days)	Increase (%/day)	mg N/ plant	kg N/ ha	% of Total N from N_2
Soybeans	26	75	8.3	260	84	25
Peanuts	42	97	8.5	221	—	25
Peas[b]	—	—	—	247	—	23

[a] Hardy and Havelka (1975).
[b] T. A. G. La Rue, National Research Council, Saskatoon, Saskatchewan, personal communication.

TABLE 2 Nitrogen Fertilizers: $N_2[C_2H_2]$ Fixation and Yield[a]

Form[b]	Application		$N_2[C_2H_2]$ (% of Control)	
	Locus[c]	Age (days)	Fixation	Yield
NO_3^-	S	0	43	98
NH_4^+	S	0	51	104
Urea	S	0	48	109
Protein	S	0	112	123
NO_3^-	S	40–50	50	106
NH_4^+	S	40–50	37	106
Urea	S	40–50	44	97
Urea	F	50–71	29	100
Urea	F	78–99	53	100

[a] Hardy *et al.* (1973a).
[b] 135 kg N/ha of indicated form.
[c] S, soil; F, foliar.

fertilizer nitrogen. The response to soybean meal used as a fertilizer may be an exception since it does not inhibit N_2 fixation and has produced yield increases of about 20 percent. However, it can be concluded that soybeans grown on agricultural soils of high fertility, which is the case for most domestic production, are nonresponsive to conventional forms of fertilizer nitrogen (Hardy *et al.*, 1973a).

Opportunities for research in this area are numerous (Hardy and Havelka, 1974) and include (a) development of nitrogen-responsive legume varieties by breeding or use of growth regulators, (b) development of bacterial strains or plant varieties that utilize fertilizer nitrogen without inhibition of N_2 fixation, (c) discovery of novel forms of fertilizer nitrogen that are utilized but do not inhibit N_2 fixation, and (d) new cultural practices that enable legumes to utilize nitrogen fertilizer without inhibition of N_2 fixation. Each of these alternatives is being actively explored in various laboratories, but practical developments that increase nitrogen input and yield through the use of nitrogen fertilizer remain in the future.

POSSIBLE FACTORS LIMITING N_2 FIXATION

Figure 6 is a schematic representation of the symbiotic N_2-fixing system and its relationship to the rest of the plant. It is based on the known biochemistry of N_2 fixation. Nitrogenase, the N_2-fixing

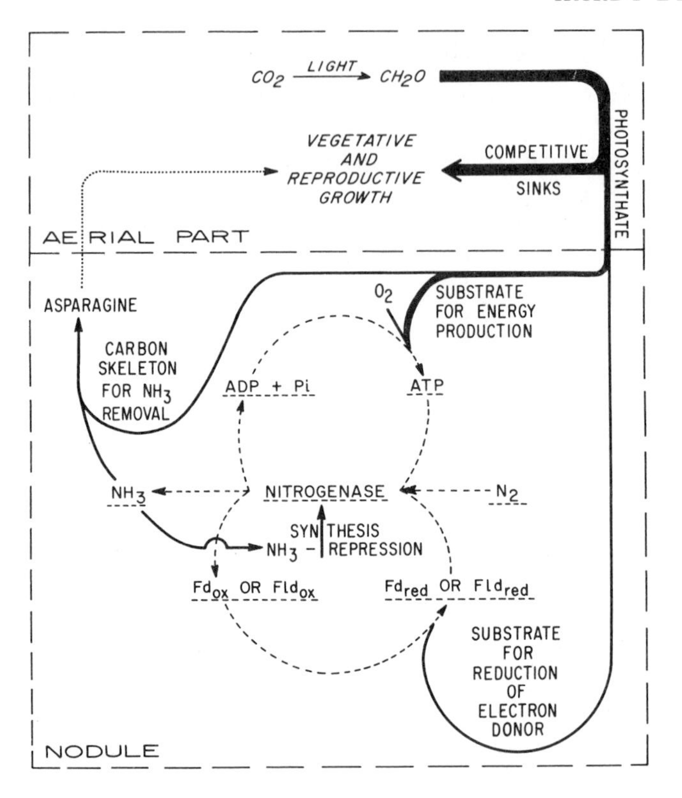

FIGURE 6 Requirements and products of N_2 fixation by a nodule and relationship of these requirements and products to the aerial part of the plant. It is suggested that photosynthate is the limiting factor for symbiotic N_2 fixation (Hardy and Havelka, 1975).

enzyme, converts N_2 to NH_3 and utilizes ATP and reduced ferredoxin and/or flavodoxin; although the exact *in vivo* energy requirements are not established, the value is between 6 and 20 molecules of ATP per molecule of N_2 fixed. The product NH_3 can repress nitrogenase but is normally translocated to the aerial part of the plant after incorporation into a carbon skeleton. The low K_m* of N_2 in the nitrogenase reaction eliminates N_2 from consideration as a limiting factor. The temperature–activity relationship of the nodules suggests limitations in other factors besides nitrogenase, and, in fact, nitrogenase may be present in

*K_m, Michaelis constant, is the concentration of substrate at which the velocity of enzyme action is half maximal.

excess. The factor common to production of energy, reductant, and carbon skeleton is photosynthate. This suggested to us that photosynthate may be the limiting factor for N_2 fixation. We used CO_2 enrichment as an experimental aproach to test this hypothesis (Hardy and Havelka, 1973, 1974, 1975).

EXPERIMENTAL SYSTEM FOR CO_2 FERTILIZATION

The objective of our experimental system for CO_2 fertilization was to provide a controlled CO_2 enrichment within the canopy of a field situation without altering the temperature or light intensity. Since photorespiration is markedly influenced by temperature and light, it was of utmost importance that these parameters be maintained as ambient in order for our results to be relevant to the crop as grown in the field. Previous work with long-term CO_2 fertilization has employed closed chambers with resultant alterations in temperature and light intensity. One of our colleagues, Dr. J. E. Carnahan, proposed the use of open-top enclosures as shown in Figure 7 to achieve our objective. With these enclosures of base dimensions of 1 m × 4 m, the introduction of about 0.5 kg CO_2 per hour during the day into the forced-air stream produced concentrations of 800 to 1,200 ppm CO_2 within the crop canopy.

N_2 FIXATION BY CO_2-FERTILIZED SOYBEANS

The profile of N_2-fixing activity of soybeans enriched with 800–1,200 ppm CO_2 during the day from 38 days of age to maturity is compared with air-grown plants in Figure 8. The control profile is similar to those found previously for field-grown plants (Figure 5), demonstrating that the open-top enclosure did not alter the N_2-fixing characteristics. On the other hand, there is a dramatic increase in N_2 fixation produced by CO_2 enrichment, with a substantial effect observed even at the time of first measurement, 1 week after initiation of CO_2-enrichment. The maximum N_2-fixing activity of CO_2-enriched plants was more than three times that of control and occurred at a somewhat later date. The age at cessation of exponential phase was extended 8 days to 83 days of age, thereby providing substantially more nitrogen from N_2 during the demanding pod-filling stage of reproductive growth. A remarkable result is the fixation of more N_2 by CO_2-enriched plants in 1 week from 87 to 94 days of age than by control plants during the complete growth cycle. Total N_2 fixed per plant was 843 mg by CO_2-enriched, versus 167 mg by control.

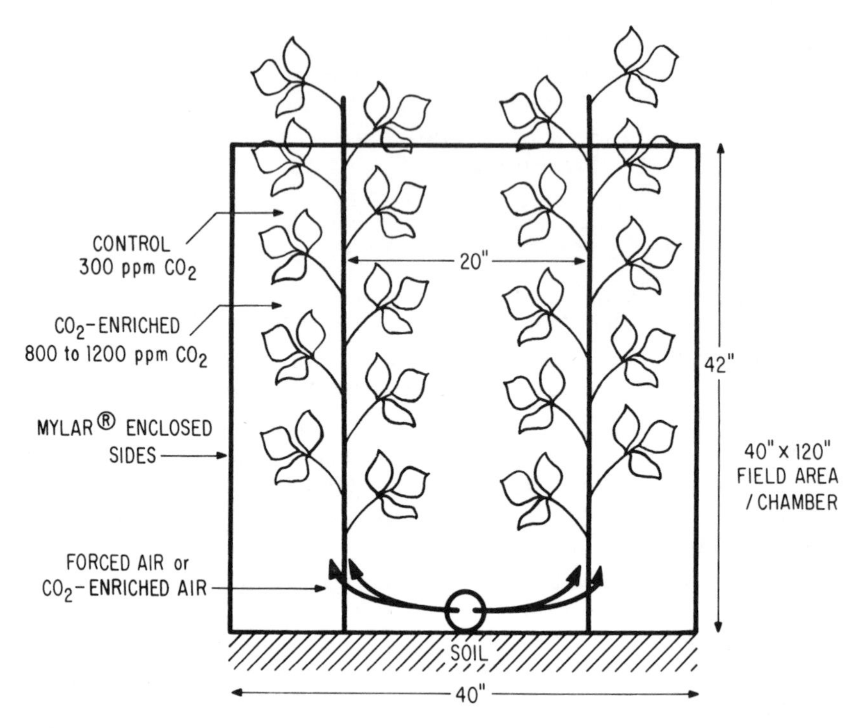

FIGURE 7 Schematic of open-top enclosures used for CO_2 enrichment of field-grown soybeans (Hardy and Havelka, 1975).

This multifold increase in N_2 fixation by CO_2 enrichment can be assigned to three factors: (a) delay in loss of exponential phase, (b) increase in nodule fresh weight by an average of twofold to a maximum of 8.97 g fresh weight/plant at 87 days of age versus 3.48 g at 77 days for control, and (c) doubling of the specific N_2-fixing activity of the nodule by CO_2-enrichment with a maximum of 5.36 mg N_2 fixed/g nodule fresh weight/day at 64 days of age versus 3.29 g at 73 days for control. This increase in specific activity is attributed to a more complete saturation of the suggested excess nitrogenase, and this proposal is supported by the demonstration of a 70 percent increase in specific N_2-fixing activity of nodules after only 6 hours of CO_2-enrichment.

The dry weight of the mature plants collected at 101 days of age was increased by CO_2-enrichment (Table 3). For example, reproductive tissue consisting of pods and seeds was 1.78 times control. The nitrogen content was determined by Kjeldahl analysis. Percent nitrogen of pods and seeds and seeds alone was unchanged, while that of vegetative tissue was decreased slightly. Total nitrogen was increased to an average of 1.009 g/plant for CO_2-enriched plants versus 0.648 g/plant for control.

Finally, the number of plants per area at maturity was increased 11 percent by CO_2-enrichment. We attribute this increased density to the improved photosynthetic competitiveness of shaded plants growing in a CO_2-enriched environment. Since crop production is based on a unit area of land rather than on a plant, we have expressed total nitrogen, N_2 fixation, and nitrogen from soil on an area basis. Nitrogen from soil is the calculated difference between total nitrogen and N_2 fixation. Over 400 kg N_2/ha was fixed by CO_2-enriched plants, giving a total nitrogen content of >500 kg N/ha versus fixation of 76 kg N/ha and total nitrogen of <300 kg N/ha by controls. Measured N_2 fixation accounts for only 26 percent of the total nitrogen of control plants but for 83 percent of CO_2-enriched plants, making these plants almost self-sufficient with respect to meeting their nitrogen needs. The contribution required from soil nitrogen decreased 65 percent with CO_2 enrichment.

These observations on the effect of CO_2-enrichment on nitrogen input are interpreted as effects of photosynthate on N_2 fixation. We suggest that elevated pCO_2 increased net photosynthesis, and we attribute the major part of this increase to a decrease in O_2 inhibition of photosynthesis, commonly called photorespiration. The result is to make more photosynthate available to the nodule for N_2 fixation. However, CO_2 enrichment also increases carbon assimilation in plants, such as corn, with low or no photorespiration (Wittwer, 1970), and therefore part of the effect of elevated CO_2 in our soybean experiments may relate to other than decreased photorespiration. Further support for the conclusion that N_2 fixation is source- rather than sink-limited is provided by the direct relationship between several factors that influence source production of photosynthate availability of the photosynthate to the nodule and N_2 fixation (Table 4).

It is concluded from these crop physiology experiments with soybean that seeking ways to increase total photosynthate so that more is available to the nodule without decreasing that available to other tissue is a most attractive research goal. In fact, enhancing nitrogen assimilation through enhanced symbiotic N_2 fixation becomes primarily a problem in carbon assimilation. Decreasing photorespiration, a process that occurs in all legumes, is one possible approach for increasing photosynthate.

Among the various proposals for the mechanism of photorespiration,

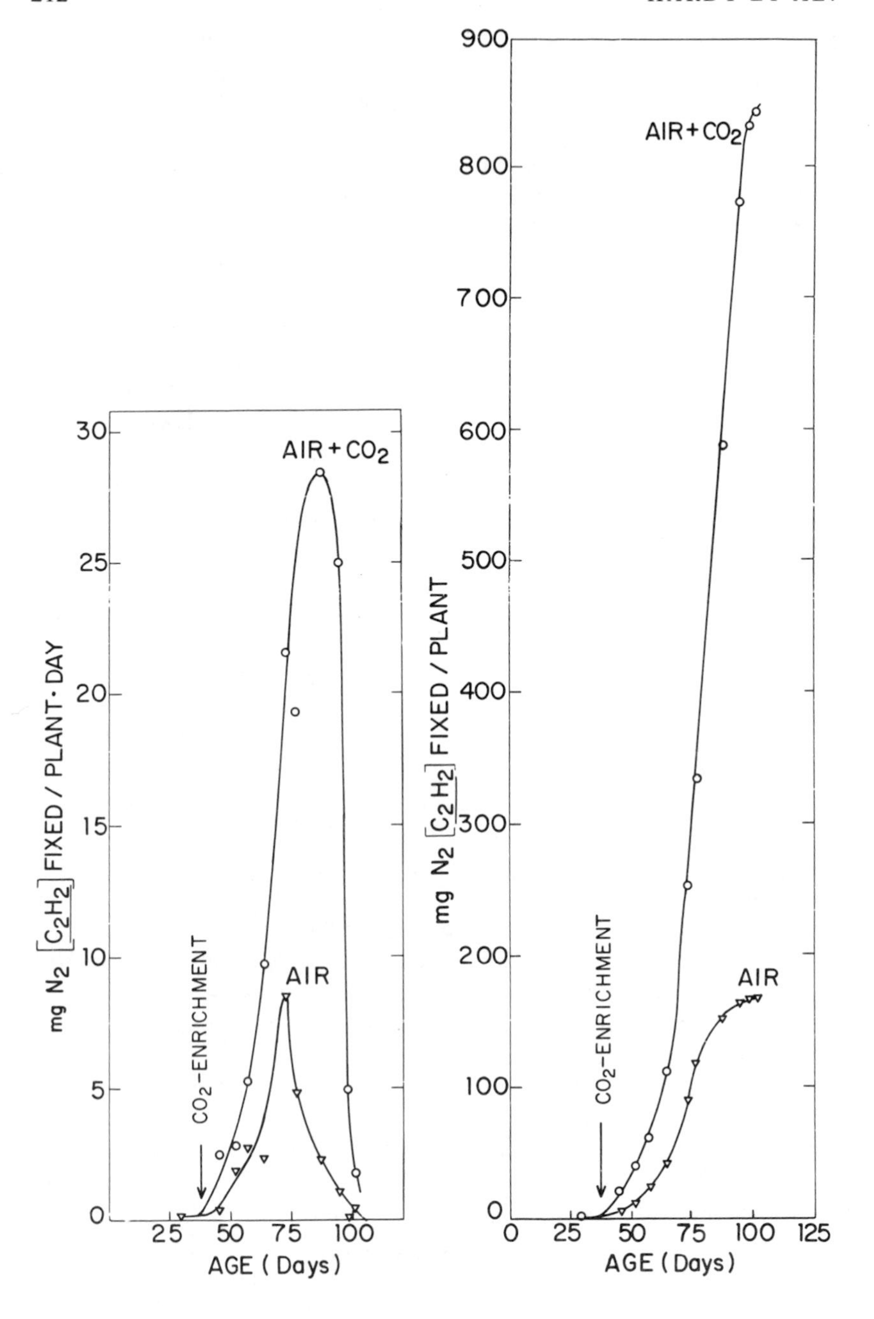

AIR + CO_2
AIR
CO_2-ENRICHMENT
mg N_2 [C_2H_2] FIXED / PLANT · DAY
AGE (Days)
0
5
10
15
20
25
30
25
50
75
100
AIR + CO_2
AIR
CO_2-ENRICHMENT
mg N_2 [C_2H_2] FIXED / PLANT
AGE (Days)
0
100
200
300
400
500
600
700
800
900
0
25
50
75
100
125

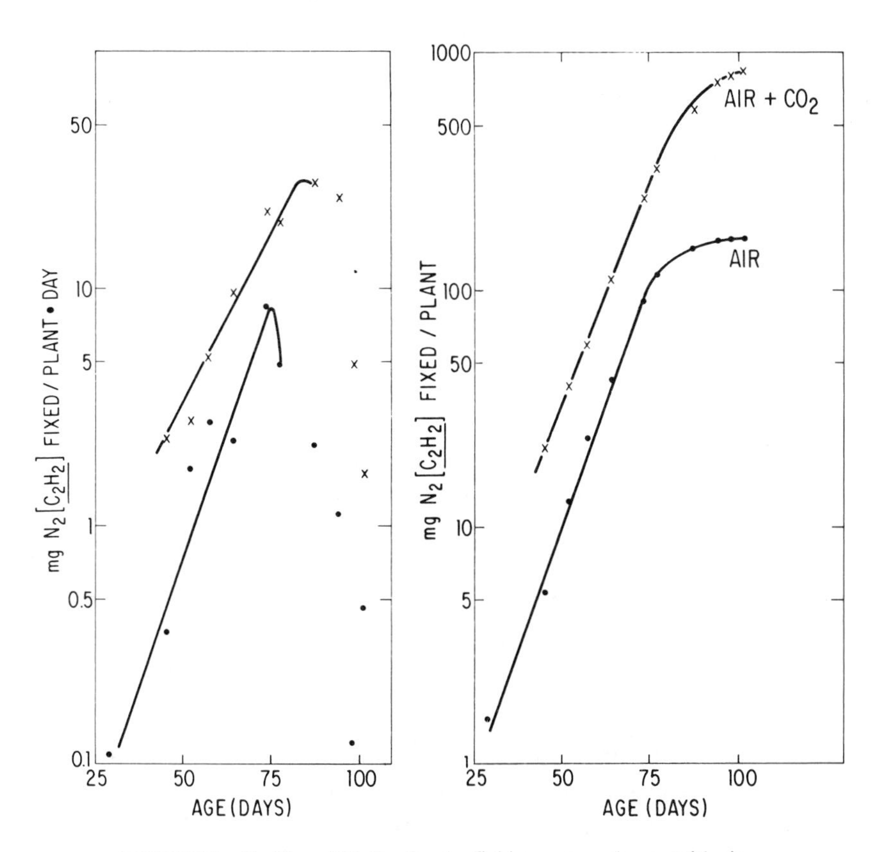

FIGURE 8 Profiles of N_2 fixation by field-grown soybeans with air or CO_2-enriched air. The CO_2-enriched air contained 800–1200 ppm CO_2 and was provided from about 40 days of age to maturity during the day (Hardy and Havelka, 1975).

TABLE 3 Effect of CO_2 Enrichment on Mass, Percent N, N Content, N_2 Fixation, N Input, and Number of Plants[a]

	Air	CO_2-Enriched Air
Mass[b] (g dry wt/plant)		
Nodules	0.73	0.71
Roots	1.2	1.8
Stems	5.6	7.1
Leaves	4.3	5.4
Pods and seeds	8.8	15.7
Total	20.6	30.7
% N[b]		
Nodules	3.1	4.1
Roots	1.1	1.1
Stems	1.5	1.3
Leaves	3.7	3.2
Pods	4.3	4.4
Seeds	6.5	6.5
N content (g N/plant)		
Nodules	0.017	0.028
Roots	0.013	0.020
Stems	0.083	0.089
Leaves	0.16	0.18
Pods	0.38	0.70
Total	0.648	1.009
$N_2[C_2H_2]$ fixed (g N/plant)	0.167	0.843
Number		
Plants/ha	455,000	505,000
Pods/plant	23	34
N Input (kg N/ha)		
Total	295	511
$N_2[C_2H_2]$ fixed	76	427
N from soil[c]	219	84

[a] Hardy and Havelka (1975).
[b] Determined at maturity at 101 days of age for six replicates of three plants per sample for each of air and CO_2-enriched air samples.
[c] Difference between $N_2[C_2H_2]$ fixed and total N.

the dual function of the CO_2-fixing enzyme ribulose diphosphate carboxylase, as both a monooxygenase and carboxylase, appears to be most attractive (Figure 9). Evidence that supports ribulose diphosphate carboxylase as the primary site of the CO_2–O_2 antagonism common to photorespiration includes the already noted dual activity of the enzyme, the similarity of the K_m of CO_2 as an inhibitor of the

TABLE 4 Direct Relationship between Alteration in N_2 Fixation and Photosynthate Availability Produced by Several Factors[a]

Factors	Increased N_2 Fixation	Decreased N_2 Fixation
Light	Day	Night
	Supplemental light	Shading
Source Size	Additional foliage	Defoliation
	Low plant density	High plant density
		Lodging
Competitive Sinks	Pod removal	Pod filling
CO_2	CO_2 enrichment	CO_2 depletion (?)
O_2	O_2 depletion (?)	O_2 enrichment (?)
Photosynthate translocation	—	Girdling

[a] Hardy and Havelka (1975).

oxygenase reaction, the similarity of the K_m of O_2 as a substrate and the K_i of O_2 as an inhibitor of the carboxylase reaction, and the conversion of ribulose 1,5-diphosphate to glycolate phosphate by O_2 and ribulose diphosphate carboxylase in a dark reaction (Chollet, *et al.*, 1974). The dissimilar responses of the O_2 and CO_2 reaction of ribulose diphosphate carboxylase to temperature, pH, and enzyme storage (Andrews *et al.*, 1973a,b) suggest that there may be exploitable differences between the interaction of O_2 and CO_2 with the enzyme. Obviously, we need to learn much more about the mechanistic details of this unique O_2 and CO_2 reaction catalyzed by ribulose diphosphate carboxylase; furthermore, the unusually large concentration of this enzyme in plant tissue in spite of its low turnover number has not been adequately explained.

In terms of photorespiration, our CO_2-enrichment experiments changed the CO_2/O_2 ratio and thereby favored the carboxylase over the oxygenase reaction (Figure 10). It is, of course, impractical to consider CO_2 enrichment for field crops. However, mutants or synthetic growth regulators may be found that favor the carboxylase over the oxygenase activity at ambient CO_2 and O_2 levels and thereby yield similar effects on nitrogen input without CO_2 enrichment.

A possibly significant corollary of these observations is that extension of N_2 fixation to crops with photorespiration such as wheat and rice may not result in greater yields, since improvement in carbon assimilation by the source may be fundamental to enhancing N_2 fixation in legumes or extending N_2 fixation to photosynthetically inefficient crops.

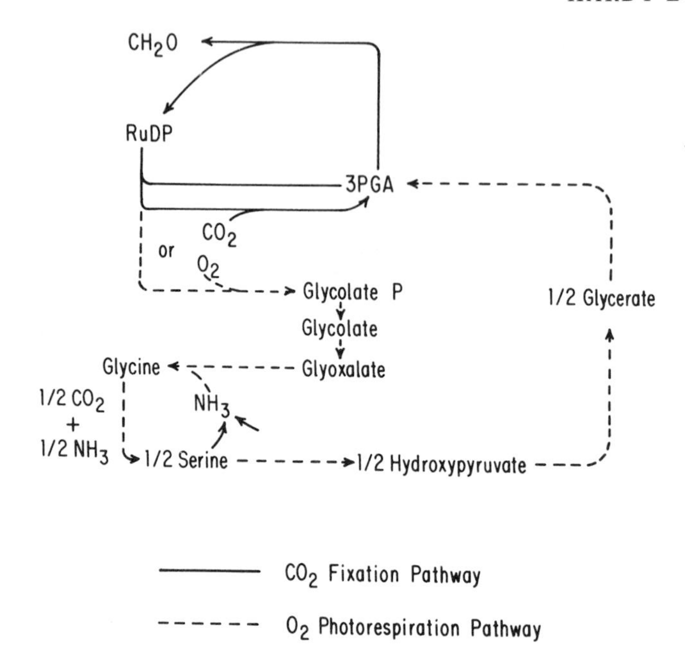

FIGURE 9 The CO_2 fixation and O_2 photorespiration pathways of inefficient plants (e.g., wheat, soybeans, rice, and forages). The initial O_2 and CO_2 competition is attributed to their reaction with ribulose 1,5-diphosphate carboxylase.

METHODS OF ASSESSING PHOTORESPIRATION

In order to select genetic material or chemical growth-regulating agents that reduce photorespiration, a valid high-capacity test system is a necessity, and several methods have been proposed and are used. These include CO_2 compensation point, zero CO_2 intercept of the CO_2 response curve, O_2 suppression of net photosynthesis, efflux of pulse-labeled ^{14}C during a light period, and postillumination burst of CO_2. The most used method is compensation point or a modification that determines rate of senescence in an environment containing sub-compensation-point concentrations of CO_2. The failure of surveys of collections of wheat and soybean varieties to find even one with a lower compensation point may suggest the absence of varieties with reduced photorespiration. Alternatively, the failure may suggest that the compensation point is not a satisfactory method; Zelitch (1973) has pointed out some of its limitations. Photorespiration is a rapid response and methods utilizing release of pulse-labeled ^{14}C during a postlabeling

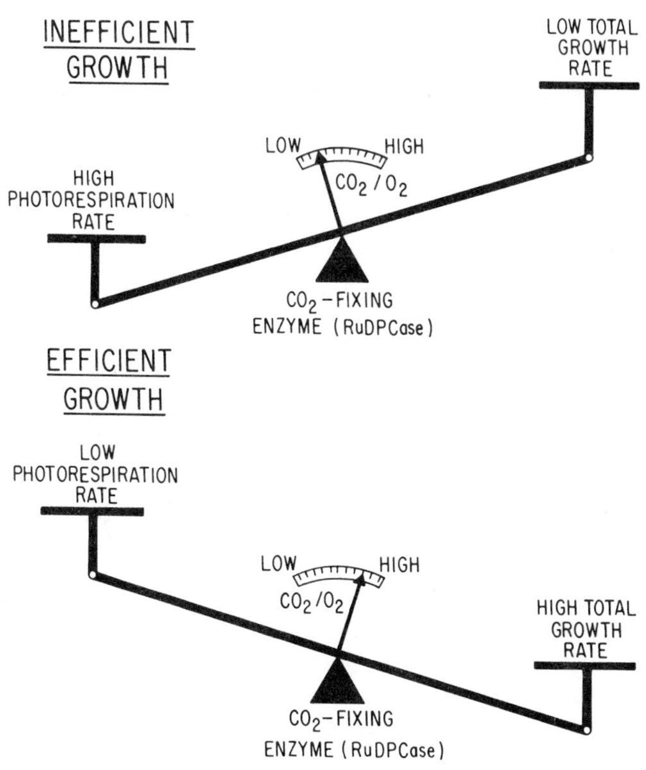

FIGURE 10 The CO_2/O_2 ratio via ribulose 1,5-diphosphate carboxylase controls growth efficiency in C_3 plants such as soybeans.

light period of 15 to 30 minutes may be in error because of the long period, which permits recycling of carbon. Postillumination burst of CO_2 or uptake of O_2 permits measurement with possibly the least perturbation of the leaf and its environment. However, this measurement may be difficult to use at a high capacity and has not been used for large-scale screening. In summary, perhaps the greatest limitation in this area is the test system.

OXYGEN CONCENTRATION

Besides optimization of carbon and nitrogen assimilation in crop plants for high total growth, it is necessary to distribute the assimilates between vegetative and reproductive growth so as to maximize the Harvest Index and thereby increase protein yield. However, knowledge of the physiological and biochemical systems that regulate Harvest Index is still almost totally lacking, just as concluded by Wardlaw (1968) in a review: "The greatest lack in these studies is an understanding of the mechanism involved in assimilate distribution." Stoy (1969) said at a Symposium on Physiological Aspects of Crop Yield, ". . . it is to be hoped that within a relatively few years we will be able to understand much better the way in which different crop plants absorb, distribute and store the energy they derive from the sun and how to utilize this information to produce more food for mankind." Improvement of Harvest Index has been the random product of empirical selection rather than the designed product of plant physiological and biochemical developments.

Recent exploratory work in our laboratory on the effect of oxygen concentrations on plant growth has identified an effect of oxygen on the balance between reproductive and vegetative growth, suggesting that an oxygen-mediated process may be a major key to controlling harvest index (Quebedeaux and Hardy, 1972, 1973a,b, 1974; Hardy and Quebedeaux, 1974). This section will describe the experiments that support this hypothesis.

Numerous measurements of photosynthetic rate of leaves under subatmospheric concentrations of O_2 demonstrate an increase of 50–100 percent for the so-called C_3, or inefficient, plants, but not the so-called C_4, or efficient, plants (Zelitch, 1971). Short-term experiments comparing vegetative growth of beans and corn at 21 percent and 2.5 percent O_2 (Bjorkman *et al.*, 1966) demonstrate a doubling in growth rate at low O_2 concentrations for bean with no appreciable change for corn.

In 1971 we set out to determine the long-term effect of O_2 concentrations on vegetative and reproductive growth. In all of these experiments, unless otherwise indicated, the aerial portion of the plant was exposed to the indicated gas phase from early vegetative growth stage to maturity. Reduced O_2 concentrations were routinely produced by dilution of air with N_2 and re-addition of CO_2 to a concentration of 300 ppm.

As shown in Figure 11, soybean plants grown in 5 percent versus 21 percent O_2 from 14 to 83 days of age had increased vegetative growth in agreement with the earlier observations with short-term experiments. Leaves were thicker, stems were larger and roots were markedly increased in size. However, reproductive growth was arrested under 5 percent O_2, with only minimal development of pods and seeds.

We have used various O_2 concentrations to titrate this effect on reproductive growth, and Figure 11 and Table 5 record typical results

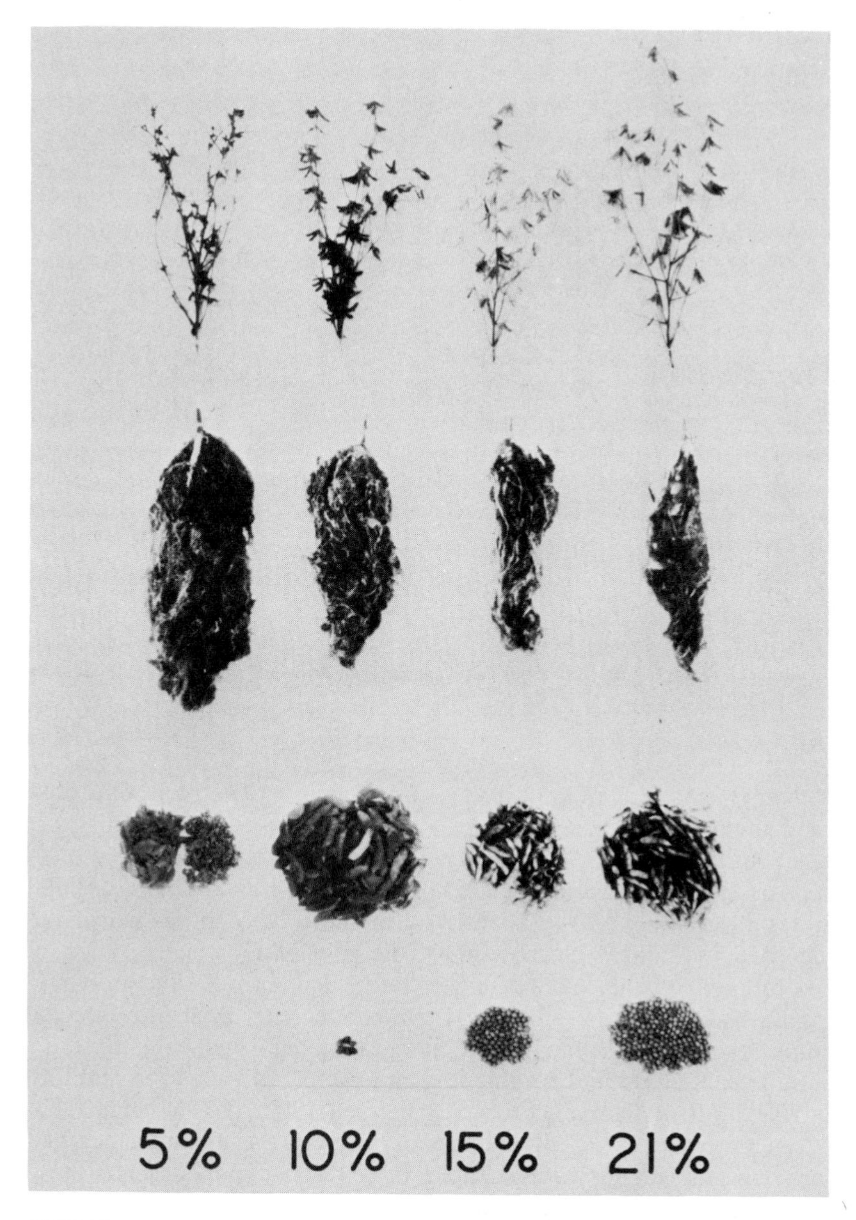

FIGURE 11 Photograph of vegetative and reproductive parts of mature soybean plants in which the aerial part was exposed to 5, 10, 15, or 21 percent O_2 from early emergence to senescence. Note the increased vegetative growth and decreased reproductive growth at subambient O_2 concentrations (Quebedeaux and Hardy, 1973a).

TABLE 5 O_2 Concentration and Soybean Vegetative and Reproductive Growth[a,b]

Growth	Percent O_2 5	10	15	21
Vegetative growth (gm dry wt/plant)				
Leaves	44.2	20.7	8.2	9.2
Stems	25.5	14.4	3.6	5.0
Roots	18.3	14.8	3.4	2.9
Reproductive growth (gm dry wt/plant)				
Pod wt	7.6	12.9	5.8	8.8
Seed wt	0	1.1	12.8	21.5
Reproductive growth/ total growth	0.08	0.22	0.55	0.64
No. developed pods	0	11	64	105
No. undeveloped pods	804	221	35	6
No. developed seeds	0	12	125	235

[a] Quebedeaux and Hardy (1973a).
[b] Measured at 83 days of age with aerial portion of the plant exposed to indicated gas phase from 14–83 days.

with soybeans grown at 5, 10, 15, and 21 percent O_2. Note that seed yield at even 15 percent O_2 or three-quarters of the concentration in air, was reduced almost 50 percent. Thus, normal reproductive growth appears to require a concentration of O_2 that approaches or equals that in air. The ratio of reproductive growth to vegetative growth as recorded in Table 5 clearly shows the role of O_2 concentration in controlling the balance between vegetative and reproductive growth.

One may ask if the effect of O_2 is common to all seed-producing crops or unique to soybeans. We have examined wheat as an example of another C_3 crop and sorghum as an example of a C_4 crop and find similar results with respect to reproductive growth and the balance between reproductive growth and vegetative growth as already described for soybeans (Figure 12 and Table 6). A recent report extends these observations to rice (Akita and Tanaka, 1973).

In order to further describe this oxygen system, we have determined, among other things, the effect of pO_2 and pCO_2 concentration, growth stage of the plant at exposure, and part of the plant exposed. Most of these results have been obtained with soybeans but are presumably

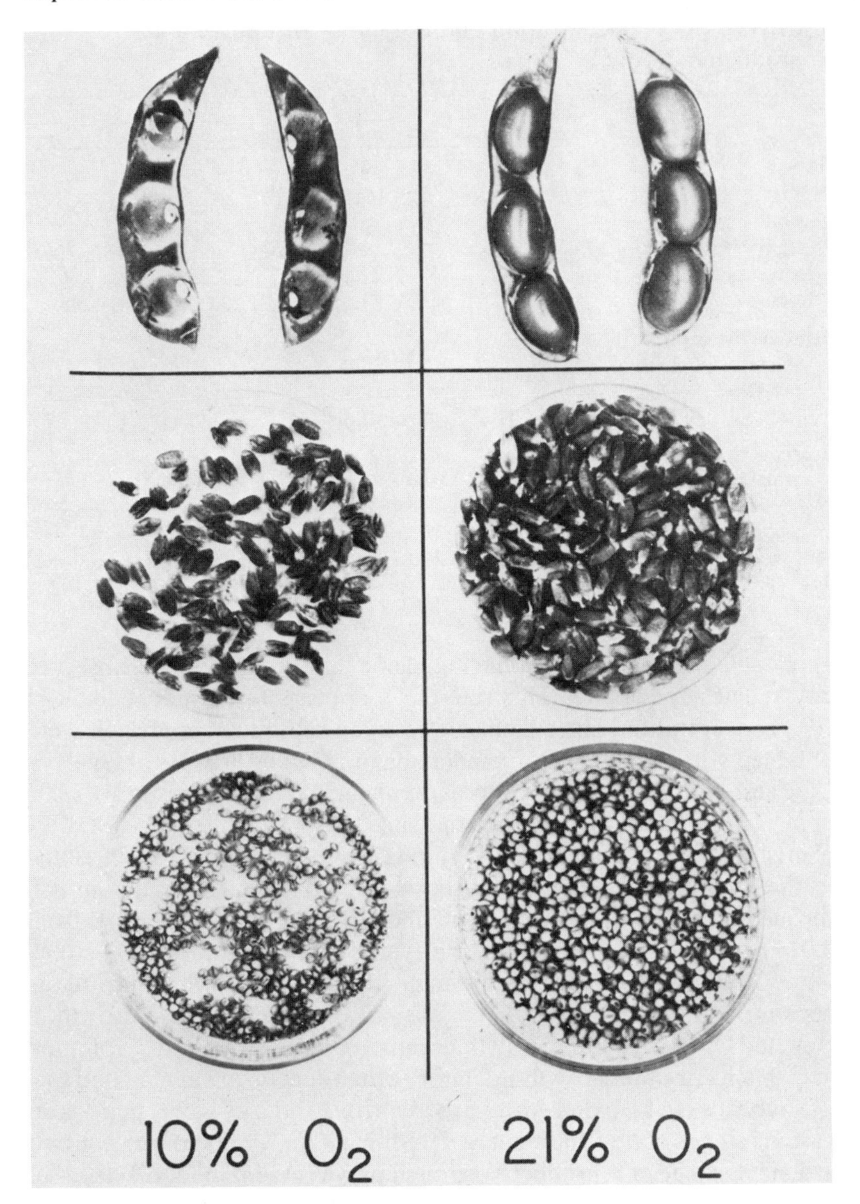

FIGURE 12 Photographs of seed from soybeans (*top*), wheat (*center*), and sorghum (*bottom*) in which the aerial part was exposed to 10 or 21 percent O_2 (Hardy and Quebedeaux, 1974).

TABLE 6 O_2 Concentration and Sorghum Vegetative and Reproductive Growth[a,b]

Growth	Percent O_2 5	10	15	21
Vegetative growth (gm dry wt/plant)				
Leaves and stems	57.5	47.6	53.2	66.7
Roots	38.2	41.1	30.9	43.2
Reproductive growth (gm dry wt/plant)				
Seed Head	5.9	8.5	9.1	16.8
Seeds	0	1.6	1.7	7.7
Reproductive growth/ total growth	0.06	0.10	0.11	0.18

[a] Quebedeaux and Hardy (1973a).
[b] Measured at 168 days of age with aerial portion of the plant exposed to indicated gas phase 67–168 days.

applicable to other seed-producing plants, since we have not observed any species specificity. This role of O_2 appears to be independent of CO_2 concentration, since similar effects on reproductive growth were obtained with subambient O_2 and ambient pCO_2 or depressed levels of CO_2 and with the CO_2/O_2 ratio as in air or CO_2 enriched up to 2,000 ppm. These results, as well as the similarity of results for C_3 and C_4 plants, indicate that the role of O_2 in reproductive growth is unrelated to the O_2 inhibition of photosynthesis in C_3 plants. The optimum pO_2 for maximum reproductive growth and seed yield was measured from 0.015 to 0.60 atm with ambient or elevated pCO_2 and is ≥ 0.21 at < 0.40 atm. Conclusions regarding optimum pO_2 are somewhat hazardous because of the competing effects of improved reproductive growth at elevated O_2 and inhibited growth because of increased photorespiration in C_3 plants or other growth-inhibitory effects occurring at elevated O_2.

Soybeans were exposed to 5 percent O_2 for 2–3-week periods at various stages of development to determine if the O_2 response occurred at a single stage of reproductive growth or was essential for all stages of reproductive growth. A 20-day exposure during the early vegetative stage inhibits pod but not seed development, while a 15-day exposure during either flowering or various stages of pod filling arrests only seed development. The arrest of reproductive growth by 5 percent O_2 is reversible for exposures of ≤ 3 days and irreversible for exposures of ≥ 10 days.

Split shoots of soybean have been used to explore the localization of the O_2 effect on reproductive growth. Exposure of one part of a split shoot to 5 percent O_2 and the other to air showed that the O_2 effect on reproductive growth is localized to the part of the plant exposed.

A number of conclusions can be made and a number of hypotheses suggested with respect to this work on O_2 effects on reproductive and vegetative growth. Some of these are listed below:

1. O_2 concentration controls total growth in most crops and reproductive growth in all crops tested.
2. Root and vegetative growth—O_2 limitation increases in inefficient (C_3) plants, with greatest effect on root.
3. Reproductive growth—O_2 plays an essential role for both C_3 and C_4 plants:

    O_2 affects all stages of reproductive growth.

    Seed development more sensitive than pods.

    Early exposure arrests pod while later exposure arrests seed development.

    The effect is localized to the part exposed.

    The effect is irreversible except for short exposures of ≤ 3 days.

    The effect of O_2 is independent of pCO_2.
4. O_2 concentration for maximal reproductive growth approaches or exceeds 21 percent.
5. Physiological and Biochemical Effect of O_2 on Reproductive Growth:

    Fertilization is normal.

    Translocation is altered by source–sink relationship.

    Altered level or activity of endogenous hormones may mediate O_2 effect.

    An enzyme with low affinity for O_2 may be involved.

    A physical versus a chemical phenomenon is also a possibility.
6. Evolutionary implication—plants reproduced by seeds may have evolved only after the concentration of O_2 in the atmosphere reached or surpassed 10 percent.
7. Altitude effects—seed yield at high altitudes may be limited by amount of O_2.
8. Crop yield—understanding the role of O_2 may provide information to optimize harvest index; i.e., maximize seed yield of cereals and vegetative growth of forage and root crops.

Of greatest importance to improvement of crop productivity is the clearly defined effect of O_2 concentration on the balance between reproductive and vegetative growth (Figure 13). Exploitation of this O_2 role could lead to maximization of harvest indices with distribution in favor of reproductive growth for seed-producing crops and vegetative growth for forage and root crops. The reproductive growth requirement for O_2 concentrations that approach or exceed 21 percent at sea level suggest that development of crops with high seed yield at high altitudes may indeed be possible with oxygen limitations. Alternatively, seed-producing crops that are endogenous to this area may have overcome this requirement for high absolute concentrations of O_2 for normal reproductive growth. Finally, seed-producing plants may not have evolved until the O_2 concentration exceeded 10 percent.

In summary, we have described a new role of O_2 in controlling sink activity of reproductive structures. The work described has been restricted to the gross observational level on whole plant growth. It is

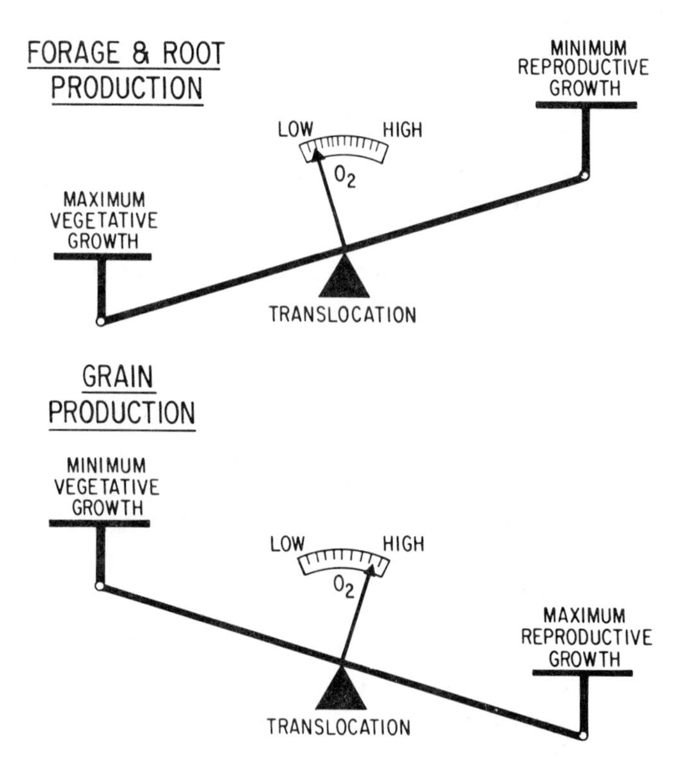

FIGURE 13 Oxygen concentration via an effect on translocation controls the balance between vegetative and reproductive growth.

hoped that further studies will lead to a chemical or physical explanation for this new role of O_2 and that it can be exploited to permit control of harvest indices.

CONCLUSIONS

Research with three of the gases of air—N_2, CO_2, and O_2—is providing definition of opportunities for the improvement of nitrogen and carbon assimilation and distribution of assimilates between reproductive and vegetative growth. New routes to couple N_2 more effectively to plants may arise from the current explorations in biological and abiological N_2 fixation. The demonstration of an almost doubled nitrogen input with almost all of the nitrogen from N in field-grown soybeans provided with CO_2-enriched air identifies the need for improved carbon nutrition as a prerequisite for increased N_2 utilization (Figure 14), while the effect

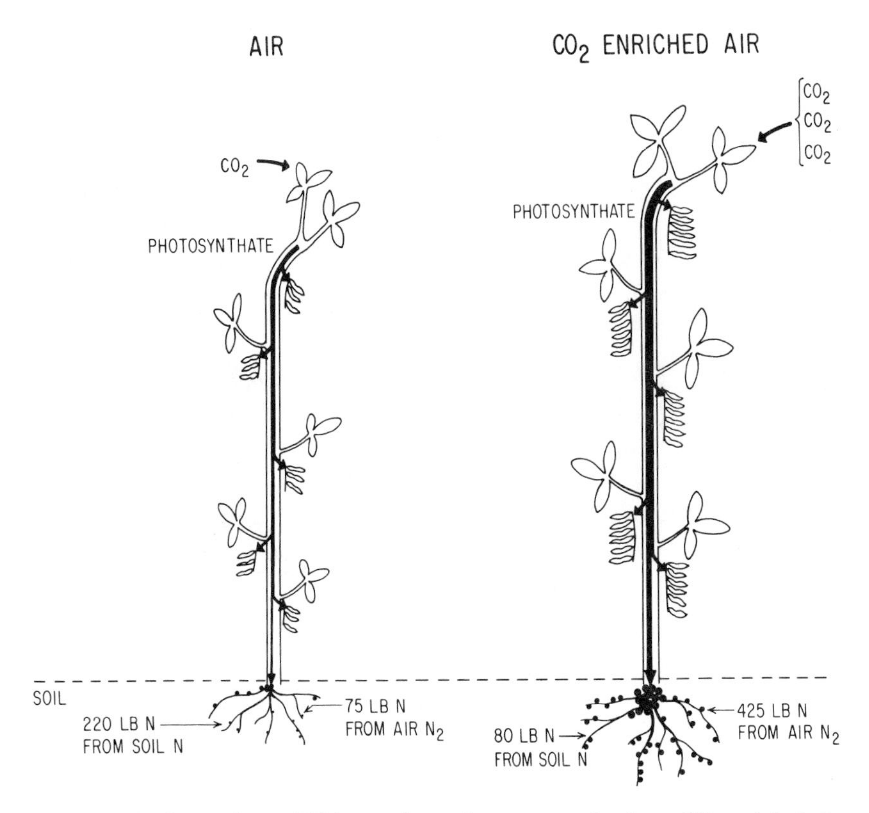

FIGURE 14 Comparison of N inputs for soybeans grown in air or CO_2-enriched air (Hardy and Havelka, 1974).

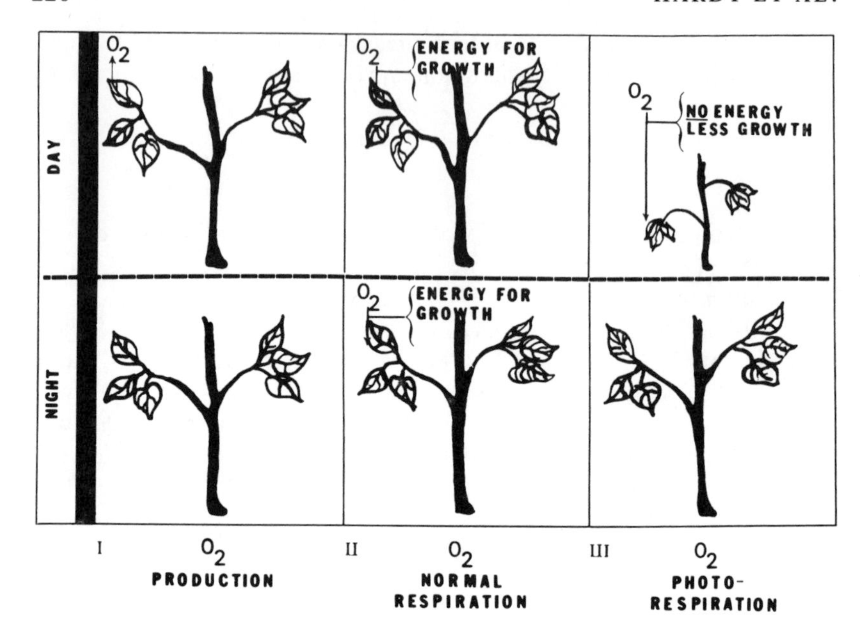

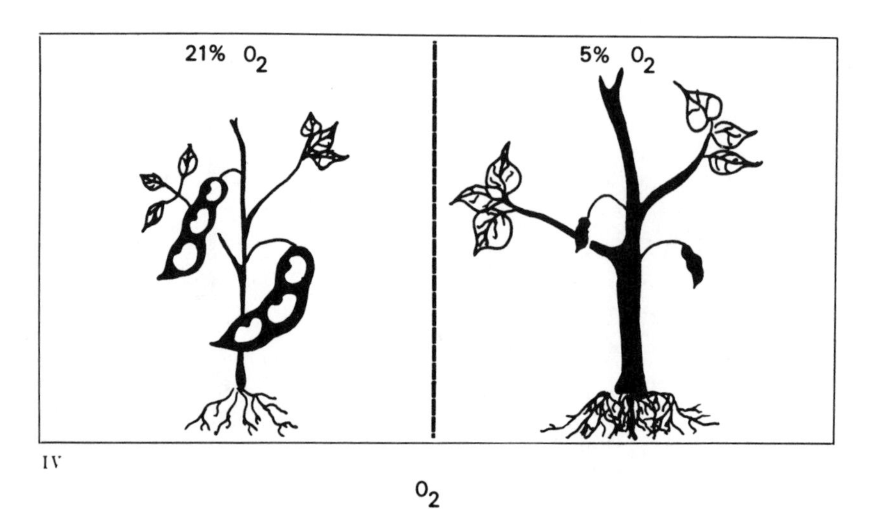

FIGURE 15 Four relationships between plants and oxygen indicating (*III*) growth reduction due to photorespiration and (*IV*) control of the distribution between reproductive and vegetative growth by O_2, concentration as well as (*I*) photosynthetic O_2 evolution and (*II*) mitochondrial respiration (Hardy and Quebedeaux, 1974).

of O_2 on reproductive growth provides the first example of a factor that clearly dictates assimilate distribution (Figure 15). Successful exploitation of these opportunities through biochemical and physiological research to develop new varieties or synthetic plant growth regulators could provide major improvements in quantity of yield and protein production.

REFERENCES

Akita, S., and I. Tanaka. 1973. Proc. Crop Sci. Soc. (Japan) 42:18.

Andrews, T. J., G. H. Lorimer, and N. E. Tolbert. 1973a. Biochemistry 12:11.

Andrews, T. J., G. H. Lorimer, and N. E. Tolbert. 1973b. Biochemistry 12:18.

Anonymous. 1971. Chem. Week 108(17):49.

Anonymous. 1972. U.S. Agr. Res. 20(10):6.

Bjorkman, O., W. M. Hiesey, M. A. Nobs, F. Nicholson, and R. W. Hart. 1966. Yb Carnegie Inst. Wash. 66:228.

Borlaug, N. 1971. Bull. At. Sci. 27(6):6.

Burns, R. C., and R. W. F. Hardy. 1974. Nitrogen Fixation in Bacteria and Higher Plants. Springer Verlag, New York.

Chollet, R., W. L. Ogren, and G. Bowes. 1974. What's New Plant Physiol. 6(1):1.

Dixon, R. A., and J. R. Postgate. 1972. Nature 237:102.

Dobereiner, J., J. M. Day, and P. J. Dart. 1973. Soil. Biol. Biochem. 5:157.

Galston, A. W. 1972. Nat. Hist. 81:24.

Hardy, R. W. F., and R. C. Burns. 1973. Iron-Sulfur Proteins, Vol. 1 (W. Lovenberg, ed.). Academic Press, New York, p. 65.

Hardy, R. W. F., and U. D. Havelka. 1973. Plant Physiol. 48(Suppl.):35.

Hardy, R. W. F., and U. D. Havelka. 1974. Crops Soils 26(5):10.

Hardy, R. W. F., and U. D. Havelka. 1975. In IBP Synthesis Volumes on Nitrogen Fixation and the Biosphere. Cambridge Univerity Press, New York.

Hardy, R. W. F., and B. Quebedeaux. 1974. Crops Soils 26(4):10.

Hardy, R. W. F., R. D. Holsten, E. K. Jackson, and R. C. Burns. 1968. Plant Physiol. 43:1185.

Hardy, R. W. F., R. C. Burns, R. R. Hebert, R. D. Holsten, and E. K. Jackson, 1971. Plant Soil. Spec. Vol.:561.

Hardy, R. W. F., R. C. Burns, and R. D. Holsten. 1973a. Soil Biol. Biochem. 5:47.

Hardy, R. W. F., R. C. Burns, and G. W. Parshall. 1973b. *In* Inorganic Biochemistry, Vol. 2 (G. I. Eichorn, ed.). Elsevier, New York. p. 745.

Hardy, R. W. F., R. C. Burns, J. T. Stasny, and G. W. Parshall. 1975. *In* IBP Synthesis Volumes on Nitrogen Fixation and the Biosphere. Cambridge University Press, New York.

Holsten, R. D., R. C. Burns, R. W. F. Hardy, and R. R. Hebert. 1971. Nature 232:173.

Phillips, D. A., J. G. Torrey, and R. H Burris. 1971. Science 174:169.

Pimentel, D., L. E. Hurd, A. C. Bellotti, M. J. Forster, I. N. Oka, O. D. Sholes, and R. J. Whitman. 1973. Science 182:443.

Quebedeaux, B., and R. W. F. Hardy. 1972. Agron. Abstr., p. 36.

Quebedeaux, B., and R. W. F. Hardy. 1973a. Nature 243:477.

Quebedeaux, B., and R. W. F. Hardy. 1973b. Agron. Abstr., p. 370.

Quebedeaux, B., and R. W. F. Hardy. 1974. Plant Physiol. Abstr., p. 62.

Schrauzer, G. N., G. Schlessinger, and P. A. Doemeny. 1971. J. Am. Chem. Soc. 93:1803.

Schrauzer, G. N., G. W. Kiefer, K. Tano, and P. A. Doemeny. 1974. J. Am. Chem. Soc. 96:641.

Shilov, A. E., N. T. Denisov, N. O. Efimov, N. Shuvalov, N. I. Shuvalova, A. K. Shilova. 1971. Nature 231:460.

Stoy, U. 1969. *In* Physiological Aspects of Crop Yield (J. D. Eastin, F. A. Haskins, C. Y. Sullivan, and C. W. M. Van Bavel, ed.). American Society of Agronomy, Madison, Wisc., p. 185.

Trinick, M. J. 1973. Nature 244:459.

Van Overbeek, J. 1969. *In* Int. Bot. Congr., Seattle, August.

Wardlaw, I. F. 1968. Bot. Rev. 34:79.

Wittwer, S. H. 1970. Trans. Am. Soc. Agr. Eng. 13:249.

Wittwer, S. H., and W. Robb. 1964. Econ. Bot. 18:34.

Zelitch, I. 1971. Photosynthesis, Photorespiration, and Plant Productivity." Academic Press, New York.

Zelitch, I. 1973. Curr. Adv. Plant Sci. 3(5):44.

DISCUSSION

DR. JENSEN: I don't find it surprising that the plants have evolved so that their best reproduction takes place at 21 percent oxygen. What I would like to ask the speaker is, if you crossed such lines and took the hybrid populations to a higher altitude, let's say with 15 percent oxygen, selected for best adaptation, and then backcrossed to the best 21-percent-adapted lines, repeating this process, would you speculate on what might be the situation with such adapted 15 percent lines when grown at the 21 percent altitude level where, in effect, they would be under luxury oxygen conditions?

DR. HARDY: I think that this would be an interesting area to examine. Dr. Quebedeaux and I have requested seeds from high-altitude areas and plan to study them under defined oxygen pressures to determine if they have developed systems to overcome the requirement for high concentrations of oxygen. However, there are no data yet available to answer this question. The distribution of crops grown at high altitude may also be relevant to this question.

DR. TOLBERT: Dr. Hardy's presentation here is backed up by information on the carboxylase–oxygenase reaction. Our recent research deals with whether these two activities, carboxylation or oxygenation, can be regulated, and, of course, that is what Dupont and other chemical companies would like to do by chemical modifiers

To answer the question of whether the plant is regulating these two reactions, we have found that there are naturally occurring effectors that are fine-tuning these activities. There is also, of course, the gross competition between CO_2 and oxygen. At the level of fine-tuning of the enzyme, there are compounds that stimulate the carboxylase and inhibit the oxygenase, and vice-versa. Let me give you two examples: Fructose diphosphate is a known inhibitor of the carboxylase, as if it slowed down photosynthesis when too much sugar or starch were being made. Concurrently, fructose diphosphate stimulates the oxygenase, which is photorespiration. An opposite effect is obtained from ribose-5-phosphate, which is a compound present before the carboxylation reaction. It is a stimulator of the carboxylase and it inhibits the oxygenase. These results suggest that there is some control over the enzyme between CO_2 fixation and photorespiration. Further, it suggests that we should be able to magnify these reactions either through plant breeding or by chemicals.

DR. HARDY: Are the levels that influence the activity close to the physiological level that occur in a plant?

DR. TOLBERT: Ribose-5-phosphate and fructose diphosphate at one millimolar are effectors of the carboxylase–oxygenase reactions. Those are physiological levels.

DR. DIECKERT: Can you tell something about the oxygen effect on the coconut seed? The cells of the endosperm of the coconut that adjoins the seed coat look like liver cells as far as the population of mitochondria is concerned. Now, as you work in toward the central cavity, which in a coconut you would expect to become oxygen-starved, these things disappear. As a matter of fact, they look as though they are degrading; the mitochondria are just going to pot, the same way you would expect them to if they were under starvation conditions in an animal.

Now, I have a hunch that, if you were to look at the developing seed coat as well, you would find that mitochondria are involved in the production of the chemical energy necessary to grow it and not chloroplasts.

DR. HARDY: We may be looking at a unique chemical reaction; alternatively, we may be looking at a physical process. You are suggesting the latter with high oxygen consumption at the seed coat endosperm interphase, thereby producing oxygen depletion towards the interior.

DR. DIECKERT: The seed is not the same kind of an organism that the rest of the plant is in this respect. It is a mitochondria-based system, and the rest of the plant you are worrying about is a photosynthetic device. One requires oxygen and the other doesn't.

DR. HARDY: However, they both contain the same type of mitochondria, with cytochrome oxidase as the O_2 acceptor. Cytochrome oxidase has high affinity for oxygen. Your proposal requires a high rate of oxygen consumption and a poor rate of diffusion to deplete the oxygen to a level where it would be inadequate for normal dark respiration.

DR. DIECKERT: There may be a family of tissues that you have to get the oxygen to in a seed. In the case of a coconut, there is this enormous husk, and inside that there is a hermetically sealed can, so to speak, in the form of endocarp, and then inside of that is presumably the transporting tissue, that is, the seed coat, and then inside of that is this endosperm that is growing.

I don't know how the oxygen is getting to it, but it probably comes in with the fluids, so the partial pressure in those fluids may be an important parameter.

DR. HARDY: A possibly relevant but more complex problem occurs in the nodule. A N_2-fixing nodule has a high rate of oxygen consumption and has developed a transport system that facilitates the diffusion of oxygen within the nodule. The molecule responsible for this transport is leghemoglobin. The nodule also contains nitrogenase, which is extremely oxygen-sensitive. The transport system must therefore transfer oxygen to the site of respiratory activity for energy generation but exclude oxygen from nitrogenase.

DR. DIECKERT: The population density of the mitochondria is also going to be a parameter here, I think, that is important. In the case of this layer of cells in coconut endosperm, the cell is just packed with mitochondria, with obvious highly developed cristae. In other parts of the plant the mitochondria are there but they don't look anything like heart muscle mitochondria.

DR. HARDY: An interesting experiment would measure the effect of elevated oxygen on the coconut, then, because that may tell you whether you can get better distribution into the coconut. The retention of mitochondria should be improved if oxygen deprivation is the limiting factor.

DR. DIECKERT: Maybe.

DR. CANVIN: I think Dr. Dieckert has raised an important point here in terms of developing seed: that the principal energy source for its expansion as a sink comes from dark respiration. Now, there is another part of the plant that Dr. Hardy carefully kept at 21 percent oxygen, and this is the root.

It is similar in that you have long-distance transport to it and the development of this sink is primarily dependent upon dark respiration.

The same situation then may be found if you keep the roots in these different oxygen concentrations and keep the leaves in the low oxygen.

DR. QUEBEDEAUX: We have compared the effects of growing the entire plant in 5 percent O_2 with the effects of exposing only the aerial portion to 5 percent O_2. In this system as described here the results on reproductive development are the same. I think the effect you are referring to in roots occurs at much lower concentrations of O_2, and the effects reported here for seed development may not be directly related.

DR. VARNER: In the CO_2 enrichment studies, how much of the effect is due to CO_2 concentration change in the roots?

DR. HARDY: The effect of carbon dioxide concentration in the root area on nitrogen fixation has been examined, as well as the requirement for rhizobial growth. There are effects on N_2 fixation but at carbon dioxide levels that are orders of magnitude higher than used in the experiments described here. Five to ten percent carbon dioxide in the root area is reported to affect nodulation and nitrogen fixation. Certainly the 800 to 1,200 parts per million with which we are surrounding the aerial portion is not going to produce root effects that require 50,000 ppm. Carbon dioxide is necessary for rhizobial growth, but only 200 to 300 ppm.

DONALD BOULTER

Biochemistry of Protein Synthesis in Seeds

The biochemistry of protein synthesis in seeds covers a variety of topics. As many of these are dealt with by other speakers, this paper is confined strictly to the biochemistry of protein synthesis on the ribosome, except where reference to related topics is necessary for comprehension. Several well-documented reviews have appeared on plant protein biosynthesis, and these should be consulted for literature references, which have not been quoted extensively here. They are Mans (1967); Allende (1970); Boulter (1970); Boulter *et al.* (1972); Zalik and Jones (1973). The paper is written from the standpoint that the ultimate goal of the workshop is to attempt to devise new approaches to plant breeding, rather than as a review suited to molecular biologists. Furthermore, I have been encouraged to be thought-provoking, even speculative; at least, I hope to generate discussion.

THE MECHANISM OF BIOSYNTHESIS

Seeds of legumes and cereals contain several thousand different proteins. Different proteins have different functions, e.g., structural, storage, recognition, control, transport, protection, but by far the largest number are enzymes.

Irrespective of their function, all classes of protein appear to be synthesized by the same basic mechanism, i.e., on polysomes by an energy-requiring multienzyme process. Polysomes consist of messenger RNA, the base sequence of which specifies the amino acid

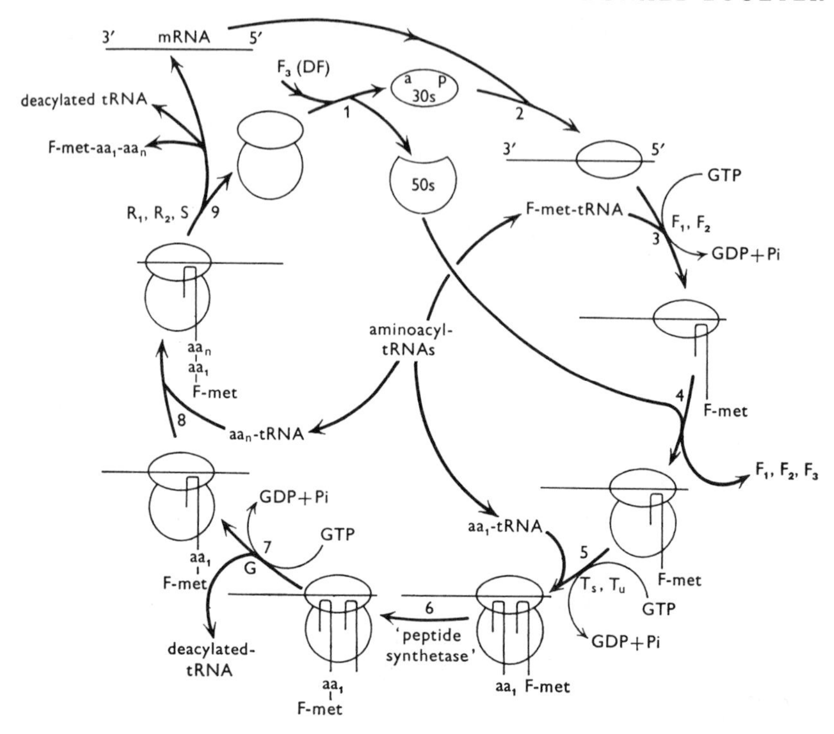

sequence of the protein to be synthesized, and several ribosomes that read the message simultaneously.

THE MECHANISM

The biochemistry of protein synthesis was first elucidated with cell-free systems of 70S ribosomes from microorganisms (Lucas-Lenard and Lipmann, 1971; see Figure 1). Table 1 sets out the various components of this system, all of which, except mRNA and the release factors, have been directly shown to be involved in plants; mRNA and release factors are also known to occur from indirect evidence (see, for example, Shih and Kaesberg, 1973; Wells and Beevers, 1973).

The mechanism of protein synthesis on 80S plant cytoplasmic ribosomes and 70S microbial ribosomes differs in some small details—for instance, in initiation. In the former, initiator met-$tRNA_F$ is not formylated, and if plants prove to be like animals, met-$tRNA_F$ attaches to the small subunit prior to mRNA, rather than vice versa as in microorganisms (but see Weeks *et al.*, 1972).

◀ FIGURE 1 Mechanism of protein synthesis on the 70S microbial ribosome.

Initiation

1,2: Dissociation of monomer into native subunits requires initiation factor F_3 (dissociation factor DF), which also promotes binding of mRNA to 30S subunit. 'a' and 'p' indicate relative positions of aminoacyl- and peptidyl-sites respectively.

3: Formation of initiation complex requires factors F_1 and F_2, GTP and f-met-tRNA. The latter is shown entering directly into the 'p' site, alternatively it may enter the 'a' site and then move to the 'p' site.

4: Functional ribosome formed by addition of 50S subunit; initiation factors released.

Chain elongation

5: Transfer factors T_s and T_u and GTP required for binding of aminoacyl-tRNA in 'a' site.

6: Peptide bond formation requires peptide synthetase (peptidyl-transferase) thought to be a function of the ribosome.

7: GTP and transfer factor G are required for displacement of discharged tRNA and translocation of peptidyl-tRNA to the 'p' site. Although often referred to as the 'translocase,' factor G appears to promote the release (uncoupled from translocation) of the deacylated tRNA molecule from the 'a' site, while translocation *per se* seems to be a function of the 50S subunit (Roufa, Skogerson & Leder, 1970; Tanaka, Lin & Okuyama, 1971; however, cf. Lucas-Lenard & Haenni, 1969).

8: Chain elongation proceeds by repetition of steps 5, 6 and 7.

Chain termination

9: Requires release factors R_1, R_2 and S. A specific peptidyl-tRNA hydrolase may also be required to give the free polypeptide (Cuzin *et al.* 1967; Vogel, Zamir & Elson, 1968; de Groot, Panet & Lapidot, 1968). Ribosome may be released in form of subunits, but in absence of F_3 these will spontaneously recombine. [Reproduced from Boulter *et al.* (1972) by kind permission of the Cambridge University Press.]

THE SUBCELLULAR SITE OF SYNTHESIS

In plants, protein synthesis occurs on 80S cytoplasmic ribosomes, both free and attached to membranes; on mitochondrial 80S ribosomes*; and on 70S chloroplast ribosomes. Mitochondria and chloroplasts have their own protein-synthesizing machinery, the mechanism of which is essentially the same as that given in Figure 1. However, minor differences do exist: for example, the initiator tRNA is formylated.

Protein synthesis on 80S cytoplasmic ribosomes is normally inhibited by cycloheximide, although there are exceptions; it is not inhibited by other inhibitors, such as D-threo-chloramphenicol. In

* Mitochondrial ribosomes from a number of higher plants have sedimentation values of 78–80S.

TABLE 1 Mechanism of Protein Synthesis

Components[a]	Potential Limitations
Building Blocks	
20 protein amino acids	Amounts of.
tRNAS	Amounts of (number of copies of genes; transcription rate, etc.). Possible need to match tRNA fraction to messenger RNAs (see Aviv *et al*, 1971).
Aminoacyl-tRNA synthetases; ATP + Mg^{2+}	Activity of. Environmental and physiological Mg^{2+} deficiency.
"Workbench"	
mRNA [codons same as in microorganisms; limited work done (Basilio *et al.*, 1966)]	Amounts and type [number of copies of genes for mRNA; attachment of RNA polymerase (sigma factors, rho factors); rate of transcription; rate of transport of mRNA from nucleus to cytoplasm; half-life (i.e. how many times read)].
Ribosomes	How many[b] (number of copies of genes for, etc.; trace element deficiency).
Initiating factors[c]	Rate of attachment of mRNA and methionyl tRNA to ribosomes; specificity of mRNA recognition (see Partington *et al.*, 1973).
Mg^{2+}, K^{3+}	Adverse ionic strength, pH values, and redox potential.
GTP	
Elongation factors	Rate of translation.
Termination factors	Rate of release of protein (recycling of ribosomes[d]).
(Timing of the onset and the overall length of the period of protein synthesis of a particular protein[e])	
Post-translation	
Enzymes responsible for changing residues, e.g., hydroxyproline from proline.	Amounts and specificity of enzymes—location.
Enzymes responsible for covalent attachment of nonprotein groups, e.g., sugars to form glycoproteins	Amounts and specificity of enzymes—location.
Transport of protein (ER production, ER vesicles, Golgi, etc.)	Factors affecting biochemistry of transport.
Storage capacity (aleurone grains or protein bodies)	Factors affecting the production of membranes, etc.

[a] All these components have been demonstrated either directly or indirectly in plants.

[b] For example, in legumes, there is a vast increase in the protein-synthesizing machinery consisting of membrane-bound ribosomes at the onset of storage protein synthesis. This is accompanied by large increases in DNA (ploidy?) (Öpik, 1965; Smith, 1973). Probably the whole genome is replicated

contrast, protein synthesis on mitochondrial and chloroplast ribosomes is not inhibited by cycloheximide but is inhibited by D-threo-chloramphenicol. Chloroplasts and mitochondria, while possessing some genetic information, are not autonomous organelles. Tentative estimates of the number of different proteins coded for are more than 1,000 for the chloroplast and about 25 for the mitochondrion. Because of difficulties with the use of inhibitors and *in vitro* systems, direct identification of the proteins synthesized by these organelles has not progressed very far (see Boulter *et al.*, 1972; Gray and Kekwick, 1973).

In seeds, the major proteins are storage proteins, which are synthesized on the rough endoplasmic reticulum.

IDENTIFICATION OF POSSIBLE LIMITATIONS OR CONSTRAINTS

GENERAL CONSIDERATIONS

Plant breeders are interested in the final product of this concerted process: that is, in the amount and quality (essential amino acid composition, for example) of the protein synthesized. For an optimum rate of protein synthesis to occur, each component of the synthesizing system, representing a unit of protein-synthesizing capacity, must be present in the correct proportions. However, it does not necessarily follow that a potential increase in the rate of any step would increase the overall rate of protein synthesis. This would be true only if this step were the limiting reaction (pacemaker) of the process. Increasing the number of methionine codons in the template of a legume storage protein deficient in methionine would be ineffective if the concentration of methionyl-tRNA then limited protein synthesis; it is possible that there are mechanisms that control the process of protein synthesis overall.

In cells that are very active in protein synthesis, the numbers of polysomes per cell are greater (see, for example, Payne and Boulter, 1969), and it is possible, therefore, that the total amount of polysomes, or the number of units of synthesizing capacity, could be limiting the amounts of protein synthesized in some instances.

---

as the simplest way of ensuring the production of the extra copies of the genes necessary to assemble the protein synthesizing machinery.

[c] Factor = protein.

[d] Ribosomes consist of large and small subunits that dissociate on completion of a cycle of synthesis.

[e] For example, the major storage proteins of *Vicia faba* are synthesized at different rates and at different times during seed development (Wright and Boulter, 1972).

While it is clear that a range of genetic, physiological, and environmental factors affect the biochemistry of protein synthesis in seeds, there is little understanding of the relative importance of these factors in different crop lines.

IMPROVING SEED PROTEIN QUANTITATIVELY

The Need To Improve Assay Methods

In order to find out whether particular steps in the overall process of the protein synthesis are limiting in particular instances, it will be necessary to improve the accuracy and reproducibility of some of the assays now available. Thus, in spite of the fact that it has been possible to identify most of the components of protein synthesis, assays using developing seed systems still lack quantitative accuracy in some cases: In isolating the enzymes or cell-free systems involved, damage is caused by hydrodynamic sheer, and plant cell vacuoles break under these conditions, releasing acids, tannins, and other substances that may inactivate enzymes.

Furthermore, proteases and nucleases are often activated (Payne and Boulter, 1974), which results in further damage to both enzymes and subcellular components (Anderson and Key, 1971; Schreier and Staehelin, 1973a). Results obtained *in vitro* under these conditions may reflect the extent of the damage to the systems during isolation rather than a real difference *in vivo*; qualitative ambiguities can also arise because of the presence in fractions of structural proteins needed for the reconstitution of damaged ribosomes.

The criteria that a satisfactory assay must fulfill are as follows:

1. It must be dependent on the exogenous components to be tested.
2. It must be efficient so as to allow quantitative analyses and conclusions.
3. Methods for preparation of components must be of general application.

Use of the Improved Assays

As satisfactory assays become available, these could be used either to further our basic understanding of protein synthesis by identifying possible constraints to the process, or for screening in breeding programs.

In order to achieve the first objective, it is necessary to use the improved assay methods to investigate high- and low-protein lines and normal and improved protein quality lines of various crop plants. In the case of some crops—e.g., wheat, barley, maize, rice, and soya—some of these lines already exist. For other crops—e.g., cowpeas—world germ plasm collections have been assembled and a start made on screening them for these qualities; methods for screening for high-protein and good-protein-quality lines already exist (see Table 2), although they may need to be adapted for use with a particular crop plant for optimum results.

Once assay methods suitable for basic investigations have been established, these should then be modified so as to be satisfactory for use where appropriate in screening programs, since such programs might identify useful lines for breeding—e.g., ones that have an excellent through-put in part of the overall process, even though the final amount of protein synthesized might be small. However, only fast, simple, inexpensive methods, capable of dealing with large numbers of samples, are appropriate in screening programs. At this stage, therefore, it is not possible to predict how much information derived from the basic investigations suggested above will be employed in practical breeding situations. Nevertheless, the history of science lends confidence to the suggestion that some of this information will be of use.

Specific Areas in Which Further Work Is Needed

As can be seen from Table 1, the components of the assay systems, which need to be improved, are understood. The problem is that of

TABLE 2 Screening Procedures for Total "Crude" Protein and the Common Limiting Amino Acids

Component	Screening Procedure
Total protein	Automated Kjeldahl (Varley, 1966)
	IR reflectance methods (Neotec Instruments Inc.)
Cereals	
Lysine	Dye-binding (Mossberg, 1969)
	CIMMYT method (Villegas and Mertz, 1971a)
Tryptophan	CIMMYT method (Villegas and Mertz, 1971b)
Legumes	
Cystine	Herrick *et al.* (1972)
Methionine	Gehrke and Neuner (1974)

preparing them from plant material, undamaged and in physiological yield.

There are two main types of assay system: first, those in which all the enzymes and components come from the plant material under investigation (homologous) and, second, heterologous systems, where this is not so.

Heterologous Systems In an assay using a heterologous system, all the components, except that under investigation, come from the basic system (e.g., rabbit reticulocyte). The assayed component is isolated from the different lines to be compared. The advantage of using a heterologous system is that the components of the basic system have already been purified and the system well characterized. Two examples of such systems are the *Escherichia coli* (Lucas-Lenard and Lipmann, 1971) and the rabbit reticulocyte systems (Sargent, 1967; Schreier and Staehelin, 1973a). The only plant system in a comparable state of development is that of wheat germ (Allende and Bravo, 1966; Allende, 1970; Leis and Keller, 1970; Marcus *et al.*, 1970a,b; Tarrago *et al.*, 1970; Ghosh *et al.*, 1971; Legocki and Marcus, 1970; Klein *et al.*, 1972; Lundquist *et al.*, 1972). However, a comparison of the results obtained by Weeks *et al.* (1972) in the wheat germ system with those obtained by Schreier and Staehelin (1973a,b) in the rabbit reticulocyte indicates that the former requires considerably more work.

The disadvantage of this approach is that it is not yet certain whether all steps of protein synthesis can be assayed with a physiologically meaningful result in a heterologous system, especially in mixed prokaryote and eukaryote systems. While a considerable degree of interchangeability of components among eukaryotic systems has been demonstrated (see Boulter *et al.*, 1972; Partington *et al.*, 1973), it has nevertheless not been proved conclusively that interchange of components does not have a quantitative effect. Furthermore, results obtained using viral messengers indicate that the tRNA fraction of the host may be matched to the particular messenger RNAS being read and that this relationship may involve so-called favored codons (Aviv *et al.*, 1971; Liddell, 1972).

Homologous Systems Apart from the wheat germ system, there are several other plant cell-free systems, among which are those from developing legume seeds (see, for example, Gumilevskaya *et al.*, 1971; Payne *et al.*, 1971a,b; Yarwood *et al.*, 1971a,b; Beevers and Poulson, 1972; Wells and Beevers, 1973). These, like most eukaryote fractionated systems, are inefficient relative to *in vivo* rates of protein synthesis. For this reason unfractionated cell-free extracts have been developed (see, for example, the rabbit reticulocyte lysate system of Darnbrough *et al.*, 1972; Hunt *et al.*, 1972; and Legon *et al.*, 1973). However, these systems cannot be developed from tissues that have high nuclease and protease activities, such as is the case with many plant tissues, unless conditions of isolation can be made more inhibitory to these degradative enzymes. Some RNase inhibitors, such as diethylpyrocarbonate, inhibit protein synthesis and are therefore inappropriate, and others, such as the liver inhibitor, are ineffective. Recently, Davies *et al.* (1972) have used high-ionic-strength and high-pH isolation media with some success, and Gray and Kekwick (1973) have used 0.2 m*M* vanadyl sulfate. Alternatively, plant varieties with relatively low nuclease and protease activities could be sought. A further complication is that seed storage proteins are synthesized on ribosomes attached to membranes, and the growing polypeptide chain probably passes into the lumen of the endoplasmic reticulum: i.e., there is no soluble product formed. Ribosomes can be removed from membranes, by treatment with EDTA, for instance; the effect of this treatment, however, on the protein-synthesizing capacity of the ribosomes is not known. Little work has been done on the mechanism of the attachment of plant ribosomes to membranes. The suggestion is that the attachment is via the large subunit.

In any event, considerably more work is needed before homologous plant systems from developing seeds are satisfactory. At present, they are inefficient, generally do not initiate protein synthesis, are too unstable (amino acid incorporation is of short duration), their products are not characterized, and they give quantitative and qualitative ambiguities. Table 3 sets out, with comments, some assays that need to be improved and developed using components from seeds.

IMPROVING SEED PROTEIN QUALITATIVELY

General Considerations

Proteins occur in greatly different concentrations in the seed. Furthermore, different proteins have different amino acid compositions and hence differ in their nutritional status. For example, the enzymic proteins tend to have a well-balanced amino acid composition, but they usually occur in quite small amounts. However, some enzymic proteins, for example, urease in jackbean, which, incidentally, has a well-balanced amino acid composition, may contribute as much as 1 percent of the dry weight of the seed (Bailey and Boulter, 1971).

TABLE 3 Assays Used for Different Steps of Protein Synthesis

Components	Assay System[a]	Comments
Building Blocks		
Total free amino acids	Automated ninhydrin (Evans and Boulter, 1974)	Suitable for screening.
Individual amino acids	Automated ion-exchange chromatography, GLC (Eveleigh and Winter, 1970)	See Table 2.
Individual tRNA species	Various methods (see appropriate chapters in Moldave and Grossman, 1971)	Further work required to identify minor tRNA species which may be important in translation level control.
"Workbench"	Electron microscopy	Amounts of rough ER. The protein-synthesizing efficiency may be related to number and size of protein bodies and density of protein stain.
Efficiency of Protein Synthesis		

Initiation factors Elongation factors Release factors	Various assays (see appropriate chapters in Moldave and Grossman, 1971)	—
In vitro synthesis of complete protein	Assay (see for example, Schreier and Staehelin, 1973a)	Product characterization by fingerprinting of tryptic peptides, including diagonal paper electrophoretic techniques for cysteine, methionine, and lysine peptides (Kasper, 1970). Requires exogenous mRNA (prepared by sucrose gradient or affinity chromatography (Philipson *et al.*, 1971; Pemberton *et al.*, 1972; Sheldon *et al.*, 1972; Higgins *et al.*, 1973).

[a] All, except the first assay system, are unsuitable for screening. Inhibitors, selective for specific steps in protein synthesis, may give confirmatory data: e.g., aurintricarboxylic acid inhibits initiation and not elongation in the wheat-germ system (Marcus *et al.*, 1970a).

Storage proteins, on the other hand, occur in large amounts but usually have a lower nutritional status due to limitation by an essential amino acid(s). Some seed proteins are toxic or act as antimetabolites (Liener, 1969).

Since the onset of the synthesis of different seed proteins occurs at different stages in seed development, and, since they are produced at different rates, consideration has to be given to the stage at which the components of protein synthesizing systems are isolated.

The proportions of chloroplasts, mitochondria, and free and membrane-bound ribosomes per cell change during seed development. Different organs of the seed synthesize different proteins. An organ such as the cotyledon has several different cell types. Preliminary results suggest that even cells of the same cell type synthesize different proteins, according to their location. Thus, the proteins in cells extracted from the outer cell layers of the cotyledon of beans differ from those extracted from the remainder of the cells (Wood and Cole, 1973), and this is true for some of the proteins of the protein bodies of these cell layers (Barker and Boulter, unpublished observations).

Possible Constraints on Change in the Amino Acid Composition of Proteins

The extent to which the amino acid composition of a protein may change without affecting the protein's activity varies from protein to protein. Some enzymes (for example, papain) can suffer removal of a large number of residues with no effect on activity. Other proteins, such as histones, appear to have remained virtually invariant during evolution, presumably because of functional constraints. In the case of storage proteins, while we do not know the extent to which change is possible, we can identify the following functional constraints.

In nature, these proteins are used as a source of nitrogen by the germinating seed. The amino acid composition of the protein must be such that, given the particular enzymic makeup of the germinating seed, it is broken down into residues that can be metabolized to supply the various nitrogen building blocks required. In barley, for example, Yemm and Folkes (1958) have shown that it is proline, the major amino acid of the storage protein of barley, that supplies much of the nitrogen. In *Vicia faba*, arginine plays an equivalent role (Boulter and Barber, 1963). Clearly, then, storage proteins are not just stores of amino acids; they have to fit the enzymic makeup appropriate to different plant species.

A second constraint is that the different polypeptide chains must be able to form the correct quaternary structure in the complete molecule. In the case of *V. faba*, Wright and Boulter (1974) have shown that legumin is a polymeric molecule consisting of 12 subunits. Disulfide bridges link basic and acidic subunits together, and secondary valence forces are involved in other subunit interactions. It is likely that the complex structure of storage proteins relates to their deposition (as a reserve product) during seed development and their utilization on germination. In addition, in order for the protein to enter the endoplasmic reticulum and possibly also for its transport within, other structural features may be required—e.g., with glycoproteins, specificity of glycosyltransferases. In the final analysis, rates of water uptake and other cooking qualities of seeds must not be adversely affected by changes in seed proteins.

Specific Areas in Which Further Work Is Needed

Results presented herein (Dieckert and Dieckert; Mertz) suggest that lines with a changed proportion of proteins that results in improved seed meal amino acid profiles are likely to be more common than those involving changes in RNA templates. It is important, therefore, to identify those proteins in seeds that have a reasonably nutritionally balanced amino acid profile *and* that either do occur in large amounts or can potentially do so. It is known, for example, that vicilin and legumin, the two major storage proteins of legume seeds, have different essential amino acid profiles and that their proportions differ greatly in different legumes (Boulter *et al.*, 1973). A similar argument applies to the storage proteins of cereals (Mertz, this volume). Since the "classical" methods now employed to isolate storage proteins give impure fractions, modern methods of protein fractionation (see, for example, Van Holde, 1970) should be used. After separation, the proteins should be thoroughly characterized, and, since they do not possess enzymic activity, this involves determining amino acid composition, subunit structure, N termini, and, in some cases, even the amino acid sequence. The positions of these proteins on various types of gel electrophoretograms should be determined as a basis for screening procedures for specific proteins. However, other proteins that do not occur in large amounts may also be important. These might be detected on gels by the use of radioactive tracers. For example, high-sulfoamino-acid-containing proteins of legumes could be detected on gels after *in vivo* ^{35}S feeding experiments.

The duration and rate of synthesis of individual seed proteins should also be investigated during seed development, in high- and low-protein lines and in lines with increased nutritional status. Such studies could lead to a basic understanding of the different physiologically viable ways in which seed lines of high protein content and quality arise.

CONCLUSION

In conclusion, I should like to stress that, given the people and resources, the problems and difficulties touched upon in this presentation could be resolved. Since legumes are our main high-protein crops, their relative neglect should be remedied; fortunately, there is a growing awareness of their importance, due largely to the efforts of scientists like N. E. Borlaug (Borlaug, 1973).

REFERENCES

Allende, J. E. 1970. *In* Techniques in Protein Biosynthesis, Vol. II (P. B. Campbell and J. R. Sargeant, ed.). Academic Press, New York, p. 55.
Allende, J. E., and M. Bravo. 1966. J. Biol. Chem. 241(24):5813.
Anderson, J. M., and J. L. Key. 1971. Plant Physiol. 48:801.
Aviv, H. I. Boime, and P. Leder. 1971. Proc. Nat. Acad. Sci. U.S.A. 68:2303.
Bailey, C. J., and D. Boulter. 1971. *In* Chemotaxonomy of the Leguminosae (J. B. Harborne, D. Boulter, and B. L. Turner, ed.). Academic Press, New York, p. 485.
Basilio, C., M. Bravo, and J. E. Allende. 1966. J. Biol. Chem. 241(8):1917.
Beevers, L., and R. Poulson. 1972. Plant Physiol. 49:476.
Borlaug, N. E. 1973. *In* Nutritional Improvement of Food Legumes by Breeding. Protein Advisory Group of the United Nations Systems, United Nations, New York, p. 7.
Boulter, D. 1970. Annu. Rev. Plant Physiol. 21:91.
Boulter, D., and J. T. Barber. 1963. New Phytol. 62:301.
Boulter, D., R. J. Ellis, and A. Yarwood. 1972. Biol. Rev. 47:113.
Boulter, D., I. M. Evans, and E. Derbyshire. 1973. Qual. Plant. 23:239.
Darnbrough, C., T. Hunt, and R. J. Jackson. 1972. Biochem. Biophys. Res. Commun. 48:1556.
Davies, E., B. A. Larkins, and R. H. Knight. 1972. Plant Physiol. 50:581.
Evans, I. M., and D. Boulter. 1974. J. Sci. Food Agr. 25:311.
Eveleigh, J. W., and G. D. Winter. 1970. *In* Protein Sequence Determination (S. B. Needleman, ed.). Chapman & Hall Ltd., London, p. 91.
Gehrke, C. W., and T. E. Neuner. 1974. J. Assoc. Official Anal. Chem. 57:682.
Ghosh, K., A. Grishko, and H. P. Ghosh. 1971. Biochem. Biophys. Res. Commun. 42(3):462.
Gray, J. C., and R. G. O. Kekwick. 1973. FEBS Lett. 38(1):67.
Gumilevskaya, N. A., E. B. Kuvaeva, L. V. Chumikina, and V. L. Kretovich. 1971. Biochemistry Acad. Nauk SSSR 36:277.
Herrick, H. E., J. M. Lawrence, and D. R. Coahran. 1972. Anal. Biochem. 48:353.
Higgins, T. J. V., J. F. B. Mercer, and P. B. Goodwin. 1973. Nat. New Biol. 246:68.
Hunt, T., G. Vanderhoff, and I. M. London. 1972. J. Mol. Biol. 66:471.
Kasper, C. B. 1970. *In* Protein Sequence Determination (S. B. Needleman, ed.). Chapman & Hall Ltd., London, p. 137.

Klein, W. H., C. Nolan, J. M. Lazar, and J. M. Clark, Jr. 1972. Biochemistry 11(11):2009.

Legocki, A. B., and A. Marcus. 1970. J. Biol. Chem. 245:2814.

Legon, S., C. H. Darnbrough, T. Hunt, and R. J. Jackson. 1973. Biochem. Soc. Trans. 1(3):553.

Leis, J. P., and E. B. Keller. 1970. Bichem. Biophys. Res. Commun. 40(2):416.

Liddell, J. W. 1972. Ph.D. Thesis, University of Durham.

Liener, I. E. (ed.). 1969. Toxic Constituents of Plant Foodstuffs. Academic Press, New York.

Lucas-Lenard, J., and F. Lipmann. 1971. Annu. Rev. Biochem. 40:409.

Lundquist, R. E., J. M. Lazar, W. H. Klein, and J. M. Clark, Jr. 1972. Biochemistry 11(11):2014.

Mans, R. J. 1967. Annu. Rev. Plant Physiol. 18:127.

Marcus, A., J. D. Bewley, and D. P. Weeks. 1970a. Science 167:1735.

Marcus, A., D. P. Weeks, J. P. Leis, and E. B. Keller. 1970b. Protein chain initiation by methionyl-tRNA in wheat embryo. Proc. Nat. Acad. Sci. U.S.A. 67:1681.

Moldave, K., and L. Grossman (ed.). 1971. Methods in Enzymology, Vol. 20, Part C. Academic Press, New York.

Mossberg, R. 1969. *In* New Approaches to Breeding for Improved Plant Protein. International Atomic Energy Agency, Vienna, p. 151.

Neotec Instruments Inc. Rockville, Maryland, U.S.A.

Öpik, H. 1965. The form of nuclei in the storage cells of the cotyledons of germinating seeds of *Phaseolus vulgaris* L. Exp. Cell Res. 38:517.

Partington, G. A., D. J. Kemp, and G. E. Rogers. 1973. Isolation of feather keratin mRNA and its translation in a rabbit reticulocyte cell-free system. Nature New Biol. 246:33.

Payne, E. S., D. Boulter, A. Brownrigg, D. Lonsdale, A. Yarwood, and J. N. Yarwood. 1971a. Phytochemistry 10:2293.

Payne, E. S., A. Brownrigg, A. Yarwood, and D. Boulter. 1971b. Phytochemistry 10:2299.

Payne, P. I., and D. Boulter. 1969. Free and membrane-bound ribosomes of the cotyledons of *Vicia faba* (L.). Planta 84:263.

Payne, P. I., and D. Boulter. 1974. Planta 117:251.

Pemberton, R. E., D. Housman, H. F. Lodish, and C. Baglioni. 1972. Nat. New Biol. 235:99.

Philipson, L., R. Wall, G. Glickman and J. E. Darnell. 1971. Proc. Nat. Acad. Sci. U.S.A. 68:2806.

Sargent, J. R. 1967. *In* Techniques in Protein Biosynthesis (P.N. Campbell and J. R. Sargent, ed.), Vol. 1. Academic Press, New York.

Schreier, M., and T. Staehelin. 1973a. J. Mol. Biol. 73:329.

Schreier, M., and T. Staehelin. 1973b. Nat. New Biol. 242:35.

Sheldon, R., C. Jurale, and J. Katers. 1972. Proc. Nat. Acad. Sci. U.S.A. 69:417.

Shih, D. S., and P. Kaesberg. 1973. Proc. Nat. Acad. Sci. U.S.A. 70:1799.

Smith, D. L. 1973. Ann. Bot. 37:795.

Stulberg, M. P., and G. D. Novelli. 1962. *In* Methods in Enzymology, vol. 5 (S. P. Colowick and N. O. Kaplan, ed.). Academic Press, New York, p. 703.

Tao, K. L., and T. C. Hall. 1971. Biochem. J. 125:975.

Tarrago, A., O. Monasterio, and J. E. Allende. 1970. Biochem. Biophys. Res. Commun. 41(3):765.

Van Holde, K. E. 1970. *In* Protein Sequence Determination (S. B. Needleman, ed.). Chapman & Hall Ltd., London, p. 4.

Varley, J. A. 1966. Analyst 91:119.
Villegas, E., and Mertz, E. T. 1971a. International Maize and Wheat Improvement Center, Mexico, Research Bulletin No. 20, p. 12.
Villegas, E., and Mertz, E. T. 1971b. International Maize and Wheat Improvement Center, Mexico, Research Bulletin No. 20, p. 4.
Weeks, D. P., D. P. S. Verma, S. N. Seal, and A. Marcus. 1972. Nature 236:167.
Wells, G. N., and L. Beevers. 1973. Plant Sci. Lett. 1:281.
Wood, D. R., and C. V. Cole. 1973. *In* Nutritional Improvement of Food Legumes by Breeding. Protein Advisory Group of the United Nations System, United Nations, New York, p. 325.
Wright, D. J., and D. Boulter. 1972. Planta 105:60.
Wright, D. J., and D. Boulter. 1974. Biochem. J. 141:413.
Yarwood, A., D. Boulter, and J. N. Yarwood. 1971a. Biochem. Biophys. Res. Commun. 44:353.
Yarwood, A., E. S. Payne, J. N. Yarwood, and D. Boulter. 1971b. Phytochemistry 10:2305.
Yemm, B. F., and E. M. Folkes. 1958. New Phytol. 57:106.
Zalik, S., and B. L. Jones. 1973. Annu. Rev. Plant Physiol. 24:47.

DISCUSSION

DR. MUNRO: This is an interesting survey delineating the numerous points of regulation of protein synthesis. Part of it is genetically controlled and part is dependent on substrate—namely, amino acids—and on energy.

In the case of the animal studies, which are very extensive, there is quite a lot of evidence about the factor of amino acid supply. Perhaps the most striking example comes from Australia, where you can control wool growth with sulfur amino acid supplements to the sheep. The type of wool changes, and you get, with supplements of methionine, a protein in the wool, increasing in amount, that has 25 percent cysteine in it. Every fourth residue is a cysteine residue, and this is specifically stimulated by the supply of sulfur. (See A. Broad, J. M. Gillespie, and P. J. Reis. 1970. Austr. J. Biol. Sci. 23:140.)

The question is, you may be able to manipulate biosynthetic pathways and pools of free amino acids, but is there any evidence that these can be controlling factors not only for the rate of synthesis overall but also for specific proteins, let us say with high sulfur content?

DR. BOULTER: To my knowledge, we don't actually have any information that suggests that that can happen, although I don't, of course, rule it out.

We know, as Dr. Miflin pointed out, that the activity of some of the enzymes can be regulated by end-product inhibition, but what we don't know is what the effect of that is in terms of either the amounts of protein or the actual amino acid composition of the protein.

We ourselves have done one or two preliminary experiments in tissue slice situations with broad bean where we can actually feed sulfur—$^{35}SO_4$ and get it incorporated quite actively in a tissue slice into the sulfur amino acids of the protein. We have tried adding, in this artificial situation, methionine and cysteine to see whether we could then increase and change. That was a negative result. In fact, we didn't change the actual composition of the protein of these developing seed cotyledon slices.

But, of course, there are many arguments against this kind of experiment. It is done in a tissue slice, it is done with unphysiological concentrations of the added amino acid, and so on. But I don't know that there is any clear-cut information.

Now, there have been a few experiments done with sulfur fertilization and its effect upon both the actual amount and type of protein, and there are a few reports that say it has an effect and there are other reports that say that it doesn't.

DR. MIFLIN: We have started feeding lysine to developing grain, and we have got some preliminary results that, when we repeat them, we may decide are essentially negative.

There are a lot of reasons why they may be negative. There are 20 amino acids, and although one of these may be limiting nutritionally, it may not be the same amino acid that is limiting biosynthetically. But it is certainly encouraging to hear that with the wool system you can get these results.

I think this is something that we shall have an answer on in a foreseeable period of time.

DR. MUNRO: I think the evidence is fairly clear on the point that the specific protein is being favored by a substrate that is rate-limiting.

DR. RABSON: I was most interested in your suggestion of converting the protein within the grain or seed from one that occupies a low percentage of the total protein but that has a high nutritional value, to one that might be much higher in content of the overall. Do you have any sort of strategy to offer on how we might do this?

DR. BOULTER: I don't know how feasible this is as a screening procedure; it may be limited. If, for example, legumin has desirable nutritional features in *Phaseolus*, there are several approaches to screening for it.

The first approach is to identify its position on SDS gels. It has two subunits with molecular weights of about 20,000 and one with a weight of 35,000, and the positions of these are known. So you could then screen for these using SDS gel electrophoresis.

How easy is it, though, to use gel electrophoresis for screening? By the use of large electrophoresis boxes and semiautomation, a couple of technicians could do a couple of hundred samples daily.

Legumin is insoluble at certain ionic strength conditions at pH 4.7, whereas vicilin is soluble. Therefore, an alternative approach would be to arrange to extract meal under these conditions. The legumin stays behind and the vicilin is solubilized. You then determine total nitrogen on the residue, which will contain proteins plus legumin, but since the latter is by far the most prominent protein, it is a reasonable measure of it.

Coming back to the problem of the prolamin in cereals, an alcohol extraction would solubilize these. Probably your method is better, but it is a strategy similar to the one suggested here.

DR. JENSEN: I have comments on protein variation and screening systems in breeding. We have just completed UDI protein analysis on about a thousand wheat lines. The range in protein found was from slightly over 7 percent to 19 percent. These are relatively homozygous lines from which we also had yield data: The best wheats were yielding about 90 bushels per acre. But, if we look at the high percentage protein wheats, the yields at 19 percent and 18 percent may be only 1.6 or 2 bushels per acre. If we continue to the 17 and 16 percent levels the yields will rise slightly, but it is only below 15 percent or so that yield levels become respectable.

The plant breeder is interested in breeding strategies. We have looked at such things as the total grain yield, the percent of protein, and the amount of protein produced per acre. It turns out that the best strategy to develop a high-performance, high-protein wheat is to look for high grain yield. The next best strategy is to screen for the largest total amount of protein produced per acre. The poorest strategy, by itself, is to select on the basis of the protein percentage of the grain. We look at these in combination, of course—all three together.

One way to employ a screening system on segregating material is to use a reduced population per acre of 100,000 to 500,000 plants instead of a normal density of 500,000 to 750,000. Planting in rows with walk space between allows one to see each plant. Since protein level in plants is invisible in the absence of a genetic marker, we apply first a selection screen for other important characters that have to be in the wheat in order for it to be used commercially. Later, we screen this selected group for protein.

DR. COFFMAN: I have a practical question about how we should manage nitrogen when we are screening lines for protein content and yield.

Most of the rice in Asia is grown with less than optimal rates of nitrogen. Now, if I understood Dr. Canvin correctly earlier, he would recommend that we have an abundant, or at least adequate, supply of nitrogen for the plants when we are growing them in the field. Would this preclude evaluating for any genetic variation for amino acid supply?

DR. BOULTER: Do you screen under the optimal conditions, and then hope that that screen, if you produce something, will be applicable to other conditions?

The answer, as I would see it, is that if you want to see the maximum potential for protein synthesis, then you should screen under nonlimiting conditions of nitrogen. There is no guarantee that such a plant, when you put it under the less than optimal conditions of the field that you want it to grow in, will do as well as some other plant that perhaps doesn't have as great a potential but that, in fact, can realize its potential more easily under difficult conditions.

DR. HAGEMAN: I would like to say that nitrate reductase is one way around this dilemma. I think at the seedling stage you could use nitrate reductase to detect those that have the potential to use high amounts of nitrogen. But again, you will have to check that out if you want to prove it.

DR. JOHNSTON: Dr. C. Roy Adair started a program at ARS, Stuttgart, Arkansas, some 22 or 23 years ago on breeding for protein content in rice. In relation to the point that Dr. Coffman brought out, over the years we have used various levels of nitrogen fertilization. I was just reviewing some earlier work that we did there in protein determinations. We classed a 5½ percent protein sample as low-protein, and a 7½ percent protein as high-protein. That is in brown rice.

But with higher levels of nitrogen fertilization, those levels for the low-protein varieties have moved up so that 6 or 6½ percent will be our normal low-protein varieties. In one case this past year, without any nitrogen applied, the low-protein variety was about 6.2 percent; the high-protein was about 8.2 percent.

As we pushed the level of nitrogen up—and all nitrogen was applied just prior to first flood when the plants were about 3 or 4 weeks old, and all the nitrogen put down as top dressing and then immediately flooded to take it into the soil—the levels increased some at about 80 pounds of nitrogen per acre. At 160 pounds per acre of nitrogen the high-protein varieties went on up to about 10.2 percent and the low-protein up to about 7.2 percent, so we still have three percentage points' difference.

These so-called high-protein varieties are from crosses with an ADT-3 parent variety from India from years ago, so a lot of these different selections go back to that one source. They have followed through with the differential in protein at various nitrogen levels over the years.

DR. CANVIN: I would like to re-emphasize that in terms of our understanding, and the possibility of beneficial advances, we need to get every different kind of mutant and every different kind of composition we can. I want to elaborate slightly by reading from this paper (G. Rakow and D. I. McGregor. 1973. J. Am. Oil Chem. Soc. 50:400) again on Neal Jensen's point about the principle of "what you see is what you get."

I really don't think the analogy is that good, because control of fatty acid composition is by an entirely different mechanism from control of the proportion of protein composition in a seed. But it is one of the closest analogies we have, because it is occurring in a seed and it is very precise. Rakow and McGregor are trying to develop plants with different linolenic acid contents. They say:

> However, only a few of these selections continue to transmit their modified fatty acid composition through further generations. Many reverted to the original Oro type of fatty acid composition in the next generation. Some took an additional generation. The cause of this reversion is unknown. It added to the selection work the requirement of repeated cultivation of selected plants through several self-pollinated generations in order to determine the stability of the fatty acid composition, and greatly increased the number of plants to be grown and analyzed, thus increasing the need for space-saving growing conditions and a rapid and simple analysis technique.

I think you can initially screen the M2 but you have to take that M2 and grow it up as a plant and do the M3, the M4, until you get ones that are stabilized. But I think you can do an initial screening without getting homozygosity.

III

GENETIC REGULATORY MECHANISMS

STERLING B. HENDRICKS

Genetic Regulatory Mechanisms: Introduction

The papers in this section will address the area of molecular biology—in other words, controls exercised over protein synthesis at this level. I want to point out one or two things—namely, that cells are extremely precisely programmed for protein synthesis, both with respect to amount and with respect to the time at which the synthesis takes place.

Moreover, it is part of the general consensus at the present time that all cells of a particular organism carry all of the genetic information, and that in the case of differentiation of cells, this is suppressed in some manner, and that it can be activated in some cells. The degree to which this is shown in plants and animals differs greatly. In the case of the plant, a specialized cell can in many cases be used to regenerate the whole plant. In the case of the animal, that is ofttimes a much more difficult process. For instance, to take a nerve cell and regenerate the whole animal would be an entirely different proposition than regenerating a plant from a single cell.

There are one or two other distinctions that I want to make. One is that most of the basic knowledge that goes into the control of protein synthesis has been developed from the prokaryotes. Nevertheless, the principles that are involved in these processes are supposedly the same throughout all types of organisms. This distinction should be borne in mind.

The only way in which one might consider its application to the higher plant is to draw some distinctions between the process in such plants and the process in the prokaryotes. There are only three of these that I would like to hit upon. One is the fact that the information is carried on a number of chromosomes in the higher plant, while it is carried on a single chromosome in the prokaryotes. The fact that it is a polychromosome in the plant might not really make any difference.

The fact that you are using diploid tissue in the plant, however, brings up the question of the importance of alleles or allelic phenomena in the plant, as contrasted with the prokaryotes.

And finally, while the compounds and the methods of control might differ from one to the other, the principles are supposedly the same for the two. This might be particularly important in the question of the appearance of histones and their possible function in the higher plant as contrasted with their function in the prokaryotes, where the function for the histone might be carried by some other type of molecule. So you might have great differences in the type of molecules effecting control, but the principle nevertheless remains the same.

GEOFFREY ZUBAY

Some Aspects of the Regulation of Gene Expression

In 1961 Jacob and Monod proposed a most useful hypothesis for gene regulation known as the operon model. Experiments based on this model have greatly expanded our understanding of regulatory processes, and it seems to be an opportune time to review and update some of the basic concepts and terminology in this field. I have attempted to do this here in a simple and direct style so that it may be used by both expert and nonexpert. To this I have added a description of the crucial observations made on the operon we understand the best, the *lac* operon. The latter is intended to give some perspective to the experimental approaches available in this rapidly moving field.

MECHANISMS FOR REGULATING GENE EXPRESSION

Genes become transcribed into RNA molecules of three types, messenger RNAS, transfer RNAS, and ribosomal RNAS. Gene expression is regulated by factors that control the rate of transcription and, in the case of messenger RNAS, also by factors that control the rate of translation. It is clear that most transcription is controlled by modulation of the rate of initiation rather than by propagation or termination. Transcription control factors function in one of two ways, either by reacting with the DNA or by reacting with the RNA polymerase.

CONTROLS ACTING DIRECTLY ON THE DNA

The initiation locus for transcription contains a binding site for RNA polymerase and frequently a binding site(s) for one or more regulator proteins. Regulator proteins that function by binding to the DNA either inhibit or augment polymerase interaction with its binding site.

Negative Control Systems

In *negative control* systems the regulator protein *inhibits* polymerase interaction. Some negative control regulator proteins bind strongly by themselves and are called *repressors*. The *lac* operon is regulated in part by such a repressor protein. This operon is not expressed as long as the *i* gene repressor (a highly specific repressor encoded by the *i* gene) binds to a segment of the initiation locus. In order for the *lac* operon to be expressed, the repressor must be removed from the initiation locus. Spontaneous dissociation of the repressor is very small because of the very low dissociation constant for this complex—around 10^{-11} mole liter^{-1} (Zubay and Lederman, 1969). Appreciable dissociation requires direct interaction of repressor with a small molecule modulator known as the *antirepressor*. (We prefer the more expressive term *antirepressor* to the commonly used term *inducer*.) Genes that are controlled in this way are said to be *inducible*. Binding of the antirepressor to the repressor is believed to produce an allosteric transition that drastically lowers the affinity of the repressor for the initiation site on the DNA. For the *lac* operon a derivative of lactose or various synthetic derivatives such as isopropyl-β-D-thiogalactoside (IPTG) serve as antirepressors.

A negative control regulator protein that must combine with one or more molecules in order to function should be referred to as an *aporepressor*. An aporepressor is converted to active repressor by combination with a *corepressor*. Genes that are controlled in this way are said to be *repressible*. The *trp* operon consists of a cluster of five structural genes involved in the biosynthesis of tryptophan. This operon is repressed at one end by the binding of an aporepressor–corepressor complex. The aporepressor is a protein encoded by a distantly located *trp* R gene, and the corepressor is tryptophan (see Rose *et al.*, 1973).

The selective advantages of the two control systems just described seem clear. The *lac* operon that encodes enzymes for metabolizing lactose is silent in the absence of lactose. If insufficient lactose is present, the enzymes would serve no purpose. The *trp* operon is inactive at high levels of tryptophan. As long as there is an adequate supply of tryptophan there is no need for the enzymes responsible for the biosynthesis of tryptophan. The scheme of control seen here is often referred to as *feedback repression*. Let us now turn from a consideration of negative control systems to positive control systems.

Positive Control Systems

In *positive control* systems the regulator protein facilitates polymerase interaction. Inherent in the concept of the positive control system is the notion that the RNA polymerase molecule by itself binds poorly to the initiation locus. Affinity of the polymerase for the various initiation loci in the cell is probably modulated by the sequence of base pairs in each initiation locus. Clearly, this can be designed for different initiation loci so that the polymerase affinity would be very high, very low, or somewhere in between. The binding of the positive control regulator protein at the initiation locus facilitates the strong binding of the RNA polymerase to an adjacent site in the initiation locus. A positive control protein that binds strongly without a cofactor is called an *activator*. The λ phage Q gene product (see Ptashne, 1971) is probably such a protein and its function will be discussed below. A small molecule modulator that prevents the binding of the activator should be called an *antiactivator*. No antiactivator is known for the λ Q protein. A positive control protein which cannot function by itself is called an *apoactivator* and the necessary attendant molecule(s) known as *coactivator*(s). The catabolite gene activator protein, CAP, is an example of an apoactivator and CAMP is the corresponding coactivator. The value of this particular activation system, which stimulates a large number of catabolite-sensitive genes in *E. coli,* will be taken up below.

Generalizations

Based on our current limited knowledge several generalizations regarding the mechanism of action of gene regulator proteins seem warranted. These generalizations will be stated as a series of provisional rules in this section.

- *Repressors bind to the double helix without significant disruption of the base-paired structure*. Only minor conformational changes such as slight bends or change in pitch of the DNA double helix may and probably do occur so that the binding sites of the DNA and repressor can interact optimally.
- In a simple negative control situation, the repressor binding site may be located on either side of the polymerase binding site with

respect to the direction of transcription. The two sites are in close proximity or even overlapping so that *steric hinderance prevents the simultaneous binding of repressor and polymerase*. In a complex negative control situation a repressor can interfere with the binding of an activator by a similar mechanism.

• *Repressors are highly symmetrical molecules made of an even number of identical subunits*. As such, they possess numerous dyad axes. The DNA double helix also possesses a dyad axis due to the antiparallel arrangement of the two chains. Alignment of the protein and the DNA so that their symmetry axes are coincident would seem to be a logical mode of interaction. This interaction would be augmented if the sequence of bases in one strand of the DNA were identical in reverse fashion to the sequence basis in the other strand at least in those regions that make direct contact with repressor. Thus *DNA repressor binding sites should have dyad axes of symmetry*.

• *Activators facilitate the binding of polymerase to the initiation locus by a local disruption of the double helix structure*. This disruption should involve some unwinding of the double helix structure and, very probably, breakage of a few hydrogen-bonded base pairs. The activator-induced disruption should facilitate further disruption of the double helix structure, which is required for the polymerase to achieve a stable binding configuration prior to initiation of transcription. There is little evidence to support the idea yet, but it would not be surprising if activator binding sites showed a dyad axis of symmetry.

• *The activator binding site must be sufficiently far removed from the polymerase site that the two proteins can bind simultaneously*. Once the initiation complex has been achieved and the first phosphodiester linkage made, elongation is a relatively rapid process, occurring at rates of 30–50 bases per second. *The activator binding site is usually located on the side of the DNA polymerase binding site away from the direction of transcription*, so the elongation of RNA chains is not delayed after initiation.

CONTROLS OPERATING DIRECTLY ON THE POLYMERASE

Whereas most transcription level controls influence the polymerase indirectly by acting on the DNA, there are some controls that operate directly on the polymerase molecule itself. Changes in the RNA polymerase entail either covalent modification of the existing polymerase or addition of factors that appear to selectively alter the affinity of the polymerase for various initiation sites. The physiologic significance of the examples that can be given for this type of control is not well established. In T_4 bacteriophage infection, there is considerable chemical modification of the *E. coli* polymerase that is believed to favor its transcription of certain phage genes (Schachner *et al.*, 1971). Modification of the host polymerase also occurs during sporulation in *B. subtilis* (Losick *et al.*, 1970). Under conditions of amino acid starvation in normal *E. coli* cells, the concentration of guanosine tetraphosphate (ppGpp) rises dramatically (Cashel and Gallant, 1969). *In vitro*, this complex has been shown to complex noncovalently with RNA polymerase, thereby inhibiting about half of the initiations that begin with guanosine nucleotides (Cashel, 1970). Some of these examples will be discussed more fully below.

GENE EXPRESSION OPERATING AT THE POSTTRANSCRIPTIONAL LEVEL

I shall discuss posttranscriptional control of gene expression very briefly, since gene level control is the main interest here. In prokaryotes like *E. coli* the messenger lifetime is a few minutes at most. This, plus the fact that there is no separation of nucleus and cytoplasm, has made it relatively difficult to study posttranscriptional control in *E. coli*. Consequently, most of the observations relating to posttranscriptional control come from studies on eukaryotes. Even here the mechanisms are poorly understood. There are three possibilities for posttranscriptional control that should be carefully weighed.

Selective Processing of Synthesized RNA

In eukaryotes most RNA is made in considerably larger sizes than found in the cytoplasm. These RNA molecules must be tailored before entering the cytoplasm. Similarly, late in the infection of certain cells by the animal viruses SV40 (Aloni, 1973) and adenovirus (Lennart Philipson, Wallenberg Laboratory, Uppsala University, Uppsala, Sweden, personal communication), there is considerably more RNA transcribed than enters the cytoplasm. The RNA that is not translated is either degraded while in the nucleus or immediately after it enters the cytoplasm.

Variable Messenger Lifetime

In eukaryotes there is evidence that messengers of different types have quite different lifetimes. It seems likely that lifetime is controlled by base sequences at one or both ends of the messenger. In the dramatic instances of cells that become dominated by one messenger type, a combination of circumstances is involved, including the selective transcription of a very limited portion of the DNA and extreme messenger stability. Examples of such messengers would be globin, ovalbumin, and silk fibroin messenger.

Stimulation or Inhibition of Translation

Once the messenger is in the cytoplasm, there is the possibility of controlling gene expression by inhibiting or stimulating translation. As in transcription, the initiation site is believed to be the most likely site of control, and a variety of controls acting on the initiation factors in a more or less selective fashion have been suggested in both prokaryotes and eukaryotes.

PATTERNS OF CONTROLS IN DIFFERENT LIVING SYSTEMS

Thus far we have considered gene regulating controls from the simplest point of view—as though genes or operons are regulated by only one type of control. In fact, many genes are regulated by more than one control, and controls frequently interact in complex ways. The way in which controls are organized in complex situations depends upon the particular needs of the cellular or viral system. In this section we shall briefly consider the pattern of controls seen in the bacterial cell, the bacteriophage, and the differentiated eukaryotic cell.

ORGANIZATION OF CONTROLS IN THE *E. coli* BACTERIAL CELL

The single *E. coli* chromosome contains enough DNA for about 3,000 genes. Every bacterial cell is totipotent and responds rapidly to changes in environmental conditions. This requires all the gene-regulated responses to be readily reversible in either the *on* or *off* direction. A system delicately poised for rapid response probably contains sufficient quantities of most gene regulator proteins at all times. At any given time only a small fraction (about 5–10 percent) of the total genes are turned on, and this is determined not by the concentrations of regulator proteins, which are usually present in sufficient concentrations, but by the concentrations of small molecule modulators (i.e., coactivators, corepressors, antirepressors and activators as well as those small molecule modulators that interact directly with the RNA polymerase).

Many genes are subject to multiple controls, and in such cases organization into a hierarchy is usually evident. This can best be illustrated by considering a couple of fairly well understood systems, the *lac* operon and the *ara* operon.

In the *lac* operon there is evidence for three control elements. The small molecule modulators for these controls are lactose, CAMP, and ppGpp. Lactose (or, rather, a derivative of lactose known as allolactose) is the antirepressor for the highly specific *i* gene repressor, which affects the activity of only the *lac* operon. This repressor binding site partially overlaps the polymerase binding site as discussed below. Somewhat farther removed from the structural genes is a binding site for CAP. CAP in combination with CAMP is a general activator for many catabolite-sensitive genes, probably hundreds, in *E. coli.* In order for CAP to activate the *lac* operon, the *i* gene repressor must also be removed. Finally, ppGpp stimulates the expression of the *lac* operon by binding to the polymerase. The concentration of ppGpp is inversely proportional to the gross rate of cellular RNA synthesis. Binding of ppGpp to RNA polymerase has a stimulatory, inhibitory, or negligible effect, depending upon the gene or operon (Yang *et al.*, 1974). For the *lac* operon, the effect is stimulatory at least *in vitro*.

The *ara* operon has an even more complex control system. It provides an example of a regulator protein that can act as both a repressor or an apoactivator. The *ara* operon contains a cluster of three structural genes involved in the catabolism of L-arabinose (Englesberg *et al.*, 1969). This cluster of genes is bound on one side by an initiation locus. The initiation locus probably contains four binding sites: one negative control site for the *ara* C (*o* locus) specific regulator protein, followed by one positive control site for the same protein (*a* locus), followed by a CAP binding site (*c* locus), followed by the polymerase binding site (*p* locus). The structural genes follow the polymerase binding site. The evidence for the two binding sites for the *ara* C protein and their relative order is solidly based on genetic and biochemical studies of whole cells by Englesberg and his coworkers (1969). Evidence for a CAP-binding site is inferred from the CAP requirement for *ara* operon expression both *in vivo* and *in vitro*. Location of the polymerase-binding site is consistent with the rules we have proposed above—that all activator binding sites should be located on the side of the polymerase binding site away from the direction of transcription. Location of the CAP binding site between the *ara* C activator site and the polymerase binding site is inferred from biochemical data that indicate that a large excess of *ara* C protein partly overcomes the requirement for CAMP–CAP (Yang and Zubay, 1973). This is an interesting fact in itself, but more direct evidence is required to certify this assignment.

Activation of the *ara* operon probably requires the following sequence of events. In the absence of L-arabinose, the *ara* C protein is tightly bound to the *o* locus. In the presence of sufficient amounts of L-arabinose, complex formation with the bound *ara* C protein takes place, resulting in its release from the *o* locus. A conformation is adopted by the *ara* C protein when complexed to L-arabinose that enables it to bind strongly to the *a* locus, where it serves as an activator. The *c* binding site for CAMP–CAP can now be occupied, and this in turn enables the polymerase to bind to the adjacent *p* site in a configuration suitable for initiation of transcription. It should be noted that the *c* site in the *ara* operon appears to have a lower affinity for the CAMP–CAP complex than the *c* site in the *lac* operon, as the former requires prior activation by the binding of the *ara* C protein. These different affinities for CAP must be controlled by a different sequence of bases at the two loci. Finally, ppGpp also stimulates the *ara* operon in the same way and to about the same extent as it does the *lac* operon.

ORGANIZATION OF CONTROLS IN DNA BACTERIOPHAGE CONTAINING 30 TO 160 GENES

In the DNA bacteriophages that multiply and ultimately produce mature viruses and cell lysis, controls are introduced in a sequential and irreversible manner. We will briefly consider some of the controls for three phages: λ, T_4 and T_7.

λ *Bacteriophage*

The λ bacteriophage encodes four regulator proteins, C_I, N, tof, and Q (Ptashne, 1971). It is of interest that no small molecule modulators such as coactivators or corepressors have been found for λ. Rather, the controls seem to be brought into action by the sequential appearance of the various regulator proteins. This bacteriophage can exist in an active or an inactive form in *E. coli*. In the inactive form the λ chromosome is integrated into the host chromosome. This inactive state is maintained by the presence of the C_I repressor, which binds to the so-called early right and early left initiation loci. This repressor is synthesized in small amounts when λ is in the prophage state and represents the only λ gene that is active at this time. Lack of sufficient C_I protein or its inactivation leads to early right and early left transcription from λ. This is the first irreversible step leading to λ replication and cell death. The early right and early left transcriptional units are quite short. Early left transcription leads to synthesis of the regulator protein N, and early right transcription leads to the synthesis of the regulator protein tof. The N protein may be a typical activator or an atypical antiterminator. At any rate, once sufficient N protein has been produced, transcription continues from the early left and early right cistrons to points considerably beyond the original terminator points. The extended transcription from the right transcriptional unit results in the synthesis of Q protein. This protein appears to be an activator that turns on late right transcription of λ. In the meantime, the buildup of the tof gene product resulting from early right transcription eventually turns off early right and early left transcription, presumably by acting as a repressor.

T Bacteriophages

The T bacteriophages cannot adopt an inactive lysogenic state like λ, so infection inevitably lead to vegetative phage replication and ultimate cell lysis. In the case of T_4 infection, modification of the host polymerase is a key factor in the regulation of transcription, whereas with T_7, replacement of the host polymerase by a phage-encoded enzyme is a key factor in the regulation.

T_4 infection begins with the transcription of 25 "pre-early" genes by *E. coli* polymerase, mainly from the *l* strand (see Zubay and Marmur, 1973). Products of early transcription shut down host transcription and translation. Some small polypeptide products of "pre-early" synthesis become associated with host polymerase. Other products lead to modification of the subunits of host polymerase, including changes in size, adenylation, and phosphorylation. Further transcription results principally from the phage *r* strand by the modified host polymerase. Presumably, the various modifications of the *E. coli* polymerase change its affinity for the different initiation sites so that the appropriate genes are transcribed when they are needed.

T_7 infection begins with transcription of about 20 percent of the DNA from the end of one strand of *E. coli* polymerase (see Zubay and Marmur, 1973). This transcription leads to the synthesis of a phage polymerase that transcribes the remaining 80 percent of the same DNA strand. The *E. coli* polymerase cannot transcribe the late genes of the T_7 phage.

ORGANIZATION OF CONTROLS IN A DIFFERENTIATED MAMMALIAN CELL CONTAINING 50,000 OR MORE GENES

One can only speculate about controls of gene regulation in mammalian cells since not one case is thoroughly understood. In the case of differentiated mammalian cells, there are two types of gene control that should be considered. There are the relatively irreversible series of changes that have led from the totipotent embryonic cell to the differentiated cell in which only a limited number of genes can be turned on. This process of differentiation probably involves introduction of new control elements in a sequential and irreversible manner much as one sees during bacteriophage development. Finally, when one has achieved the metastable differentiated state, it seems likely that, within the limited spectrum of potentially active genes remaining, there exists a hierarchy of reversible positive and negative controls as in *E. coli*. The cyclic nucleotides (CGMP and CAMP) and the steroid hormones (e.g., estradiol, aldosterone, progesterone, dihydrotestosterone, and cortisol) are strongly implicated as small molecule modulators of gene expression both during and after differentiation. Using such modulators with their related control proteins in various combinations could help explain the great variety of gene control patterns seen in various differentiated cells.

CONTROL OF *lac* OPERON EXPRESSION

Let us turn now from these general considerations to a consideration of the genetic and biochemical approaches that have led to our current detailed understanding of *lac* operon control. These investigations should serve as a model for similar investigations on other systems. Again, the treatment will be brief, and only some of those observations that were historically crucial will be reviewed.

Jacob and Monod (1961) proposed the operon model for control of the *lac* genes. This model was based on genetic and biochemical studies on whole cells. The genetic map of the *lac* region was shown to contain three contiguous structural genes, *z*, *y*, and *a*, with an adjacent control locus known as an operator (*o*).

i — — — — — *o* *z* *y* *a*

Another control locus or gene called *i* is located close to the *o* locus, as indicated in the diagram. The three structural genes code for different proteins involved in lactose metabolism. In particular, the *z* gene codes for the enzyme β-galactosidase (β-gal), which hydrolyzes lactose to its component monosaccharides.

Expression of the *z*, *y*, and *a* genes is very low unless lactose or a synthetic analog of lactose such as isopropyl-β-D-thiogalactoside is present in the growth medium. Under these conditions one sees a thousandfold increase of *lac* operon proteins in the cell. In the absence of lactose or IPTG, large quantities of the *lac* proteins can also occur by mutation of o to o^c or i^+ to i^-. This suggests that o and i are involved in regulation. In partial diploids, it was found that o^c is *cis* dominant to o^+. Thus the o locus affects only those genes to which it is in direct apposition. In contrast, i^+ is dominant to i^- in the *cis* or *trans* position. The results of these two dominance tests led to the suggestion that the i gene makes a diffusible product (repressor) that normally binds to the o locus (operator), preventing expression of the operon. The small molecule lactose or IPTG antagonizes this repressor–operator interaction by binding to the repressor, thereby permitting expression of the operon.

In 1964 Jacob *et al.* (1964), through further genetic studies, found a site near the operator required for expression of the operon. This site was presumed to be the place where RNA polymerase initiates transcription. It was named the promoter. I prefer to use the term *promoter* for the combination of sites, activator sites, and polymerase binding sites required for optimal transcription. Subsequently, by careful genetic studies, Ippen *et al.* (1968) mapped the promoter on the side of the operator distal to the *z* gene.

Further understanding and testing of the operon control required that cell-free biochemical techniques be used to dissect out of the cell each control factor, study it in isolation, and demonstrate the effects which occur when the control factors are recombined in the purified state. This analytical phase of the problem began in 1966 with the isolation of repressor by Gilbert and Müller-Hill (1966). They used radioactive IPTG as a probe for isolating repressor from whole cells. This was done by fractionating whole cells by standard techniques of protein purification, showing at each step which fraction contained the IPTG binding material, and purifying it further. At the same time, through subtle genetic maneuvering, mutants were isolated that produce much higher quantities of repressor, thus simplifying the task of purification. This pioneering work made possible direct chemical studies on repressor. It was ultimately established that repressor is a stable tetramer composed of four identical subunits, each with a binding site for IPTG (Beyreuther and Klemm, 1970; Riggs and Bourgeois, 1970).

A number of mutants of repressors are currently being isolated by

Konrad Beyreuther and Benno Müller-Hill of the Genetics Institute at Cologne University in the hope of discovering which parts of the repressor bind to operator and which to antirepressor. Attempts are also being made by others to crystallize repressor so that its three-dimensional structure can be determined by x-ray diffraction. Soon after they isolated repressor, Gilbert and Müller-Hill (1967) demonstrated that repressor binds to operator and that this binding is weakened by antirepressor. These measurements were carried out with λ*lac* DNA, a 30 × 10^6 dalton DNA molecule with the *lac* operon inserted. Radioactive ^{35}S-labeled repressor was mixed with the DNA and the mixture sedimented on a gradient. Some protein was found to sediment with the DNA. The DNA sediments much more rapidly than unbound repressor, so any radioactivity sedimenting with the DNA must be bound to it. The proof that the repressor binding was significant came from controls that showed that the binding was eliminated in the presence of IPTG or by substituting DNA with an o^c mutation of o.

At the same time that these studies were in progress, our laboratory (Zubay *et al.*, 1967) was attempting to develop a cell-free system in which it would be possible to demonstrate the inhibiting effect of repressor on synthesis. This system uses the same λ*lac* DNA that Gilbert and Müller-Hill used in their repressor binding studies but contains in addition a crude cell-free extract of *E. coli* known as an S-30 and all the small molecules and substrates necessary for RNA and protein synthesis. This system can make β-galactosidase (β-gal) in sufficient quantities to be detected by a standard colorimetric assay. When the S-30 is prepared from an i^- strain, the amount of β-gal synthesized is directly proportional to the amount of λ*lac* DNA added to the system. Substantially less DNA is made when S-30 is prepared from an i^+ strain. Proof that this is due to the inhibiting effect of repressor comes from addition of IPTG to the cell-free system. IPTG increases the β-gal yield in the system containing repressor but has no effect on the system lacking repressor. Numerous quantitative relationships could be established in the cell-free system which would be difficult or impossible to establish in whole cells (Zubay and Chambers, 1971). For example, it was shown by varying repressor concentration that one repressor molecule is sufficient for inhibiting gene expression. It was also shown that in the presence of excess repressor, gene expression is directly proportional to the square of the TIPG concentration. The latter fact suggests that derepression requires two antirepressor molecules. From this information and symmetry considerations we suggested that the repressor contains two functional operator binding sites, each sensitive to only two of the antirepressor binding sites. Other symmetry arguments lead to the prediction that the operator binding site should have a dyad, as discussed above. Further studies on nucleotide sequence of the repressor binding site, described below, lend strong support to this prediction.

Other observations in the cell-free synthetic system provided strong support for the idea that repressor blocks initiation of transcription rather than elongation (Zubay *et al.*, 1970). Zubay *et al.* started the cell-free incubation in the presence of repressor but in the absence of IPTG. At a later time, rifampicin and IPTG were added simultaneously. If an initiation complex is made in the presence of repressor, then it should not be inhibited by subsequent addition of rifampicin, and some β-gal synthesis should occur. In fact, no β-gal synthesis occurs in this experiment, supporting the idea that repressor blocks initiation. It was also shown that when artificial situations were created *in vitro* so that initiation of transcription occurred outside of the normal region, considerably less IPTG was needed for induction.

The studies done in Gilbert's laboratory and my own gave strong support to the operon model of gene regulation, but they said little about another more obscure phenomenon. This phenomenon was known as *catabolite repression*, and it has various manifestations. If cells are grown on glucose, the *lac* operon is very poorly expressed even at high concentrations of lactose. Glucose has a similar repressive effect on a wide variety of genes that are involved in catabolism. Glucose is the simplest biochemically manipulable source of carbon for both biosynthesis and energy production. If sufficient glucose is available, other more complex carbon sources are not utilized, and there appears to be a regulatory mechanism repressing the synthesis of enzymes that process other carbon sources.

The phenomenon called catabolite repression had been observed at the turn of the century, but it was only in 1965 that a beginning was made in understanding the molecular basis of this repression. At this time Makman and Sutherland (1965) found that addition of glucose to growing cells leads to a sudden and drastic decline in the intracellular cAMP concentration. Subsequently, in 1968, Perlman and Pastan (1968) and Ullman and Monod (1968) found that addition of cAMP to growing cells reverses the repressive effect of glucose on the *lac* operon. These experiments suggested that cAMP might be the active agent required by catabolite-sensitive genes and that catabolite repression was due to a lowering of the cAMP level. This idea was verified by Chambers and me in 1969 (Chambers and Zubay, 1969); we found that addition of cAMP to the DNA-directed cell-free system for β-gal synthesis increased the yield of β-gal synthesized twentyfold. Other studies indicated that this was due to increased transcription rather than to increased translation, showing that the stimulation effect was occurring at the gene level.

The next problem to be considered was the site of action of CAMP. Experiments were designed to specifically answer this question. These experiments were based on the hypothesis that CAMP affected most catabolite-sensitive genes and that the expression of such genes depended upon a good supply of CAMP and an intact receptor for the CAMP. Jon Beckwith of the Harvard Medical School designed a selection technique for isolating mutant *E. coli* cells with either defect. He reasoned that if he simultaneously selected for the ara^-, lac^- phenotype, he would be most likely to obtain single mutants in one of the components required for the CAMP response. In fact, mutants of two types were obtained; some were phenotypically reverted to ara^+, lac^+ by adding CAMP to the growth media and some were not. It seemed likely that the former had defects in the ability to make CAMP and the latter had defects in the target site for CAMP action. Mutants of the latter type were used to make an S-30 for cell-free synthesis and were shown to be unaffected by CAMP. Addition of small amounts of extract from normal S-30 brought back the CAMP stimulation. This stimulation effect was used to purify a single protein from normal extracts that cause the CAMP response (Zubay *et al.*, 1970). The protein known as CAP has been characterized as a dimer containing identical subunits (Riggs *et al.*, 1971). CAP binds CAMP, and it also binds to DNA in the presence of CAMP. The binding of CAP to single-stranded DNA is about five times stronger than that to double-stranded DNA, suggesting that CAP binding to double-stranded DNA involves some local disruption of the normal base-paired DNA structure.

Eron *et al.* (1971) reconstituted all of the purified components involved in the activation of the *lac* operon. This mixture included λ*lac* DNA, RNA polymerase, CAP, CAMP, and the salts and substrates necessary for RNA synthesis. The radioactively labeled RNA resulting from cell-free synthesis was assayed for the content of *lac* mRNA by a selective hybridization technique. The importance of both CAMP and CAP to expression of the *lac* operon was demonstrated by the observations that the exclusion of either component from this reaction mixture resulted in a drastic lowering of the synthesized *lac* mRNA. It seemed that the components that were involved in both repression and activation of the *lac* operon were finally identified.

The next level of inquiry was already under way. Gilbert had been working for some time to isolate the repressor-binding region of DNA and determine its sequence. If fragments of λ*lac* DNA were mixed with *lac* repressor, only the fragment containing the operator region would bind repressor strongly, and this could be isolated by its adherence to nitrocellulose. Protein adheres to nitrocellulose, but DNA by itself does not. The repressor-bound DNA fragment isolated on nitrocellulose was further treated with DNase in the hope that the DNA region in direct contact with repressor would be resistant to the enzyme. The DNase-resistant fragment was subsequently transcribed into radioactive RNA, and the sequence of this 48-base double helix fragment was determined by Gilbert and Maxam (1973). About half of bases show a dyad axis of symmetry, and these have been underlined in the diagram below. The prediction from previous arguments would be only these bases are in direct contact with repressor.

$$
\begin{array}{l}
5'\text{--}\mathrm{T\,G\,G\,}\overline{\mathrm{A\,A\,T\,T\,G\,T}}\,\mathrm{G}\,\overline{\mathrm{A}}\,\mathrm{G}\,\overline{\mathrm{C\,G}}\,\mathrm{G\,A}\,\overline{\mathrm{T}}\,\mathrm{T}\,\overline{\mathrm{A\,C\,A\,A\,T\,T}}\,3' \\
3'\text{--}\mathrm{A\,C\,C\,}\underline{\mathrm{T\,T\,A\,A\,C\,A}}\,\mathrm{C}\,\underline{\mathrm{T}}\,\mathrm{C}\,\underline{\mathrm{G\,C}}\,\mathrm{C\,T}\,\underline{\mathrm{A}}\,\mathrm{A}\,\underline{\mathrm{T\,G\,T\,T\,A\,A}}\,5'
\end{array}
$$

The observations of Gilbert and Maxam were recently complemented by those of Reznikoff and coworkers (1974) who determined the entire nucleotide sequence in the *lac* control region which contains 120 base pairs. The essential DNA fragment was isolated by hybridizing two different viral DNAs that contained complementary sequences for the region in question. The single-stranded DNA was removed after hybridization by digesting with a nuclease specific for single-stranded exonuclease, and the resulting double helix fragment was used in transcription of a radioactive RNA chain. This RNA molecule has been sequenced and tells us indirectly the sequence of bases in the *lac* control region. The fragment extends from the end of the *i* gene to the beginning of the *z* gene (see the diagram below). These ends and beginnings can be recognized by the fact that the sequence of bases corresponds to the codons for amino acids known to be present in the C-terminal portion of the repressor polypeptide chain and the N-initial portion of the *β*-gal polypeptide chain, respectively. The repressor binding sequence can be recognized by comparison with the work of Gilbert and Maxam.

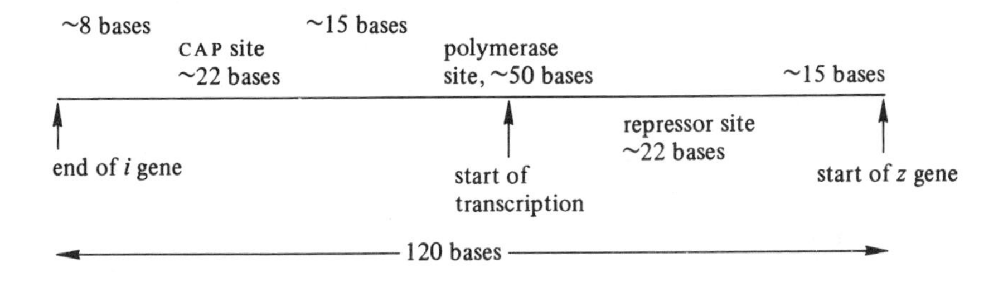

The CAP binding site is inferred from mutant studies of Beckwith *et al*. (1972), which show that the CAP-sensitive region is located in this general area, and from dyad symmetry shown by the six outermost base pairs located on the perimeters of this fragment. More direct evidence is essential and will be obtained, no doubt, by the same kind of binding technique that Gilbert used to locate the repressor site. The polymerase binding site is believed to be 40–50 base pairs in length and overlaps the operator binding site (W. Gilbert, personal communication). The initiation site for transcription occurs in approximately the middle of the polymerase binding site and about 48 base pairs away from the first codon of the *z* gene. A significant gap exists between the end of the *z* gene and the CAP binding site, between the CAP binding site and the polymerase binding site and between the repressor binding site and the first codon.

It was suggested above that activators like CAP facilitate polymerase binding by partial unwinding of the double helix. The distance between the stable binding sites for CAP and polymerase in the *lac* promoter is too great for the unwinding at the CAP binding site to facilitate direct interaction of polymerase with its binding site. Therefore we propose that polymerase enters the complex via the region immediately adjacent to the CAP–DNA complex and migrates down the DNA to the position where it binds firmly prior to initiation. The region where the polymerase first makes contact with the DNA could be called the entry site.

Work on the *lac* operon has clearly not been completed, but nevertheless a great deal has been achieved by this intensive effort. It will have to be decided how far investigations on regulation should be carried for other systems. It seems likely that similar information can be obtained on other systems with the currently available techniques.

REFERENCES

Aloni, Y. 1973. Nat. New Biol. 243:2.

Beckwith, J., T. Grodzicker, and R. Arditti. 1972. J. Mol. Biol. 69:155.

Beyreuther, K., and A. Klemm. 1970. *In* The Lactose Operon (D. Zipser and J. R. Beckwith, ed.) Cold Spring Harbor Laboratory, Cold Spring, Harbor, N.Y.

Cashel, M.. 1970. Cold Spring Harbor Symp. Quant. Biol. 35:407.

Cashel, M., and J. Gallant. 1969. Nature 221:838.

Chambers, D. A., and G. Zubay. 1969. Proc. Nat. Acad. Sci. U.S.A. 63:118.

Englesberg, E., C. Squires, and F. Meronk. 1969. Proc. Nat. Acad. Sci. U.S.A. 62:1100.

Eron, L., R. Arditti, G. Zubay, S. Connaway, and J. Beckwith. 1971. Proc. Nat. Acad. Sci. U.S.A. 68:215.

Gilbert, W., and A. Maxam. 1973. Proc. Nat. Acad. Sci. U.S.A. 70:3581.

Gilbert, W., and B. Müller-Hill. 1966. Proc. Nat. Acad. Sci. U.S.A. 56:1891.

Gilbert, W., and B. Müller-Hill. 1967. Proc. Nat. Acad. Sci. U.S.A. 58:2415.
Ippen, K., J. H. Miller, J. Scaife, and J. Beckwith. 1968. Nature 217:825.
Jacob, F., and J. Monod. 1961. J. Mol. Biol. 3:318.
Jacob, F., A. Ullman, and J. Monod. 1964. C. R. 258:3125.
Losick, R., A. L. Sonensheim, R. G. Shorenstein, and C. Hussey. 1970. Cold Spring Harbor Symp. Quant. Biol. 35:443.
Makman, R. S., and E. W. Sutherland. 1965. J. Biol. Chem. 240:1309.
Perlman, R. L., and I. Pastan. 1968. J. Biol. Chem. 243:5420.
Ptashne, M. 1971. *In* The Bacteriophage Lambda (A. D. Hershey, ed.). Cold Spring Harbor Laboratory, Cold Spring Harbor, N.Y.
Reznikoff, W., W. Barnes, J. Abelson, and R. Dickson. 1974. Personal communication.
Riggs, A. D., and S. Bourgeois. 1968. J. Mol. Biol. 34:361.
Riggs, A., G. Reiness and G. Zubay. 1971. Proc. Nat. Acad. Sci. U.S.A. 68:1222.
Rose, J. K., C. L. Squire, C. Yanofsky, H-L. Yang, and G. Zubay. 1973. Nat. New Biol. 245:133.
Schachner, M., W. Seifert, and W. Zillig. 1971. Eur. J. Biochem. 22:520.
Ullman, A., and J. Monod. 1968. FEBS Lett. 2:57.
Yang, H-L., G. Zubay, E. Urm, G. Reiness, and M. Cashel. 1974. Proc. Nat. Acad. Sci. U.S.A. 71:63.
Yang, H-L., and G. Zubay. 1973. Mol. Gen. Genet. 122:131.
Zubay, G., M. Lederman, and J. K. DeVries. 1967. Proc. Nat. Acad. Sci. U.S.A. 58:1669.
Zubay, G., D. Schwartz, and J. Beckwith. 1970. Proc. Nat. Acad. Sci. U.S.A. 66:104.
Zubay, G., D. A. Chambers, and L. C. Cheong. 1970. *In* The Lactose Operon (J. R. Beckwith and D. Zipser, ed.). Cold Spring Harbor Laboratory, p. 375.
Zubay, G., and D. A. Chambers. 1971. *In* Metabolic Regulation (H. J. Vogel, ed.). Academic Press, New York, p. 297.
Zubay, G., and M. Lederman. 1969. Proc. Nat. Acad. Sci. U.S.A. 62:550.
Zubay, G., and J. Marmur (ed.). 1973. Papers in Biochemical Genetics, 2nd ed. Holt, Rinehart and Winston, New York, p. 515.

DISCUSSION

DR. MUNRO: I want to ask about the eukaryote fate of two items. First of all, quanosine tetraphosphate has been looked for very diligently in eukaryote systems from fungi upwards, under conditions where you might expect it to be generated, and it has not emerged. However, you can make a case for uncharged tRNA, which, of course, is the signal for generation of MS_1, as a regulator coordinating a whole series of actions related to protein synthesis in the higher organisms.

The second point is the transferring of the operon concept from prokaryotes to eukaryotes, for example. If you take the histidine pathway from *Salmonella*, say, and go to *Neurospora*, where your various genetic sites are scattered over several chromosomes, what is the current view about the regulation under these conditions?

DR. ZUBAY: We have examples of broken-up operons in *E. coli*, as well. I might mention the archoperon or polyoperon, which is broken up into five clusters consisting of nine enzymes. The largest cluster is four genes. But all these genes are still under the control of one repressor molecule.

Bob Goldberger (1974. Science 183:810) has published a review article in which he mentions that Jerry Fink at Cornell has shown that it is highly likely that *Neurospora* has the same control apparatus that has been found in *Salmonella* as far as histidine system is concerned.

DR. MUNRO: What is the position of the regulator molecule, then?

DR. ZUBAY: It is just as confusing as it is in the *E. coli* system. The histidine system is one of the most involved that anybody has found. Goldberger's current idea is that this is a case of autogenous regulation, where the first enzyme in the pathway, which is also the operator enzyme or gene, appears to be the repressor protein for the operon. He has very strong evidence for that. I think that in the case of *Neurospora* one does not have such strong evidence for that, but one does have very strong evidence for the fact that the corepressor, instead of being the amino acid, is a complex between the amino acid and its tRNA, the *his*tRNA. That is the corepressor in both cases. The evidence that the repressor is the first gene product in *Neurospora* is much weaker.

I would like to make one more point, which is peripheral to your question. One does not necessarily have to find the same control elements being used. But the theme that I want to emphasize is that the basic mechanisms, the fact that we have activators and repressors and that they work in the general way that I have described, are probably going to turn out to be similar for simple and complex systems. There really are not many ways of accomplishing the turning on and turning off of genes. The same basic mechanisms will be used over and over again.

DR. DOY: Are you implying that there is now evidence in eukaryotes that there are control systems like, say, tryptophan repression or histidine repression? Are there repression systems, or must the genes be turned on rather than turned off?

DR. ZUBAY: Most of what I could say here would be speculative, but it is clear that the positive control systems used in various combinations are going to be a very popular type of mechanism. If you go all the way to highly differentiated cells that turn out only a very limited number of gene products, then it would be ridiculous to be using negative control exclusively, because this would require that you synthesize tens of thousands of different types of repressor molecule that sit all over the chromosome. However, if you have interlocking relationships, as we do with lambda, say, when you have one gene system being activated and turning on the synthesis of other positive control proteins, which affect further operons, and this goes on and on in a chain of reactions, you can see how you could lead to a situation, which you have in the red blood cell, where you really end up with only one gene turned on to any significant extent.

We certainly favor the speculation that by and large one is using positive control systems, and in various combinations, so that you can get whatever pattern you want for different types of differentiated cells. But these could be, and probably are, mixed in with negative control systems.

DR. MEANS: I might just reply to your question, Dr. Doy. In the avian oviduct there now is evidence for both positive and negative repression of the gene coding for the synthesis of ovalbumin messenger RNA. Estrogen will activate this genome by binding to the DNA and associated nonhistone chromosomal proteins, allowing the interaction of several RNA polymerase molecules. There are multiple RNA polymerase binding sites on this region, but apparently only one promoter. These two types of sites can be differentiated by using temperature sensitivity and drug resistance. One sees that there are two orders of binding, one with a dissociation constant of about 10^{-7} molar and one with a dissociation constant of about 10^{-12} molar. They can be distinguished very clearly. Estrogen results in increases in the number of polymerase molecules bound and the number of initiation sites. Progesterone and its receptor protein antagonize the estrogen effects.

The interesting fact is that estrogen and progesterone have different binding sites on the chromatin.

DR. HENDRICKS: Does someone from the plant breeding side want to comment about how far they are away from incorporating such information into their methods? The answer is no?

DR. DOY: I might point out, Mr. Chairman, that this is why we started our work, in the hope that we can start to apply these approaches. But we are a very, very great distance from application.

DR. HENDRICKS: I understand, then, that the regulatory systems in the case of the *lac* operon and the others that you spoke of, are exceedingly complex and require a number of agents to carry out conservative processes and that these processes do not require too many agents that exhaust all the material of the cell, and, that they are very apt to be cascade processes, in which there is a succession of products and that result in an amplification of the system. Would that type of system operate with the normal type of genetics that you are using in the manipulation of higher plants?

JOHN M. CLARK, JR.

Posttranscriptional Regulation of Protein Synthesis

Posttranscriptional regulation or control of protein synthesis is a relatively new area of research interest. As such, the basic concepts and literature in the area are still developing. Hence, with the exception of one recent review (Andron and Strehler, 1973), the more recent and significant findings in the general area have not been coordinated and analyzed in an overall sense. In part, this underdevelopment of the field arises from the earlier need for a more thorough knowledge of the components and events of the posttranscriptional or protein synthesis process. In part, this backwardness also arises from the rather recent development of the necessary separatory methods needed to characterize sites of posttranscriptional control within protein synthesis. In any event, this field is now one of increasing interest to a wide variety of biologists, biochemists, geneticists, and others. Many of the major advances in the field have been made by those investigating problems of developmental biology or aging. Thus, workers today can point to several sites of known or potential posttranscriptional control of protein synthesis. To date, few of these findings have been carried to the level of direct genetic implications dependent upon posttranscriptional control. Rather, the field can best be considered a source for future genetic utilization, i.e., a field that identifies sites where genetic events and considerations may yield useful future benefits.

An understanding of the posttranscriptional control mechanisms requires a stepwise consideration of protein synthesis and an examination of a wide variety of processes. Accordingly, I will consider, in turn, the various steps of protein biosynthesis and the existing evidence for control or regulation of events acting at each of these levels of protein biosynthesis.

The initial step of protein biosynthesis is the ATP-dependent activation of L-amino acids by specific aminoacyl–tRNA synthetases, followed by "charging" of these activated amino acids to amino-acid-specific transfer RNAS (reactions 1–3: AA = amino acid, Enz = aminoacyl–tRNA synthetase, PP_i = pyrophosphate, tRNA = transfer RNA).

$$\text{AA} + \text{ATP} + \text{Enz} \rightleftharpoons \text{Enz-AMP-AA} + \text{PP}_i \quad (1)$$

$$\text{Enz-AMP-AA} + \text{tRNA} \rightleftharpoons \text{tRNA-AA} + \text{AMP} + \text{Enz} \quad (2)$$

$$\text{AA} + \text{ATP} + \text{tRNA} \overset{\text{Enz}}{\rightleftharpoons} \text{tRNA-AA} + \text{AMP} + \text{PP}_i \quad (3)$$

There is currently some question about the actual existence of Enz-AMP-AA in this sequence (Loftfield, 1972) but the basic existence of a variety of amino-acid-specific aminoacyl–tRNA synthetases and tRNAS is well accepted. More specifically, it is recognized that this process is further complicated by the existence of multiple forms of both amino-acid-specific aminoacyl–tRNA synthetases and tRNAS. Specifically, any one subcellular fraction may possess two or more different aminoacyl–tRNA synthetases. This arises from the fact that any one fraction from cells may contain more than one tRNA species capable of accepting a specific amino acid. These differences are further complicated by the fact that different subcellular fractions (e.g., mitochondria, chloroplasts, cytoplasm) will exhibit further differences among their aminoacyl–tRNA synthetases and tRNAS (Barnett and Brown, 1967; Buck and Nass, 1968; Epler, 1969). Finally there are reported cases of further tissue-specific qualitative and quantitative differences in both aminoacyl–tRNA synthetases and tRNAS (Anderson and Cherry, 1969). These latter observations that different tissues and different subcellular organelles contain differences in their enzyme and tRNA populations have often been ignored by investigators in the field. This has often made it difficult to define the origin of qualitative and quantitative differences in the multiple forms of aminoacyl–tRNA synthetases and multiple or isoaccepting forms of tRNA detected in various experiments. Nevertheless, there are cases for potential control(s) of protein synthesis that arise at this level.

Much of the interest in controls at this level arises out of studies of aging. Kanabus and Cherry (1970), Bick and Strehler (1971, 1972) have examined the changes that occur in leucyl–tRNA synthetases and leucyl–tRNAs ($tRNA^{leu}$s) during the aging of soybean cotyledons. Their basic premise is to detect qualitative or quantitative changes in the populations of these components with time and then correlate these changes with known changes that occur during aging. Carrying this line of reasoning further, they could predict the site of posttranscriptional control of protein biosynthesis in that variations in the populations of synthetases would define the quantities of a specific amino acid being charged onto a specific tRNA among the various isoaccepting tRNAs in a fraction of a cell. These different rates of charging the isoaccepting tRNAs would then express themselves during the subsequent steps of protein synthesis. A corollary to this reasoning involves changes in the levels of the substrates for the aminoacy–tRNA synthetases—namely, changes in the populations of isoaccepting tRNAs. If one considers that varied charging rates for individual tRNAs will represent a form of regulation of protein synthesis, one must recognize that deficiencies of either enzyme (synthetase) or substrate (tRNA) can influence the charging rates and the eventual supply of charged tRNA (tRNA–AA).

These workers have detected quantitative differences in the levels of both leucyl–tRNA synthetases and individual isoaccepting $tRNA^{leu}$s with aging of soybean cotyledons. Further, these workers have shown selectivity by individual leucyl–tRNA synthetases for specific $tRNA^{leu}$ forms. Subsequent interpretation of these variations in levels of individual synthetases and tRNAs has been more difficult. The key result of these studies is obtained by contrasting the charging of 5-day-old and 21-day-old $tRNA^{leu}$s with 5-day-old and 21-day-old synthetases (Figure 1).

As can be seen, 5-day enzyme charges 5-day tRNAs effectively, while 5-day enzyme charges 21-day $tRNA^{leu}$ to a lower level, suggesting that one has reduced tRNA levels at 21 days. The results of other tRNA charging experiments can be interpreted to support this concept. Yet when 21-day enzyme charges 5-day $tRNA^{leu}$, less charging is obtained at equilibrium. Twenty-one-day enzyme charges 21-day $tRNA^{leu}$s to even less. With these results, the reasoning becomes more complicated. These workers initially speculate about coeluting "subspecies" of isoaccepting tRNAs to explain why a given $tRNA^{leu}$ sample will be charged at equilibrium to varying degrees by different enzymes (Bick and Strehler, 1971). A later paper speculates about the existence of repressors in this charging reaction (Bick and Strehler, 1972), and even invokes the possibility of specific hormone binding influences on the charging reactions.

Perhaps the difficulties in interpretation of these results or the inability to establish a direct regulatory role for control at the level of

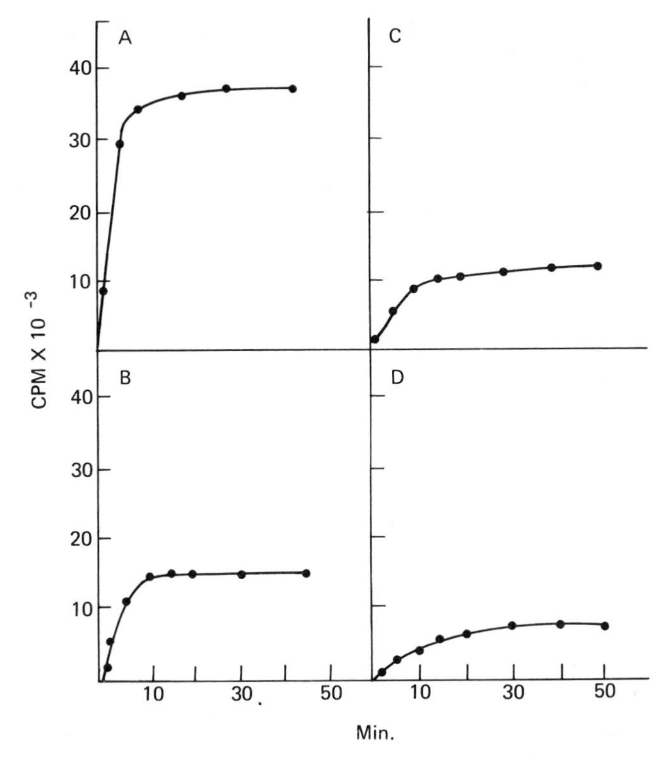

FIGURE 1 Charging kinetics of tRNA^{leu} and synthetase from 5- and 21-day-old cotyledons. 11 A_{260} units of tRNA was used in each reaction, with about 1 mg of synthetase protein. The prefix numbers refer to the age (days) of the cotyledons from which enzyme or tRNA was prepared. *A*, 5-Enz–5-RNA; *B*, 5-Enz–21-tRNA; *C*, 21-Enz–5-tRNA; *D*, 21-Enz–21-tRNA. (Bick and Strehler, 1971.)

aminoacyl–tRNA synthesis lies in the dependence of these studies on assays of charging of tRNAs. Recently, it has been shown (Schreier and Schimmel, 1972; Eldred and Schimmel, 1972) that aminoacyl–tRNA synthetases contain two activities: (a) the expected ATP-dependent ability to charge tRNAs and (b) a separate ability to hydrolyze charged amino acid from tRNAs. This latter ability is not a reversal of the expected charging reaction but rather is a separate phenomenon unaffected by AMP and PP_i levels. This latter ability also appears to be subject to potential controls: It is competitively inhibited by uncharged tRNA and responds differently to pH, Mg^{++} ion, and other tests. The existence of both these activities within synthetases can readily influence the extent of charging of tRNA observed at equilibrium and makes interpretation of such observations hazardous. Yet the very existence of these two competing reactions within aminoacyl–tRNA synthetases suggests an even more direct potential control of protein synthesis at the tRNA charging level. No doubt, future research efforts will concentrate on this topic.

The next major step in protein biosynthesis is the formation of initiation complexes between methionine-specific tRNAs, initiator sites of messenger RNAs, and ribosomes (Figure 2). The significant features of this process are that

Messenger-RNAs possess unique sites containing the trinucleotide 5′→3′ sequence AUG that serve as initiator sites.

There are specific initiator tRNAs that are charged with methionine in cytoplasmic eukaryotic systems and are eventually charged with N-formyl-methionine in prokaryotic systems and chloroplast and mitochondrial systems of higher plants and animals (eukaryotes).

Ribosomes participate in this process in a stepwise manner that first involves the smaller ribosomal subunit (30S in prokaryotes, 40S in eukaryotes) and eventually yields a full ribosome (70S in prokaryotes,

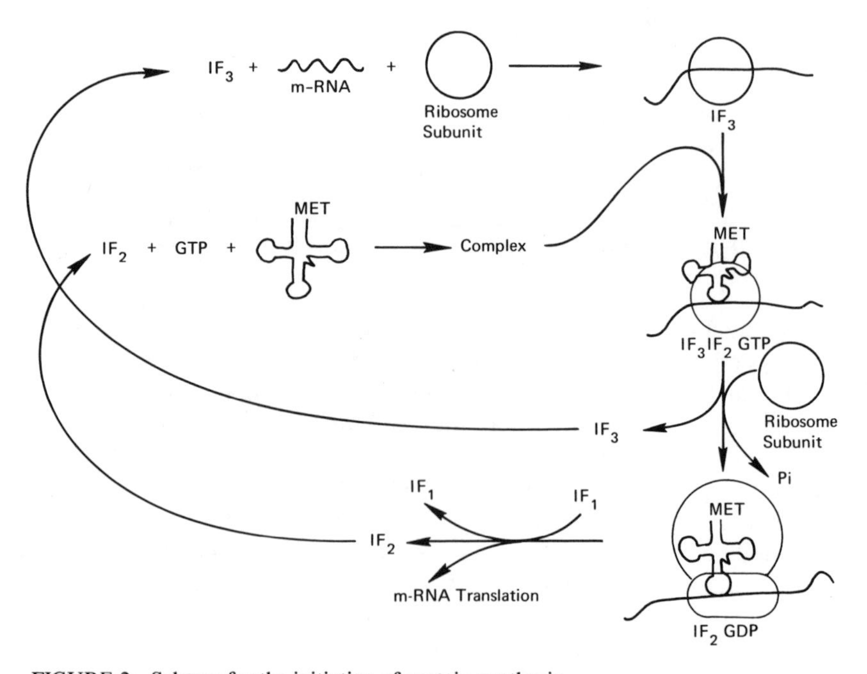

FIGURE 2 Scheme for the initiation of protein synthesis.

80S in eukaryotes) as a result of a GTP-dependent incorporation of the larger ribosomal subunit (50S in prokaryotes, 60S in eukaryotes).

The formation of the initiation complex requires the participation of at least three ribosome specific initiation factors, indicated in Figure 2 in the prokaryotic convention as IF_1, IF_2, and IF_3.

The full mechanistic details of this process are still developing and are therefore the subject of considerable debate and interest. In particular, the specific roles of IF_1, IF_2, and IF_3 are somewhat speculative, but it can generally be concluded that

IF_3 facilitates the binding of natural message to the smaller ribosomal subunit and must leave this smaller subunit in order to allow incorporation of the larger subunit.

IF_2 participates in binding the initiator tRNA and directs its binding into initiation complexes.

IF_1, in some way, enhances the action of the other two initiation factors, possibly by facilitating recycling of the factors.

Recently, these initiation factors have been strongly implicated as a site for posttranscriptional control of protein synthesis in bacterial systems. Specifically, bacterial IF_3 has been resolved into two subspecies, $IF_3\alpha$ and $IF_3\beta$, that demonstrate different selectivities for specific messenger RNAs (Revel *et al.*, 1970; Berissi *et al.*, 1971; Lee-Huang and Ochoa, 1971). Thus, one IF_3 will enhance the binding of one messenger RNA into initiation complexes, while a second IF_3 will enhance the binding of a second and different message into initiation complexes. This is truly a significant control point, for message selection defines what protein(s) will be made by the cell. The overall picture of this type of control is probably more complicated than just $IF_3\alpha$ and $IF_3\beta$. For example, recently various inhibitor, or *i*, factors have been identified that influence the potential for IF_3s to bind specific messages (Groner *et al.*, 1972).

Higher forms of life, eukaryotes, appear to have similar initiation factors necessary for initiation complex formation. It remains to be seen whether similar significant posttranscriptional control exists in the form of message-specific roles for such initiation factors at any given time in cellular development. Yet there is ample evidence that eukaryotic initiation factors demonstrate selectivity for messages derived from different stages of organism development. Specifically, Ilan and Ilan (1971) show that initiation factors are necessary for formation of initiation complexes in an *in vitro* system derived from *Tenebrio molitor*, the common mealworm. Further, these workers demonstrate that initiation factors derived from mealworm larvae will enhance formation of initiation complexes with larval messenger RNA but not with message derived from mealworm pupae. The inverse, a strict requirement for pupal initiation factors for initiation complex formation with pupal messenger RNA, is also true. The source of the ribosomes, larval versus pupal, does not influence this result. Thus, differentiation in higher organisms can be linked to selectivity at the level of the initiation factors of protein biosynthesis.

Following formation of a completed initiation complex on a 70S or 80S ribosome, protein synthesis features the protein chain elongation process catalyzed upon the ribosome by two or more "translation factors" or "elongation factors" (Figure 3).

During this process one has a stepwise association of nucleotides of the anticodon region of specific aminoacyl–tRNAs with complementary trinucleotides of messenger RNAs. This specific coding is accompanied by specific, translation-factor-mediated steps involving the formation of peptide bonds, the translocation of the messenger RNA through the two aminoacyl–tRNA binding sites of ribosomes, and the energy-dependent binding of incoming aminoacyl–tRNAs. The key sites for potential control of this process involve the coding interaction of the aminoacyl–tRNAs with the trinucleotide codons or triplets of the messenger RNA. It is known that specific amino acids are often coded

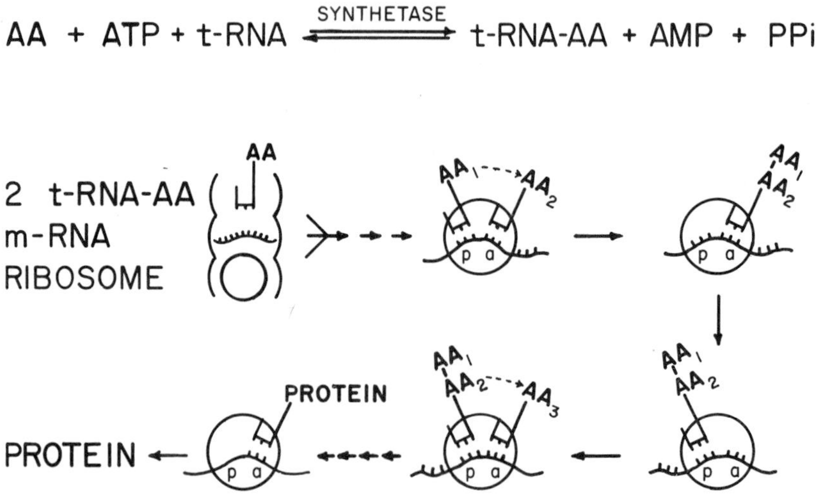

$$\text{AA} + \text{ATP} + \text{t-RNA} \xrightleftharpoons{\text{SYNTHETASE}} \text{t-RNA-AA} + \text{AMP} + \text{PPi}$$

2 t-RNA-AA
m-RNA
RIBOSOME

PROTEIN

FIGURE 3 Scheme for the translation of messenger RNA.

for by a variety of specific codons. The following examples illustrate this point.

RNA Codons for Leucine	RNA Codons for Serine
CUA	UCA
CUG	UCG
CUC	UCC
CUU	UCU
UUA	AGC
UUG	AGU

This "degeneracy" with the genetic code is a site for potential regulation of protein synthesis. For example, one can envision two different messenger RNAS, each coding for leucine in the second amino acid so that

– – AUG–CUA–UUU– . . . $\xrightarrow{\text{translation}}$ Met–Leu–Phe– . . .

– – AUG–UUA–AAA– . . . $\xrightarrow{\text{translation}}$ Met–Leu–Lys– . . .

These different codons for leucine are recognized by different iso-accepting tRNA^{leu}s. Thus, alteration (for example, lowering) of the pool sizes of either of these tRNA^{leu} forms could result in rate-limiting synthesis of that particular protein—in short, a translational control mechanism.

Much of the interest in this line of reasoning has occurred among investigators of developmental biology and aging processes. Figure 4 shows how alterations in the levels of isoaccepting tRNAS, most likely in a state of general deficiency of leucine to charge the various tRNA^{leu} forms, can influence the synthesis of specific proteins.

Such reasoning has prompted a wide search for alterations in the relative levels of isoaccepting tRNAS. This search has been aided by the recent development of a variety of ion exchange media that can resolve various isoaccepting tRNAS. Many specific changes in the levels of isoaccepting tRNAS are now known as a result of developmental stages (Lee and Ingram, 1967; Molinaro and Mozzi, 1969; Bagshaw *et al.*, 1970; Dewitt, 1971; White *et al.*, 1973b), viral infections (Sueoka and Kano-Sueoka, 1964), tumor growth (Yang and Comb, 1968), and aging (Anderson and Cherry, 1969; Kanabus and Cherry, 1970; Bick and Strehler, 1971, 1972). All of these findings suffer from the question of whether tRNA differences are the cause of posttranscriptional control on the effect of yet other control mechanisms. Specifically, no *in vivo* experiments show alterations in tRNA levels working at the coding level

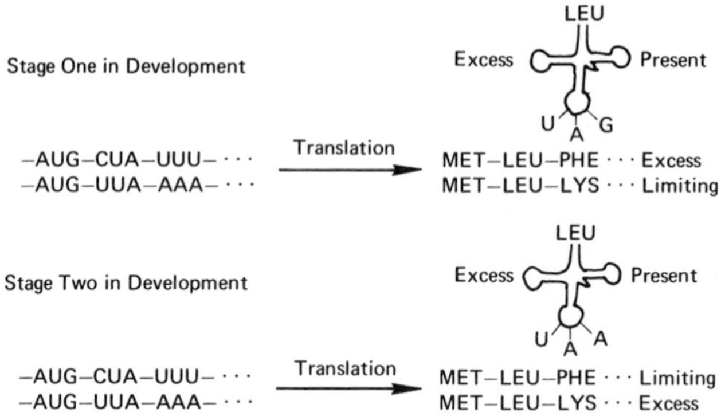

Stage One in Development

LEU
Excess Present
U A G

–AUG–CUA–UUU– · · ·
–AUG–UUA–AAA– · · ·
Translation
MET–LEU–PHE · · · Excess
MET–LEU–LYS · · · Limiting

Stage Two in Development

LEU
Excess Present
U A A

–AUG–CUA–UUU– · · ·
–AUG–UUA–AAA– · · ·
Translation
MET–LEU–PHE · · · Limiting
MET–LEU–LYS · · · Excess

FIGURE 4 A model for tRNA dependent translation control of cell development.

to function as a posttranscriptional control mechanism. An *in vitro* example of such tRNA-dependent posttranscriptional control at the coding level does exist (Anderson, 1969), but the result largely reflects the synthetic messenger RNAs employed in the assays.

In some cases, the differences in isoaccepting tRNAs reflect the action(s) of specific modifying enzymes (e.g., methylate, isopentenylate, thiomethylate) that modify specific nucleotides of a basic primary nucleotide sequence. Thus, changes in the levels of specific isoaccepting tRNAs during development or aging may reflect changes in the levels of activity of specific modifying enzymes. One study of such modifying actions on a population of isoaccepting tRNAs has been carried on sufficiently to define a new and meaningful mechanism of posttranscriptional control. Figure 5 depicts the dramatic changes that take place in the isoaccepting tRNA^{tyr}s of *Drosophila melanogaster* during its various stages of development.

The increase observed in the quantity of tRNA$_{1\gamma}^{tyr}$ during development appears to be due to the action of a specific modifying enzyme (White *et al.*, 1973b) that facilitates the formation of tRNA$_{1\gamma}^{tyr}$ (the left-hand species in Figure 5). This modified tRNA, tRNA$_{1\gamma}^{tyr}$ in its uncharged form, will specifically bind to the tryptophan pyrolase of *Drosophila* with resultant inactivation of the enzyme (Jacobson, 1971). This phenomenon is the basis for the vermilion phenotype in *Drosophila*. This vermilion phenotype can be suppressed in fruit flies containing the homozygous recursive suppressor su(s)^2 or a new induced homozygous suppressor su(s)^{le}. These suppressor mutations both map in the same location on the X chromosome and both result in the elimination of the $tRNA_{1\gamma}^{tyr}$ form of tyrosine-specific tRNA (presumably by elimination of the tRNA modifying enzyme). Most important, both suppressor mutations result in enhanced tryptophan pyrolase activity in the suppressed *Drosophila*. These data prove that

1. Specific tRNAs can serve as regulators of the activity of specific enzymes.
2. The specificity for such regulation of specific enzyme activities is dependent, in part, upon "modification" of specific tRNA structures.

We must await further experiments to characterize the nature of this significant modification. Yet these data establish that tRNAs can play a regulatory role in cells that is independent of their role in protein biosynthesis.

After completion of the protein chain elongation process by translation factors, one finally encounters the chain termination step. In this step, the completed protein chain is cleaved from the tRNA specific for the carboxyl terminal amino acid. Three messenger RNA codon combinations—UAG, UAA, and UGA—are known to trigger this process throughout nature. Bacterial geneticists have utilized the unique potentials of bacterial systems to develop classes of mutants that carry "suppressors" of this chain termination process. Such suppressor-carrying strains contain mutated tRNAs that actually override the chain termination process by inserting amino acids into protein chains at the

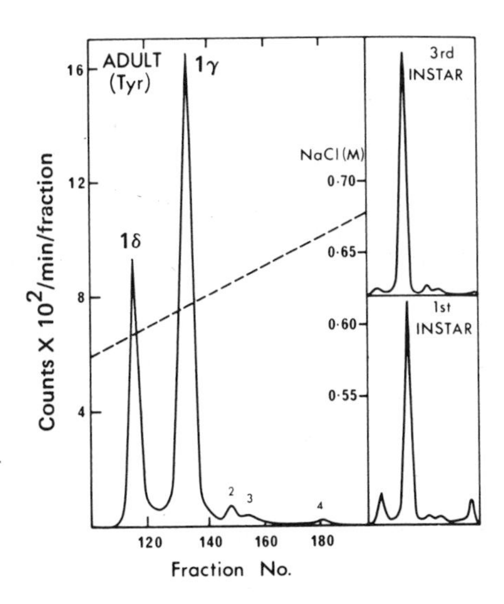

FIGURE 5 Chromatographic profile of ^{14}C-tyrosyl-$tRNA^{tyr}$s from a RPC-5 column at 37°C. Data taken from White *et al.* (1973a).

points where they would normally terminate. In the strictest sense, such suppression of chain termination is a posttranscriptional control of protein synthesis. In the practical sense, this phenomenon is of little value to future agricultural genetics, for such suppressor tRNA-carrying strains have not been found in higher forms. Further, when found in bacteria, they are really useful only as a laboratory tool to study chain termination. Such suppressor tRNA strains are particularly weak or unviable, presumably due to competition by the suppressor tRNA with the normal chain termination processes of the cell.

The various examples discussed so far represent the most thoroughly documented and investigated areas of posttranscriptional control of protein synthesis. Yet it is equally important to scan other areas of biology in the hope of pinpointing other potentially significant areas of posttranscriptional control. The remainder of this paper will concentrate on this rather speculative approach.

One potentially significant area of posttranscriptional control involves features of the three-dimensional or secondary structure of messenger RNA. Recent work with a variety of polycistronic bacterial viral messenger RNAs suggests that initiation of protein synthesis—that is, the selection of initiation sites upon a messenger RNA—involves, in part, unique secondary structures within the message (Steitz, 1969; MinJou *et al.*, 1972; Robertson *et al.*, 1973). Specifically, it is possible that sites for initiation of protein biosynthesis involve specific loops in the secondary structure of messenger RNAs. If messenger RNA secondary structure loops are an important feature in the initiation of translation of eukaryotic messages, one can envision mutations that would yield nucleotide combinations that would enhance or retard loop formation in messages. If these nucleotide substitutions or limited frame shifts result in changes in amino acids not associated with the catalytic activity of the resultant proteins, then one could achieve enhanced or reduced translation of specific cistrons without alteration of enzymatic potential. As yet, there is no evidence for such mutations, and this whole concept is rather speculative at best. Yet, as we learn more about eukaryotic messenger RNA structure and its relation to the initiation of protein synthesis, this concept may increase in importance in higher plants. Alternatively, this phenomenon may be limited to the polycistronic prokaryotic messenger RNAs.

Another speculative area of posttranscriptional control of protein biosynthesis also arises from studies of the translation of messenger RNAs of the single-stranded RNA bacterial viruses. Such viral RNAs are polycistronic messages. They also generate marked selection for the *in vitro* translation of certain of their cistrons over others. It is probable that, in part, this cistron selection represents the secondary structure of the messenger RNA burying the initiation sites of cistrons (MinJou *et al.*, 1972). Yet it is also known that certain proteins will bind to such messenger RNAS and alter the *in vitro* cistron selection process (Sugiyama and Nakada, 1967). This phenomenon may be unique to the structured polycistronic character of these bacterial viral messenger RNAS. To date, no similar activator or repressor of messenger RNA translation has been detected in higher organisms. Still, the existence of a biological analogy makes the topic worthy of thought.

The last rather speculative concern with posttranscriptional control of protein synthesis deals with ribosomes. Molecular biologists have long considered ribosomes as inert components that house the protein synthesis process. We are currently learning a great deal about the structural components of ribosomes, and it is apparent that the ribosome is more specific, if not more functional, than originally envisioned. More specifically, eukaryotic ribosomes are now known to undergo fairly specific ATP-dependent phosphorylation of ribosomal proteins (Kabat, 1970; Bitte and Kabat, 1972) and to contain an endogenous protein kinase (Kabat, 1972). Eukaryotic systems make use of many regulation instances involving phosphorylation and dephosphorylation of specific enzymes. It follows that detection of a potentially similar specific phosphorylation mechanism in eukaryotic ribosomes may represent detection of a significant posttranscriptional control of protein synthesis.

In conclusion, in an overall sense, one can see that posttranscriptional control of protein synthesis can and does occur in various systems. It is clear that in prokaryotic systems, transcriptional events serve as the major control of the eventual quantities of proteins made. With eukaryotes, the case is not as clear-cut. It is highly conceivable that posttranscriptional events may play a significant role in the control of protein biosynthesis. The field awaits further characterization before a strong link can be made between such posttranscriptional controls and practical genetic manipulations related to these principles of posttranscriptional control.

REFERENCES

Anderson, M. B., and J. H. Cherry. 1969. Proc. Nat. Acad. Sci. U.S.A. 62:202.
Anderson, W. F. 1969. Proc. Nat. Acad. Sci. U.S.A. 62:566.
Andron, L. A., II, and B. L. Strehler. 1973. Mech. Ageing Dev. 2:97.
Bagshaw, J. E., F. J. Finamore, and G. D. Novelli. 1970. Dev. Biol. 23:23.
Barnett, W. E., and D. H. Brown. 1967. Proc. Nat. Acad. Sci. U.S.A. 57:452.
Berissi, H., Y. Groner, and M. Revel. 1971. Nat. New Biol. 234:44.

Bick, M.D., and B. L. Strehler. 1971. Proc. Nat. Acad. Sci. U.S.A. 68:224.
Bick, M. D., and B. L. Strehler. 1972. Mech. Ageing Dev. 1:33.
Bitte, L., and D. Kabat. 1972. J. Biol. Chem. 247:5345.
Buck, C. A., and M. M. K. Nass. 1968. Proc. Nat. Acad. Sci. U.S.A. 60:1045.
Dewitt, W. 1971. Biochem. Biophys. Res. Commun. 42:266.
Eldred, E. W., and P. R. Schimmel. 1972. J. Biol. Chem. 247:2961.
Epler, J. L.. 1969. Biochemistry 8:2285.
Groner, Y., Y. Pollack, H. Berissi, and M. Revel. 1972. Nat. New Biol. 239:16.
Ilan, J., and J. Ilan. 1971. Dev. Biol. 25:280.
Jacobson, K. B. 1971. Nat. New Biol. 231:17.
Kabat, D. 1970. Biochemistry 9:4160.
Kabat, D. 1972. J. Biol. Chem. 247:5338.
Kanabus, J., and J. H. Cherry. 1970. Plant. Physiol. 46(Suppl.):30.
Lee, J. C., and V. L. Ingram. 1967. Science 158:1330.
Lee-Huang, S., and S. Ochoa. 1971. Nat. New Biol. 234:236.
Loftfield, R. B. 1972. Prog. Nucl. Acid Res. 12:87.
MinJou, W., G. Haegeman, M. Yaebaert, and W. Fiers. 1972. Nature 237:82.
Molinaro, M. and R. Mozzi. 1969. Exp. Cell Res. 56:163.
Revel, M., H. Aviv, Y. Groner, and Y. Pollack. 1970. FEBS Lett. 9:213.
Robertson, H. D., B. Q. Barrell, H. L. Weith, and J. E. Donelson. 1973. Nat. New Biol. 241:38.
Schreier, A. A., and P. R. Schimmel. 1972. Biochemistry 11:1582.
Steitz, J. A. 1969. Nature 224:957.
Sueoka, N., and T. Kano-Sueoka. 1964. Proc. Nat. Acad. Sci. U.S.A. 52:1535.
Sugiyama, T., and D. Nakada. 1967. Proc. Nat. Acad. Sci. U.S.A. 57:1744.
Twardzik, D. R., E. H. Grell, and K. B. Jacobson. 1971. J. Mol. Biol. 57:231.
White, B. N., G. M. Tener, J. Holden, and D. T. Suzuki. 1973a. Dev. Biol. 33:185.
White, B. N., G. M. Tener, J. Holden, and D. T. Suzuki. 1973b. J. Mol. Biol. 74:635.
Yang, S. S., and D. G. Comb. 1968. Biochem. Biophys. Res. Commun. 31:534.

DISCUSSION

DR. WILSON: A short note in the *New Scientist*, Jan. 31, 1974, p. 244, referred to the claim of K. Downey *et al.* (1973. Proc. Nat. Acad. Sci. U.S.A. 70:3400) that duck reticulocytes may be able to duplicate messenger RNA. Would Dr. Clark or Dr. Zubay comment on the possibility of this being a controlling mechanism in areas where you get a great deal of synthesis of a single protein, as in the reticulocyte or in seed?

DR. CLARK: I think that the work that has been done with the reticulocyte system has largely been that of a protein synthesis system, the so-called translation steps. And here, messages from a variety of sources have been translated, and therefore the protein that is made is the protein that has been programmed by the message. This has not been as effective a system as some of the other protein synthesis systems.

With respect to synthesis of messenger RNA within reticulocytes *per se,* this is usually not examined. If you know more about that than I do, you are welcome to speak to that. But it is not normally the place where people have looked for messenger RNA synthesis. In a sense, the mammalian red blood cell has been considered to be a highly developed differentiated thing that, after the various controls have been applied, is kind of reduced to a hemoglobin-making machine, and that is about it.

DR. ZUBAY: We are examining this question right now. I cannot give you any answer on it, but in amphibia and birds you can get nucleated reticulocytes that are immature by bleeding or by treatment with phenyl hydrazine, which makes them anemic so they turn out a large number of cells. They are very active in protein synthesis, a hundred times more active than the normal reticulocytes, and quite active in transcription. And we are investigating the question as to which messengers they are making. So all I can say is that these questions are being investigated.

DR. MERRICK: Actually, with respect to the Downey paper, I believe the report was that there was an RNA copy made, not an actual message made from the hemoglobin message. I know individuals in our laboratory have attempted to repeat it and have not been able to do so yet. In particular, my group has been interested in the initiation of protein synthesis, and, unfortunately, at least from the data that I have been able to read and have access to, it appears to be that there are very few instances in which there will be any message-specific recognition or limitation in cells. There have been several instances cited where there does appear to be recognition, perhaps, between different messages. But this certainly is a long way from saying that this definitely is a factor in the regulation of protein synthesis.

DR. CLARK: I would agree; in particular, a given cell could be said to have somewhere between 2,000 and 3,000 proteins. It would seem unrealistic to have 2,000 to 3,000 messenger-selected factors also around in order to define this process. I would suspect, therefore, that there would be, at best, a few types of factors that would demonstrate some minor degree of selection or preference. I cannot help but be impressed, though, by the case of the mealworm example, which demonstrates an all-or-nothing phenomenon. We will not know exactly what these factors are or what they do until they purify the factors.

DR. DOY: Is there any hard evidence that U–A–G is a termination signal in eukaryotes? We have done some experiments, but I do not necessarily think that is what it means.

DR. CLARK: Some of the sequences that have been looked at at the ends of plant viral nucleic acid suggest that there is a U–A–G at that particular point. Higher plant and animal suppressors are not known. So the classical suppression-chain termination experiments used with bacterial systems have not really been extended to higher systems in the absence of known populations of suppressor tRNAs *per se*.

ANTHONY R. MEANS,
SAVIO WOO, CASSIUS BORDELON,
JOHN P. COMSTOCK,
and BERT W. O'MALLEY

Hormone-Regulated Synthesis of Tissue-Specific Proteins

The ultimate response to a hormone that causes growth or differentiation of its target tissue is a change in the complement of cellular proteins. Indeed, a variety of steroids and peptide hormones have been demonstrated to stimulate protein synthesis within hours after a single injection to hormone-deficient animals or plants. Since the description of the original model for control of bacterial gene expression at the transcriptional level by Jacob and Monod (1961), it has been hypothesized that hormones stimulate protein synthesis in a similar manner. In order to demonstrate that such a control mechanism existed in eukaryotic organisms, the presence of hormone effects on mRNA production had to be demonstrated. Measurements of changes in gene transcription have indicated that several hormones do, in fact, act at this level of cell regulation. Estrogen, for example, stimulates the synthesis of rapidly labeled mRNA, and RNA polymerase activity is altered, as is the template capacity of nuclear chromatin. Analysis of these RNA products by nucleic acid hybridization and nearest neighbor base analysis also reveal marked changes in response to the steroid hormone. However, none of these studies constitutes direct proof of alterations in the transcription of specific structural genes. Indeed, obtaining direct evidence that the synthesis of mRNA is a rate-limiting step in the action of steroid hormones has continued to be a most elusive problem.

The only definitive way to prove the existence of any mRNA is to isolate an RNA fraction from the cell or tissue under investigation and demonstrate that this RNA supports the unambiguous translation of its specific protein in a cell-free protein synthesis system. Application of this procedure to studies concerning mechanism of hormone action would be made much easier if the hormone resulted in the stimulation of a specific and easily quantifiable protein in its target tissue. The presence of such proteins has made the chick oviduct a unique model in which to study the hormonal regulation of specific protein synthesis. Two steroid hormones, estrogen and progesterone, affect the chick oviduct (O'Malley *et al.*, 1969, 1971; Means and O'Malley, 1972).

Estrogen is required for the cytodifferentiation of the glandular tissue and is subsequently required for the maintenance of optimal metabolic activity. One of the salient features of optimal metabolic activity in this organ is the synthesis and secretion of egg-white proteins. One of these proteins, ovalbumin, comprises nearly 60 percent of the protein synthesized in the mature hen oviduct and a similar proportion of the oviduct protein of chicks treated with estrogen for 15–18 days. Thus, ovalbumin, because of its abundance, has been relatively easy to isolate and purify to homogeneity. Unlike estrogen, progesterone causes neither cytodifferentiation nor major changes in any metabolic event in the oviduct. However, progesterone has been shown to specifically control the synthesis of the egg-white protein, avidin. Avidin represents no more than 0.1 percent of the total egg-white protein. On the other hand, this protein, too, has been relatively easy to quantify because of its unique ability to bind biotin with high affinity (Korenman and O'Malley, 1967). In order to demonstrate whether estrogen and progesterone regulated the synthesis of ovalbumin and avidin, respectively, at the level of mRNA production, two tasks remained to be completed. It was necessary first to isolate an mRNA fraction from the oviduct and second to select a suitable protein-synthesizing system that would be capable of translating the oviduct mRNA with high fidelity.

In our earlier studies it has been demonstrated that estrogen brings about an increase in the synthesis of oviduct ribosomes within 1 day of administration to the immature chick (Means *et al.*, 1971). For at least 7 days of hormone treatment, the oviduct content of ribosomes continued to increase, but by 10 days had begun to decline. Concomitant with the increased synthesis of ribosomes are estrogen-induced changes in the distribution of ribosomes and polysomes analyzed by sucrose gradient centrifugation. A large proportion of these ribosomal particles (RNP) exist as monomers in the unstimulated oviduct. On the other hand, after 4 days of treatment with estrogen, more than 90 percent of the cytoplasmic RNP particles exist as aggregates of two or more ribosomes.

Ribonucleoprotein preparations were tested for their ability to synthesize protein in an *in vitro* cell-free protein-synthesizing system (Means *et al.*, 1971). Estrogen administration resulted in a doubling of incorporation activity within 24 hours. By 4 days of hormone treatment, polysome protein synthesis assayed *in vitro* reached a maximum; it began to decline at 7 days of estrogen treatment. The marked stimulation of incorporation activity at 4 days is in keeping with the striking increases in ribosome synthesis and conversion of monomers to polysomes noted at the same time. Again, the decline in protein synthesis *in vitro* occurs in concert with the decreased synthesis of ribosomes and a further shift in the polysome pattern.

Only 25–35 percent of the radioactive protein was released from the ribosomes following incubation in the cell-free system (Means and O'Malley, 1971). In order to demonstrate that the radioactive material released into the supernatant fluid was present as completed tissue proteins, it was necessary to characterize and identify these peptides. Thus, a specific antibody for ovalbumin was prepared and characterized. The antibody was reacted with dialyzed supernatant fluid obtained from the cell-free system. A time-dependent increase in antibody-precipitable radioactivity accounted for 25 percent of the total acid-insoluble counts present in the ribosome-free supernatant fluid (Means and O'Malley, 1971). On the other hand, no radioactive material was precipitated by the ovalbumin antibody if the cell-free reaction included polysomes from unstimulated oviducts. These data suggested that it was possible to demonstrate the synthesis of an oviduct-specific protein in a polysomal cell-free system. Although immunological competence cannot be taken as conclusive evidence for identity, it is strongly suggestive. Furthermore, the fact that the polysomes from unstimulated oviduct fail to synthesize antibody-reacting material *in vitro* strengthens the argument since the oviduct of the unstimulated chick does not produce ovalbumin *in vivo*. The implication of these studies was that mRNA for ovalbumin is associated with ribosomes only during the period in which estrogen is being administered to the animals.

Further evidence for this assumption was obtained by incubation of estrogen-stimulated chick or hen oviduct minces *in vitro* in the presence of tritiated cytidine and adenosine (Means *et al.*, 1972). This results in a labeling pattern of polyribosomes in the region of 6–15 ribosome aggregates. Ovalbumin, which has a molecular weight of 45,000 daltons, would be translated on the average by a ribosome-to-mRNA ratio of 13 to 1. Since ovalbumin appears to be the major protein synthesized under these conditions, we had reason to believe that the label would serve as a marker for its mRNA. Extraction of these polysomes by detergent treatment followed by sucrose gradients centrifugation resulted in a typical RNA profile with the bulk of radioactivity found as a broad peak with a sedimentation value of 16–18S, which corresponds closely to the expected sedimentation value for the ovalbumin mRNA.

Direct assessment of mRNA can be obtained only by demonstrating the ability of an RNA fraction to support the *de novo* synthesis of a specific protein in an *in vitro* translation system. Utilization of a homologous protein-synthesizing system was determined to be less than optimal. Several reasons for this exist. In the first place, it is difficult to demonstrate that all mRNA fragments are absent from the ribosomes one uses for protein synthesis. Second, it is difficult to effect the reinitiation of protein synthesis in a completely defined translation system consisting of 40S and 60S ribosomal subunits and the various translation initiation factors. Third, it has been very difficult to isolate and characterize the three or more protein factors that are required for initiation of protein synthesis in eukaryotic cells (Comstock *et al.*, 1972a). Therefore, it seemed desirable to use a heterologous system in our initial attempts to synthesize specific proteins from oviduct RNA. At the time our studies were initiated, only one heterologous system had been well defined with respect to its ability to translate exogenous mRNAs. This translation system was a lysate of rabbit reticulocytes and was first described by Stavnezer and Huang (1971). As will be discussed later in this chapter, the reticulocyte lysate is probably neither the easiest nor the most efficient heterologous translation system to use for most studies. As will be seen, one of the largest problems is that this system synthesizes globin from endogenous mRNA very efficiently.

Prior to addition of an 8–18S fraction of RNA isolated from chick oviduct, radiolabeled valine is incorporated almost entirely into globin chains. Addition of chick RNA results in the appearance of a ^{14}C-labeled protein peak that is coincident with authentic ovalbumin upon analysis by polyacrylamide gel electrophoresis. Further proof that the reaction product was authentic ovalbumin has been gained by several procedures (Means *et al.*, 1972): (a) interaction with a specific antiserum to purified ovalbumin; (b) solubilization of the immunoprecipitate and analysis on sodium dodecyl sulfate gels; (c) ion exchange chromatography on carboxymethyl cellulose followed by reprecipitation with antiovalbumin; and (d) the construction of peptide maps, which Rhodes *et al.* (1971) demonstrated were similar to those prepared from authentic ovalbumin. Synthesis of radioactive ovalbumin shows linear dependence on exogenous oviduct RNA preparations (Rhodes *et al.*, 1971; Rosenfeld *et al.*, 1972a). Moreover, the ovalbumin mRNA activity was specific for RNA isolated from oviduct of estrogen-stimulated chicks or from laying hens and was primarily found in the 8–17S fraction of polysomal RNA. The amount of synthesis was increased by addition of protein synthesis initiation factors prepared from rabbit reticulocytes. In addition, inhibitors of chain initiation such as edeine or aurintricarboxylic acid or of general protein synthesis such as puromycin or cyclohexamide completely block ovalbumin synthesis directed by the oviduct mRNA fraction (O'Malley *et al.*, 1974). Ribonuclease destroys the messenger activity, while deoxyribonuclease does not. Steroid hormone receptor complexes had no demonstrable effect on the translation of ovalbumin mRNA.

This system for translating the ovalbumin mRNA with fidelity allowed us to look at the hormonal regulation of the specific messenger (Means *et al.*, 1972). Oviduct from laying hens in which ovalbumin is being synthesized at its maximal rate contains the greatest amount of ovalbumin mRNA. On the other hand, there is no detectable mRNA for ovalbumin in nucleic acids extracted from the unstimulated immature oviduct of the 7-day-old chick. Stimulation of these animals with estrogen for 4, 10, or 16 days leads to increasing activity of the extractable messenger for ovalbumin synthesis. However, when chicks treated with estrogen for 16 days are subsequently withdrawn from hormone treatment for 16 days, the ovalbumin mRNA activity again becomes very low. Finally, the administration of estrogen to these animals for 1, 2, or 4 days after the 16-day withdrawal period leads once more to a progressive increase in the amount of ovalbumin messenger. These data reveal that the amounts of extractable ovalbumin mRNA from oviduct are indeed directly dependent upon estrogen stimulation.

When ovalbumin and ovalbumin mRNA are measured in the same tissue samples, a striking correlation over a period of 0–17 days can be demonstrated between ovalbumin accumulation in the stimulated oviduct and its mRNA activity (Comstock *et al.*, 1972b). This relationship required clarification, however, since during the same time of differentiation and growth there are other dramatic changes occurring, particularly in the cellular content of total nucleic acid. Therefore, in order to better study the control of ovalbumin synthesis, we used oviduct minces from chicks that had been withdrawn from estrogen for 16 days and then killed at various times following readministration of a single dose of this steroid hormone. The rate of ovalbumin synthesis was assayed by incubating oviduct slices for 2 hours in the presence of radioactive valine (Chan *et al.*, 1973). Synthesis of ovalbumin was again quantified using the specific-antibody procedure. The rate of ovalbumin synthesis was found to be time dependent, peaking at 18 hours after steroid induction, at which time the rate of synthesis begins to decline. An approximate half-life of 8–10 hours for the ovalbumin mRNA can be calculated from the decline in the rate of ovalbumin synthesis. Ovalbumin mRNA was then extracted from oviduct and quantified during this same period following a single injection of estrogen. A remarkable parallelism exists between the changes in the rate of ovalbumin synthesis and the available mRNA. Prior to injection of estrogen at zero time, very little ovalbumin mRNA was detected. Maximal induction occurred at 18 hours, and mRNA content returned to barely detectable levels at 72 hours. From the decline in the activity of translatable mRNA, it can be calculated that again the half-life of ovalbumin mRNA appears to be 8–10 hours. These studies indicate that estrogen acts at the level of gene transcription, leading to the accumulation of a specific mRNA during differentiation of the oviduct. The appearance of the message seems to be a rate-limiting factor in determining the rate and extent of synthesis of this tissue-specific protein, ovalbumin (O'Malley and Means, 1974).

Ovalbumin mRNA was chosen in our initial studies designed to investigate the hormonal regulation of tissue-specific mRNAs, since ovalbumin was present in such high concentrations in oviduct. This, of course, is a very specialized instance, and in most cases proteins are present in much lower concentration. In this regard, the egg-white protein, avidin, which is specifically induced by progesterone, represents only about 0.1 percent of the total oviduct protein. Therefore, it followed that the mRNA for this protein might also be present in considerably smaller amounts than that for ovalbumin. Indeed, extraction of total RNA from estrogen-stimulated hen oviduct proved to be less than satisfactory as a means of quantitation of avidin mRNA. When these RNA preparations were tested in the rabbit reticulocyte lysate, it was not always possible to demonstrate avidin synthesis by a specific immunoprecipitation procedure. Subsequently, it was demonstrated that reproducible results could be assured by effecting a partial purification of the mRNA fraction (Rosenfeld *et al.*, 1972b)

We were able to take advantage of the fact that most mRNAs, including the ones for avidin and ovalbumin, contain at the 3′ terminal end an extensive sequence of polyadenylate residues. The presence of this poly-A sequence was shown initially by Brawerman *et al.* (1972) to allow the mRNA to be selectively adsorbed to nitrocellulose filters. Application of this procedure to oviduct RNA results in a one-step fiftyfold purification of both avidin and ovalbumin mRNAs (Rosenfeld *et al.*, 1972b). This simple procedure allowed us to measure routinely and consistently the avidin mRNA activity that appears in oviduct in response to progesterone.

Avidin mRNA is present only in the 8–17S fraction of RNA extracted from oviduct polysomes (O'Malley *et al.*, 1972). In addition, it has been shown by sucrose gradient analysis to have an average sedimentation coefficient of 9S. This would be expected if the message were to code for a protein of approximately 15,000 daltons. This is, in fact, the molecular weight of a single subunit of avidin. Avidin mRNA activity is abolished by ribonuclease, and no avidin is synthesized when inhibitors of peptide chain initiation or elongation are present in a cell-free system.

Avidin mRNA activity is highest in oviducts of mature laying hens, where progesterone stimulation is maximal (O'Malley *et al.*, 1972). On the other hand, no activity can be demonstrated in the unstimulated immature chick or in oviducts from animals that have received multiple injections of estrogen. However, after a single injection of progesterone, avidin mRNA activity is readily detected. Good correlation between avidin mRNA activity and avidin synthesis was found following a single injection of progesterone to estrogen-stimulated chicks. Although the assay method was not sufficiently sensitive to measure the *in vivo* rate of avidin synthesis, we were able to determine total accumulation of avidin with the specific and very sensitive biotin assay. Consequently, by calculating a first derivative plot of the observed avidin accumulation, we were able to obtain the theoretical rate of avidin synthesis (O'Malley *et al.*, 1974). Avidin mRNA activity was first detected at 6 hours following a single injection of progesterone and continued to increase until approximately 24 hours. The avidin mRNA levels increased prior to the accumulation of avidin and coincident with its increased rate of synthesis. In contrast to the estrogen-mediated changes in ovalbumin mRNA, progesterone induction of avidin mRNA and avidin synthesis occurs with little or no change in net cellular RNA and protein synthesis. However, these results suggest that both estrogen and progesterone act in the oviduct to alter gene transcription in a manner that leads to the production of specific mRNAs (O'Malley and Means, 1974).

In order to continue our studies on the kinetics of synthesis and properties of the ovalbumin mRNA, a more highly purified preparation of mRNA was required. The initial purification scheme consisted of passing a total nucleic acid extract through nitrocellulose filters (Means *et al.*, 1972; Rosenfeld *et al.*, 1972b). The adsorbed RNA was recovered by elution with a pH 9.0 buffer containing detergent. The recovered RNA usually represents 1–2 percent of the total material applied to the filter and represents a 30–50-fold enrichment in mRNA for ovalbumin. Analysis of this RNA fraction on sucrose gradients reveals the presence of ribosomal RNA as well as DNA. DNA can be removed by chromatography of the filter-bound nucleic acid extract on sepharose 4 B. Under the conditions used, DNA is excluded from the column and all the detectable ovalbumin mRNA activity appears in a peak slightly preceding the 18S ribosomal RNA peak. Much of the remaining 18S ribosomal contaminant can be removed by fractionation of the RNA on sucrose gradients containing formamide, which prevents RNA aggregation. In these gradients, the ovalbumin mRNA activity runs as a sharp band at 17–18S. Finally, the 17–18S material from the formamide gradients can be purified by electrophoresis in 2 percent agarose gels containing urea. This electrophoretic procedure cleanly separates the ovalbumin mRNA activity from ribosomal RNA, and the resulting single band of ovalbumin mRNA is apparently homogeneous. Thus, a purification procedure based on adsorption of poly-A-rich mRNA and precise sizing techniques can generate reasonable quantities of highly purified ovalbumin mRNA. This purity was considered sufficient to begin studies designed to show that this steroid-hormone-induced mRNA is transcribed from a single copy (unique) DNA sequence.

The fact that ovalbumin represents such a high proportion of the total protein in the oviduct cell suggested the possibility that gene amplification might be involved in the control of the synthesis and accumulation of this protein. At present, the most sensitive and specific way to detect small numbers of mRNA-specific sequences is to utilize a complementary DNA copy of a purified mRNA (Kacian *et al.*, 1971). This was accomplished in our system by incubating the purified ovalbumin mRNA with RNA-directed DNA polymerase (reverse transcription) that had been isolated from avian myeloblastosis virus (Harris *et al.*, 1973). This reaction produced a copy of radioactively labeled DNA of high specific activity that would be the complement of the ovalbumin mRNA. The tritium-labeled DNA produced was sized on alkaline sucrose gradients and shown to contain fragments that ranged in size from 60–1,600 nucleotides, with an average length of 220 nucleotides. When the tritium-labeled DNA was reacted with excess ovalbumin mRNA, 90 percent of the labeled DNA formed a stable hybrid with the mRNA, indicating that the tritiated labeled DNA was indeed a complementary copy. A fraction of this tritiated complementary DNA containing approximately 400 nucleotides was then used in a DNA excess hybridization experiment. Whole-chick DNA was sheared to

400-nucleotide lengths and incubated with the tritiated labeled DNA at an excess of 10^7 to 1. Complementary tritiated labeled DNA hybridized to chick DNA with a $Cot_{1/2}$* of 480. Under similar conditions of second-order kinetics, single-copy or unique-sequence DNA hybridizes with a $Cot_{1/2}$ of 420. Furthermore, similar results were found when unlabeled chick liver DNA was used in place of chick oviduct DNA. Finally, identical data have been obtained using ovalbumin mRNA purified by a specific immunoabsorption technique (Palacios *et al.*, 1973) as a template for Rous sarcoma virus reverse transcriptase (Sullivan *et al.*, 1973). These data suggest that the ovalbumin gene is not amplified but rather is present only once in the chick genome. These data suggest, then, that estrogen may act at the level of transcription to stimulate production of numerous copies of a single gene. This type of hormonal regulation would lead to a high intracellular concentration of ovalbumin mRNA and subsequently of ovalbumin itself.

Tritiated complementary DNA can also be used as a specific probe for ovalbumin mRNA sequences. Whereas even the most sensitive *in vitro* translation system requires the presence of several thousand mRNA molecules per cell, the sensitivity of the complementary DNA probe should be able to detect as few as one molecule in several thousand cells. Thus, we have used the complementary DNA (cDNA) copy of the ovalbumin mRNA to calculate the exact number of mRNA copies in each oviduct tubular gland cell prior to and during hormonal simulation with diethylstilbestrol (DES). When this probe is applied to the oviduct of the mature laying hen, which should be synthesizing ovalbumin at a maximal rate, it can be calculated that approximately 80,000 copies of the ovalbumin mRNA are present in each tubular cell (Table 1). On the other hand, in the oviduct of an unstimulated 7-day-old chick less than one molecule per cell can be detected. Following 4 days of estrogen treatment, the concentration of ovalbumin mRNA increases to approximately 17,000 molecules per cell and reaches 40,000 molecules per cell by about 18 days of estrogen stimulation. When hormone treatment is discontinued, the mRNA level declines to less than 2 molecules per cell. Finally, as shown in Table 2, readministration of a single injection of estrogen results in a significant increase in ovalbumin mRNA sequences within 30 minutes, and by 4 hours there are 500 molecules per tubular cell. These data strongly suggest that, at least in the case of estrogen stimulation of ovalbumin synthesis in the chick oviduct, the hormone is acting at the level of transcription to control the relative concentration

* $Cot_{1/2}$ is half-renaturation of single-stranded DNA to double-stranded DNA.

TABLE 1 Chronic Induction of $mRNA_{ov}$ with Estrogen[a]

Hormonal State	No. Mols. $mRNA_{ov}$/Cell	% Tubular Gland Cells	No. Mols./ Tubular Cell
Unstimulated	0.2–0.5	0	—
4 days DES	5,000	30	16,670
9 days DES	15,000	50	30,000
18 days DES	30,000	75	40,000
Hen	66,000	85	77,650
Withdrawn	0.5–1.0	20	2

[a] Chicks 7 days of age received daily SC injections of estrogen (DES). Total nucleic acids were extracted and ^{3}H-cDNA was employed to determine the number of ovalbumin mRNA sequences by mRNA excess hybridization. The standard mRNA was from hen oviduct and the kinetics of hybridization revealed a $Cot_{1/2}$ of 2×10^{-2}. Withdrawn means that chicks were stimulated for 18 days with DES, and then hormone treatment was discontinued for 14 days. Percent tubular gland cells was determined by counting the cells in cross-sections of oviduct prepared at the various stages of differentiation shown.

of ovalbumin mRNA molecules; this argues against any major role for translational control in this particular steroid-hormone-regulated system.

Molecular probes such as the tritiated labeled cDNA will remain a powerful tool for monitoring transcription of specific genes. However, it is clear that such probes can be prepared only under ideal circumstances. That is, one must have a specific mRNA and it must be purified to homogeneity. Clearly, in investigating hormone-regulated transcription, such purified products are not always available. The

TABLE 2 Acute Induction of $mRNA_{OV}$ with Estrogen[a]

Hormonal State	No. Mols. $mRNA_{ov}$/Tubular Cell
Withdrawn	2
0.5 hr after DES	10
1 hr after DES	25
4 hr after DES	500
8 hr after DES	1,250
29 hr after DES	5,000

[a] Seven-day-old chicks were given daily injections of DES for 18 days. Subsequently, no hormones were administered for 14 days. At this time a single injection of DES was administered and chicks were sacrificed at various times as detailed. Ovalbumin mRNA sequences were quantitated as described in the note to Table 1.

second major problem also relates to the specificity of the hormone. If the hormone stimulates the production of one or more specific proteins, these proteins can then be isolated and purified and antibodies can be prepared against them. Thus, one has a specific radioimmunoassay to monitor and one can quantitate the specific protein synthesized in an *in vitro* translation system in response to added mRNA fractions at various stages of purification. Under such optimal conditions, any translation system can be used, and it should be possible eventually to prepare a tritiated cDNA probe to the mRNA under investigation.

As mentioned previously, when one begins to investigate the effects of any hormone on protein synthesis by target tissue and one is interested in determining whether mRNA is involved, methods must be established for assaying mRNA. The preparation of an mRNA fraction is no longer a problem. This is very similar in any tissue examined. Thus, a total nucleic acid extract followed by a simple millipore filter assay should enrich the mRNA 30–50-fold (Rosenfeld *et al.*, 1972; Brawerman *et al.*, 1972). This is going to be true for any mRNA that contains a polyadenylate sequence at a 3′ terminal end, and to date the only mRNAs from eukaryotic cells that have been demonstrated to lack this sequence are the histone mRNAs, which are transcribed from reiterated sequences in the genome (Schochetman and Perry, 1972; Adesnik and Darnell, 1972). The second problem is to decide upon a translation system for quantitating the mRNA activity. Several systems have now been described. These include a lysate of reticulocytes (Means *et al.*, 1972; Stavnezer and Huang, 1971), an S-30 preparation of ascites tumor cells (Mathews and Korner, 1970; Schutz *et al.*, 1973), the frog oocyte (Gurdon *et al.*, 1972), and, finally, an S-30 preparation derived from wheat germ (Marcus *et al.*, 1970). Each of these systems has advantages and disadvantages. The reticulocyte lysate and the frog oocyte system suffer from a common drawback. That is, one must measure a specific translation product coded for by the exogenously added mRNA because both of these translation systems have a considerable amount of endogenous mRNA activity, and even under ideal conditions only about 10 percent of the translation products will be directed by the exogenous mRNA. Other major disadvantages of the reticulocyte lysate are that it requires considerable effort to obtain it and is rather expensive. Rabbits must be purchased and subjected to an injection schedule that renders them anemic, blood has to be collected, and the lysate of the reticulocytes prepared. Second, there is a considerable amount of variation in translation efficiency from preparation to preparation. In other words, one lysate may translate exogenous mRNA very efficiently, whereas another preparation may not work nearly as well. Moreover, there seems to be no good way of predicting which preparation is going to work and which is not. On the other hand, the reticulocyte lysate is very efficient in translating mRNAs and will release better than 90 percent of all completed peptide chains.

The frog oocyte system offers the advantage of being essentially an *in vivo* situation that is extremely sensitive to low concentrations of added mRNA. However, it is extremely expensive to set up. One must purchase specialized equipments and have a continuing supply of *Xenopus laevis* or other suitable egg donors. Preparation of micro-pipettes suitable to deliver nanoliter amounts of fluid is difficult, and the injection of the egg to assure adequate survival rates is also a demanding task.

The ascites cell and wheat germ systems have similar characteristics. The major disadvantage of the ascites cell is that one has to have a continuing supply of tumor-bearing animals, the cells must be harvested and again, considerable variation in the efficiency of preparations has been noted. A rather curious and still unexplained finding in both of these systems is that a considerable proportion of the radioactive amino acid incorporated into TCA-precipitable material remains associated with the ribosomes. The efficiency of release is only 25–50 percent. The major advantage of both these systems is that they have very low levels of endogenous mRNA. Therefore, one can ask whether a hormone causes any changes in the qualitative or quantitative nature of the protein synthesized in the target tissue. This can be determined by simply making a crude mRNA preparation, adding it to the translation system, spinning out the ribosomes following incubation, and distributing the radioactive peptides released into the medium on SDS-acrylamide gels.

All things taken into consideration, the system of choice for beginning studies into mRNA translation is the wheat germ. This system is the simplest, the easiest to set up, the most fail-safe, and the most stable. This system was first described by Marcus *et al*. (1970). Briefly, preparation in our laboratory consists of grinding wheat germ with sand, adding a buffer, spinning it down at 30,000 *g*, passing the supernatant fluid through a column of Sephadex G25 and pipetting aliquots into liquid nitrogen (Roberts and Paterson, 1973). The aliquots of aqueous material form balls of known volumes that can simply be picked up with forceps and placed in containers. When stored in liquid nitrogen, such preparations lose no activity for at least 6 months.

Figure 1 shows that the wheat germ S-30 system will translate with fidelity ovalbumin from added ovalbumin mRNA. Synthesis of ovalbumin is dependent both upon the amount RNA added and upon the time

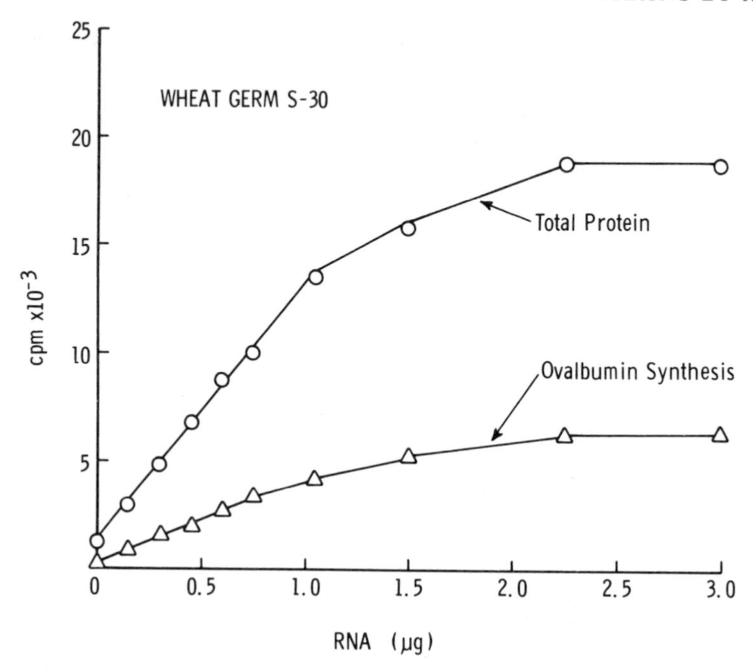

FIGURE 1 Synthesis of ovalbumin in a wheat embryo translation system primed with hen oviduct ovalbumin mRNA. Various amounts of ovalbumin mRNA (Means *et al.*, 1972; Rosenfeld *et al.*, 1972b) were added to an S-30 wheat embryo fraction (Marcus *et al.*, 1970; Roberts and Paterson, 1973) and incubated for 2 hr at 25° with ^{14}C-valine as a radioactive tracer. Total protein synthesis was determined from the total radioactivity remaining after acid precipitation and treatment of the pellet at 90°. Ovalbumin was measured by a specific antibody procedure previously documented (Means *et al.*, 1972).

of incubation. As demonstrated in Figure 2, most S-30 preparations will incorporate amino acids into ovalbumin linearly for a period of 2–3 hours. Furthermore, radioactive ovalbumin is first demonstrable on the ribosomal pellet and then in the supernatant, as would be expected. Total protein synthesis has ceased by approximately 2 hours in this experiment. These and other data suggest that the reason for the cessation of protein synthesis in the wheat germ system is a failure in the initiation of new peptide chains, not the completion of those chains that already existed on the polyribosomes. Finally, under optimal conditions, approximately 50 percent of the total protein released into the supernatant is ovalbumin (Table 3).

The general applicability of the wheat germ translation system to various mRNA preparations is shown in Figure 3. mRNA was prepared from hen oviduct, rabbit reticulocytes, and lactating rat mammary glands. mRNA fractions from each of these cell types were incubated in the S-30 system, and the completed peptide chains synthesized *de novo* on the wheat germ ribosomes were analyzed by electrophoresis on SDS-acrylamide gels. As can be seen, the patterns obtained with all three preparations of RNA were different. The protein products obtained using mRNA from reticulocytes show only a single peak that

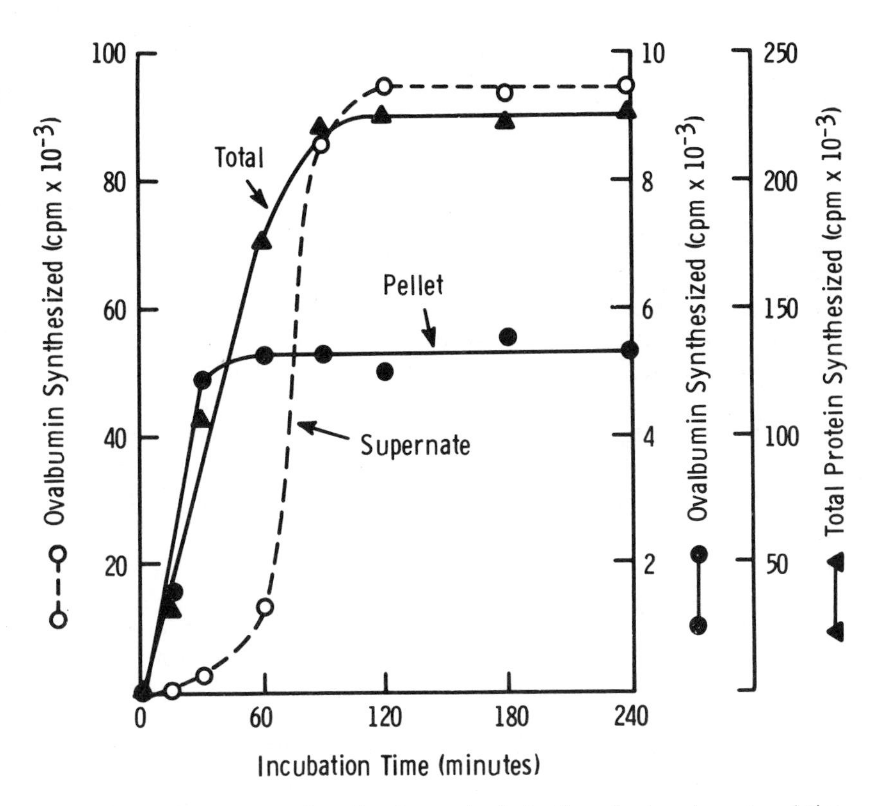

FIGURE 2 Time course of ovalbumin synthesis in the wheat embryo translation system. Aliquots of wheat embryo S-30 (Marcus *et al.*, 1970; Roberts and Paterson, 1973) were incubated with hen oviduct ovalbumin mRNA (Means *et al.*, 1972; Rosenfeld *et al.*, 1972b) at 25° for various times in the presence of ^{35}S-methionine. Small aliquots were removed for determination of total protein synthesis (▲ – ▲), and the remainder of the reactions were centrifuged at 105,000 *g*. Ovalbumin assays were then performed on both the pellet (● – ●) and the supernatant fluid (○---○).

TABLE 3 Translation of Ovalbumin mRNA in an S-30 System Derived from Wheat Embryo[a]

Fraction	Protein Synthesized		
	Total cpm	Ovalbumin cpm	% Ovalbumin
Total	15,017	4,610	30.7
Released protein	6,413	3,358	52.3
Ribosome-bound protein	7,887	1,857	23.5

[a] The S-30 wheat embryo system has been previously described (Marcus *et al.*, 1970; Roberts and Paterson, 1973), as has the preparation of partially purified ovalbumin mRNA from hen oviduct (Means *et al.*, 1972; Rosenfeld *et al.*, 1972b). Following incubation for 2 hr at 25° samples were taken for determination of total protein synthesized and ovalbumin synthesized (Means *et al.*, 1972). Replicate tubes were centrifuged at 105,000 *g* to yield a supernatant fraction (released protein) and a ribosomal pellet (ribosome-bound protein).

matches that seen in chromatographs of authentic globin. When hen oviduct RNA is translated, the proteins synthesized show a major component that migrates to a position on the gel slightly smaller than authentic ovalbumin. It should be pointed out that ovalbumin is a glycoprotein, and it is unlikely that the wheat germ system has a mechanism for attaching the carbohydrate moiety. Therefore, it is not surprising to see a slight discrepancy in the behavior of authentic and newly synthesized ovalbumin upon analysis by gel electrophoresis. Finally, a lactating rat mammary gland RNA was also translated. Again, one major component is apparent that runs slightly lighter than a marker of bovine α-casein. Casein is also a glycoprotein, so the results are not unexpected. We are also conducting studies with mRNA prepared from rat testis in attempts to determine whether follicle-stimulating hormone affects the level of testicular mRNA. Again, the wheat germ system is both sensitve and specific enough to accurately assay these mRNA preparations, even though in this system no specific product can be analyzed. Finally, it should be pointed out that the wheat germ S-30 translation system is specific for mRNA. Addition of transfer or ribosomal RNA has absolutely no effect on the very low basal level of the incorporation of amino acids into protein. Furthermore, all translation of exogenous RNA is inhibited by the addition of peptide inhibitors of peptide chain initiation or chain elongation. Taking all advantages and disadvantages into consideration, we would suggest that initial studies designed to probe the effects of hormones or other regulatory agents on mRNA investigators take advantage of this simple and efficient translation system derived from wheat embryo.

In this paper, we have attempted to present evidence supporting the hypothesis that at least two steroid hormones, estrogen and progesterone, activate specific genes and allow transcription of new species of mRNA that code for synthesis of specific proteins in the chick oviduct. Although we have direct quantitative evidence that these steroid hormones cause a net increase in the intracellular levels of specific mRNA molecules, much less is known about the mechanisms of specific gene regulation (O'Malley and Means, 1974). Steroid hormones appear to enter cells by simple passive diffusion and bind with high affinity to a specific cytoplasmic receptor protein. An activation of the receptor complex occurs at the same time as translocation to the

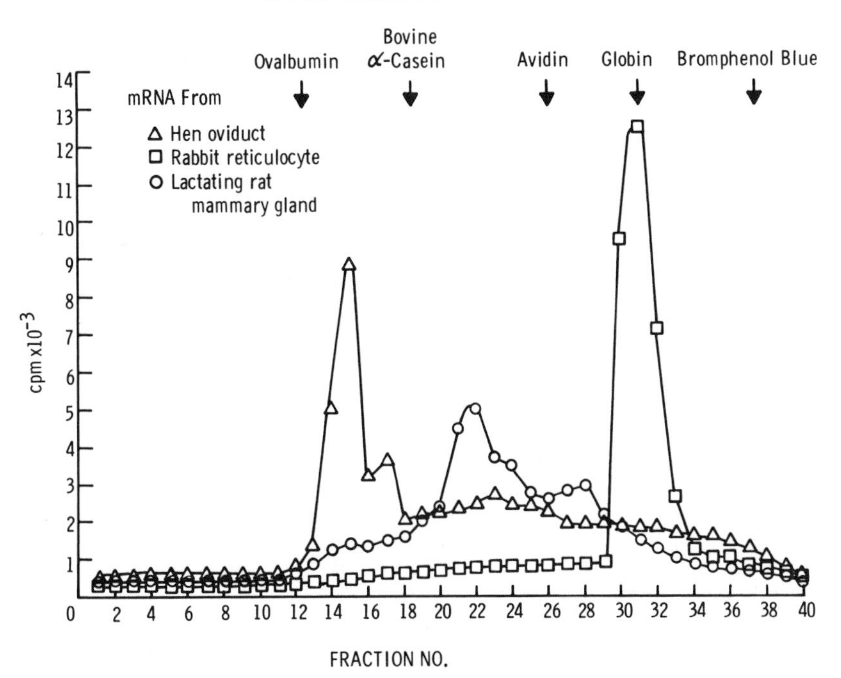

FIGURE 3 Translation of mRNA from various sources in the wheat embryo translation system. Messenger RNA fractions were prepared from hen oviduct, rabbit reticulocytes, or mammary glands of lactating rats (Means *et al.*, 1972; Rosenfeld *et al.*, 1972b). These mRNAS were incubated separately in the wheat embryo S-30 system for 2 hr at 25° with ^{35}S-methionine. Following incubation, samples were centrifuged at 105,000 *g*. The supernatant fluid was dialyzed overnight and analyzed on SDS-acrylamide electrophoresis. Gels were sliced at 1 mm, and radioactive content of each slice was determined. The arrows represent the position of authentic protein standards.

nuclear compartment and seems to be dependent upon both salt and temperature. This activated nuclear receptor can then be shown to undergo a hormone-dependent binding reaction with chromatin. This receptor binds to chromatin acceptor sites with high affinity. The acceptor sites appear to be primarily comprised of the DNA backbone, but it also seems certain that specific chromosomal nonhistone proteins quantitatively modify this binding to DNA. Therefore, careful studies of the interaction of steroid receptor complexes with isolated chromatin should be coupled with the detection of specific gene sequences transcribed from chromatin templates using endogenous RNA polymerase.

Ultimately, then, we should be able to characterize those molecules important in hormone-mediated gene activation. Indeed, the attainment of this goal may be imminent in view of the rapid progress recently made in the areas of molecular biology and hormonal regulation of specific protein synthesis. On the other hand, the enormous complexity of steroid hormone action at the molecular level must not be overlooked. One has only to consider that the model presented above deals only with induction of a specific protein. It certainly does not explain the coordinated changes in mRNAs for other structural genes, rRNA, tRNA, the synthesis of RNA polymerase, initiation and elongation factors required for translation, and the many other proteins necessary for hormone-mediated growth and differentiation. The future understanding of these processes will depend on our ability to disassemble and reconstitute the biochemical components of the transcriptional apparatus, with the ultimate goal of understanding the direct effects of hormones in a chemically defined cell-free system.

REFERENCES

Adesnik, M., and J. E. Darnell. 1972. J. Mol. Biol. 67:397.

Brawerman, G., J. Mendecki, and S. Y. Lee. 1972. Biochemistry 11:637.

Chan, L., A. R. Means, and B. W. O'Malley. 1973. Proc. Nat. Acad. Sci. U.S.A. 70:1870.

Comstock, J. P., B. W. O'Malley, and A. R. Means. 1972a. Biochemistry 11:646.

Comstock, J. P., G. C. Rosenfeld, B. W. O'Malley, and A. R. Means. 1972b. Proc. Nat. Acad. Sci. U.S.A. 69:2377.

Gurdon, J. B., C. D. Lane, H. R. Woodland, and G. Marbaix. 1972. Nat. New Biol. 236:7.

Harris, S. E., A. R. Means, W. M. Mitchell, and B. W. O'Malley. 1973. Proc. Nat. Acad. Sci. U.S.A. 70:3776.

Jacob, F., and J. Monod. 1961. J. Mol. Biol. 3:318.

Kacian, D. L., K. F. Watson, A. Burny, and S. Spiegelman. 1971. Biochim. Biophys. Acta 246:365.

Korenman, S. G., and B. W. O'Malley. 1967. Biochim. Biophys. Acta 140:174.

Marcus, A., D. P. Weeks, J. Leis, and E. B. Keller. 1970. Proc. Nat. Acad. Sci. U.S.A. 67:1681.
Mathews, M. R., and A. Korner. 1970. Eur. J. Biochem. 17:328.
Means, A. R., I. B. Abrass, and B. W. O'Malley. 1971. Biochemistry 10:1561.
Means, A. R., J. P. Comstock, G. C. Rosenfeld, and B. W. O'Malley. 1972. Proc. Nat. Acad. Sci. U.S.A. 69:1146.
Means, A. R., and B. W. O'Malley. 1971. Acta Endocrinol. Suppl. 153:318.
Means, A. R., and B. W. O'Malley. 1972. Metabolism 21:357.
O'Malley, B. W., W. L. McGuire, P. O. Kohler, and S. G. Korenman. 1969. Rec. Prog. Horm. Res. 25:105.
O'Malley, B. W., A. R. Means, and M. R. Sherman. 1971. *In* The Sex Steroids (K. W. McKerns, ed.). Appleton-Century-Crofts, New York, p. 315.
O'Malley, B. W., and A. R. Means. 1974. Science 183:610.
O'Malley, B. W., G. C. Rosenfeld, J. P. Comstock, and A. R. Means. 1972. Nat. New Biol. 240:45.
O'Malley, B. W., L. Chan, S. E. Harris, J. P. Comstock, J. M. Rosen, and A. R. Means. 1974. *In* Control of Proliferation in Animal Cells (B. Clarkson and R. Baserga, ed.). Cold Spring Harbor Laboratories, New York. p. 85.
Palacios, R., D. Sullivan, N. M. Summers, M. L. Kiely, and R. T. Schimke. 1973. J. Biol. Chem. 248:540.
Rhodes, R. E., G. S. McKnight, and R. T. Schimke. 1971. J. Biol. Chem. 246:7407.
Roberts, B. E., and B. M. Paterson. 1973. Proc. Nat. Acad. Sci. U.S.A. 70:2330.
Rosenfeld, G. C., J. P. Comstock, A. R. Means, and B. W. O'Malley. 1972a. Biochem. Biophys. Res. Commun. 46:1695.
Rosenfeld, G. C., J. P. Comstock, A. R. Means, and B. W. O'Malley. 1972b. Biochem. Biophys. Res. Commun. 47:387.
Schochetman, G., and R. P. Perry. 1972. J. Mol. Biol. 63:591.
Schutz, G., M. Beato, and P. Feigelson. 1973. Proc. Nat. Acad. Sci. U.S.A. 70:1218.
Stavnezer, J., and R. C. Huang. 1971. Nat. New Biol. 230:172.
Sullivan, D., R. Palacios, J. Stavnezer, J. M. Taylor, A. J. Faras, M. L. Kiely, N. M. Summers, J. M. Bishop, and R. T. Schimke. 1973. J. Biol. Chem. 248:7530.

DISCUSSION

DR. ZUBAY: The restimulation with estrogen led to about 1,000 molecules within 30 minutes, which would mean 30 rounds of transcription per minute, roughly, if there is only a single genome. I just wanted to mention that that is, as far as I know, about 30 times faster than any structural gene can work in bacteria.

DR. MEANS: Yes, that is true. Good point. And the only thing that we can say about that is that under maximal stimulation 18 polymerase molecules are transcribing the ovalbumin gene.

DR. ZUBAY: The other point I wanted to bring out is that no matter how good your quantitation is at the present time, I do not think it is so good that you could exclude the possibility that some of these very hot genes have thrown off only one episomal copy. In the case of immunoglobin synthesis, it has been clearly established that the messenger that makes an immunoglobin comes from two different parts of the chromosome. Either the episomal copies are sloughed off from the chromosome, so that one gets two pieces of DNA that are ligased together to make this messenger, or there has to be a recombinational event at the gene level, or the message has to be made at the gene level in two pieces and connected together.

I think the possibility of an episome being sloughed off is a very real one that we should consider in cases like the ovalbumin reticulocyte and other cases where the quantitation in the hybridization is not really good enough to exclude this possibility. I think that one could get at this by fractionating the total nuclear DNA. If the episomal possibility exists, this fragment should be present in a low-molecular-weight piece of DNA.

DR. MEANS: That is, again, an excellent point. In fact, we have someone working on the possibility that in the chick oviduct there is an episomal DNA. And remember that in the unstimulated and in the withdrawn, there were still sequences that we could detect. They were low, but they were there. The question was what these mean. We cannot rule out the possibility that some of these chicks got a little estrogen. So that is certainly a point well taken.

One of the things that we are in the process of doing now is isolating the DNA sequences for this message by coupling our purified messenger RNA to an affinity column and then passing through the unique sequence DNA, in order to pull out those sequences.

DR. MUNRO: Have you any evidence of specific initiation factors that impede the wheat germ system? It seems at the moment to be very catholic in its acceptability. Has it been tried, for example, with messenger for muscle? S. M. Heywood, D. S. Kennedy, and A. J. Bester (1974. Proc. Nat. Acad. Sci. U.S.A. 71:2928) claimed that muscle initiation factors were necessary to get actin and myosin made.

The other question relates to the general background. Stage one seemed to be differentiation of cells involving cell division. Does stage two involve cell division when you reactivate the copies?

DR. MEANS: In answer to the first question, I am still a little bit leery of specific initiation factors. There may be some factors in eukaryotic cells that are helpers, that are not initiation factors 1, 2, and 3, but are helper molecules for various classes of message. But that there are message-specific ones, no, I do not think so. We cannot repeat Stu Heywood's results. We have tried and tried and tried, and I know other people who have, as well. I cannot say that they are wrong. I am simply saying that it is not possible for us in our laboratory to reproduce such results. There are others. I am sure Donald T. Wigle and Alan E. Smith at the Imperial Cancer Research Fund Laboratory, London, and Stephen C. Marker, U.S. Army Medical Research Institute for Infectious Diseases, Frederick, Md., and several other people have talked about the possibility of message specific initiation factors. We now have initiation factors 1 and 2 purified fairly well from oviduct. Initiation factor 1 has been purified to homogeneity by Dr. Comstock in our laboratory. This is the first demonstration of homogeneic initiation factor 1—to my knowledge, at least. Initiation factor 2 is very pure, and Dr. John Ficunding, who has joined me from Robert Traut's and John Hershey's laboratory (University of California at Davis) has done this by his phosphorylation technique that he described for *E. coli.*

We can add these back to the wheat germ system and they do not do anything. They just sort of sit there. We do not have purified M-3. The only purified M-3 we have is from reticulocyte that H. French Anderson (NIH, Bethesda, Md.) was kind enough to give us, and those experiments were done yesterday, and I do not know the results.

The second question that you asked had to do with DNA synthesis during primary and secondary stimulation. Certainly, during primary stimulation, DNA synthesis must occur; the mitotic index is increased some 100-fold from the resting primitive epithelial cell. There is a paper by Susan Socher and Bert O'Malley (1973. Develop. Biol. 30:411) that documents this.

During the restimulation, these cells are apparently quiescent. It is really very analogous during the early stimulation of a withdrawn animal to the rat uterine system or the androgen–prostate system. For the first 24 hours, there is no detectable increase in the incorporation of thymidine into DNA. It apparently is the same thing that happens when you give estrogen to the rat uterus. I am not exactly sure what that is. And so there is a stimulation, a very rapid stimulation in a lot of things.

We have been able to show by looking at specific initiation sites for RNA polymerase on the chromatin from these animals that very rapidly after you give estrogen back to these withdrawn animals you get approximately a fourfold increase in the number of specific initiation sites. This would suggest that it takes a very minimal rearrangement, and that this can happen very quickly, in order to begin expressing some of the genes that were temporarily repressed at the time that you withdrew the hormone.

Dr. Zubay can talk about repression and depression in this system. We still cannot talk about them, unfortunately, in ours.

DR. MERRICK: You mentioned that your initial CDNA probe was rather small based on the finding that people have been able to find poly-A stretches upwards of 150 or more nucleotides. Can you make any estimate of how much of the ovalbumin sequence you are actually incorporating beyond what you would normally anticipate for the bases involved only with poly-A, which presumably would not be specific for ovalbumin, but would be found with any message?

DR. MEANS: First of all, ovalbumin, as it is isolated in our laboratory, runs under several electrophoretic conditions as a single band and will back-hybridize 100 percent to the CDNA probe. The mRNA has approximately 70 adenines attached to the 3′ terminus. Our average product size from the reverse transcriptase action is about 400 nucleotides. Since the coding portion of ovalbumin in RNA requires 1,161 nucleotides, we would suggest that we are synthesizing about one-fourth of the total DNA molecule. We also have evidence that at the 5′ terminal end of ovalbumin there is some nontranslatable material. At this time we do not know what it is. However, Chuck

Liarakos, who has just finished his Ph.D. with us, is now doing a joint project with us and Harris Busch of the Baylor College of Medicine, Houston, to try to sequence this 5′ end to see what this might be.

DR. MARCUS: You did not talk much about the turn-on of ribosomal RNA that is concomitant at least with the first stimulation. First of all, does it turn on? Do you get that same kind of induction with your second stimulation? With regard to receptor estrogen, can you mimic the process *in vitro* and make ovalbumin messenger? Does it do the same thing for ribosomal RNA, too?

DR. MEANS: Well, yes, it does during primary stimulation by estrogen. Except avidin, I cannot think of a single molecular event that occurs in the oviduct that is not stimulated by estrogen. This is one of the problems, and this is one of the reasons that working with withdrawn–restimulated animals is much better.

One day of primary stimulation results in a concomitant increase in the number of ribosomes and the synthesis of ribosomal RNA in addition to the synthesis of message for ovalbumin. In the withdrawn–restimulated case within the same 1-day period there is no evidence of the synthesis of new ribosomes or ribosomal RNA. So we would say that the latter case is analogous to the rat uterine system with estrogen. However, the uterine system has recently been re-evaluated. It was thought that the initial RNA response was a stimulation of ribosomal RNA. That very clearly is not the case. There is an activation of RNA polymerase II that occurs first. This response is inhibitable by actinomycin D but not by cycloheximide. However, the subsequent increases in protein synthesis and RNA synthesis are inhibited by cycloheximide, suggesting that there is the very early synthesis of a few RNA species that must be necessary to produce some proteins, which then potentiate and amplify the effects of estrogen.

DR. DOY: You commented on the use of embryo from wheat. The whole plant field is plagued by the fact that one system does certain things, that if you go to another plant, well, too bad. Would you comment on whether you feel one could use embryos from other sources?

DR. MEANS: My collaborator, Dr. Cassius Bordelon, has also tried rice and barley embryos. The answer is yes.

DR. DOY: I am very relieved to hear it, because this could open up enormous potential. Since you mention barley embryo, I should mention that Sparrow and Shepherd (personal communication) of the Waite Institute, Adelaide, Australia, have regenerated very nice-looking homozygous diploid barley plants from haploid tissue cultures from an embryo.

J. E. VARNER, D. FLINT, *and* R. MITRA

Characterization of Protein Metabolism in Cereal Grain

PROTEIN SYNTHESIS IN THE ALEURONE LAYERS

The mobilization of the endosperm reserves of cereal grains during germination and early growth of the seedlings has received much attention (Yomo and Varner, 1971; Jones, 1973a) because this mobilization is initiated and controlled by gibberellins from the embryo. The target cells for these gibberellins are the aleurone layer cells. The response of these cells to gibberellins includes a proliferation of rough endoplasmic reticulum (Jones, 1969a; Vigil and Ruddat, 1973); swelling and dissolution of the protein bodies (Jones, 1969b); release of inorganic ions from the phytate globoids of the protein bodies and from the cells (Jones, 1973b); release of sucrose (Chrispeels *et al.*, 1973); release of existing phosphatase, peroxidase, esterase (Ashford and Jacobsen, 1974) and β-glucanase (Jones, 1972); and the synthesis and secretion of α-amylase (Filner and Varner, 1967), protease (Jacobsen and Varner, 1967), β-glucanase, and ribonuclease (Bennett and Chrispeels, 1972).

Although the marked changes in levels of specific enzymes are often observed, it is not generally appreciated to what extent gibberellins interrupt whatever was happening in the aleurone cells and redirect the

---

This work was initiated while J. E. Varner and D. Flint were at the MSU/AEC Plant Research Laboratory in East Lansing, Michigan, and was supported in part by the United States Atomic Energy Commission under contract no. AT(11-1)1338 and in part by the National Science Foundation (GB-33944).

cells toward an almost exclusive occupation with hydrolase synthesis and secretion. Of 12 proteins released from aleurone layers after treatment with gibberellic acid (GA), 10 become labeled if the layers are incubated in radioactive amino acids (Jacobsen and Knox, 1974). The other two bands may represent release from cell walls of proteins already secreted before the gibberellin treatment. Thus, most, or perhaps all, of the secreted proteins are synthesized in response to the added gibberellic acid. We have further examined by double labeling and SDS gel electrophoresis the control by gibberellic acid and abscisic acid of the synthesis of proteins by isolated aleurone layers. We also describe in this paper a new method for studying the biosynthesis of amino acids during the germination and early growth of seedlings.

MATERIALS AND METHODS

Aleurone layers from barley (*Hordeum vulgare* L. Cv. Himalaya) were prepared and incubated as previously described (Chrispeels and Varner, 1967) with the following exceptions: (a) the buffer solution used for incubation was 20 m*M* sodium succinate, 20 m*M* $CaCl_2$, pH 5.0; and (b) the layers were stripped after 2.5 days' imbibition. Two-hour periods were used for labeling. The double-labeling procedure was as follows: Layers were stripped and incubated (6 layers per flask) in incubation buffer for approximately 12 hours. Pairs of flasks were selected, the hormone added to one and the other left untreated as a control. After treatment for the specified time, 6 μCi of ^{14}C-leucine was added to the flask containing hormone and 40 μCi of ^{3}H-leucine was added to the control. Incubation continued for 2 hours after the addition of label. After the final incubation, the aleurone layers were removed from solution, combined, rinsed twice with water, and ground with mortar and pestle in 1.5 ml 0.2 *M* NaCl and 1.0 m*M* $KBrO_3$. The homogenate was centrifuged at 12,000 *g* for 10 minutes, the supernatant removed and saved, the pellet resuspended in 0.5 ml of the grinding solution, and centrifugation was repeated. The final supernatant was combined with that from the first centrifugation; this mixture contains what is referred to here as the salt-soluble proteins. The pellet was resuspended and extracted with a solution containing 8.0 *M* urea, 2 percent sodium dodecyl sulfate (Pierce Chemical Co., Rockford, Ill.), 5 percent 2-mercaptoethanol, 0.156 *M* tris (pH 6.8) overnight at room temperature. The supernatant from this final extraction contains what we refer to here as the salt-insoluble proteins. The salt-soluble protein mixture was diluted with 1 ml of the final extraction solution described above. The salt-soluble and salt-insoluble proteins were run on SDS gels using the discontinuous buffer system of Laemmli (1970). The gels were stained with coomassie brilliant blue, sliced into 1-mm-thick layers, digested in 0.6 ml NCS solubilizer (Amersham Searle, Arlington Heights, Ill.), and counted in a three-channel scintillation counter.

RESULTS AND DISCUSSION

The double-labeling technique, in conjunction with SDS gel electrophoresis, clearly shows the dramatic shift in the kinds of proteins being synthesized by aleurone layers in response to exogenous gibberellic acid. In both salt-soluble and salt-insoluble proteins, the shift is detectable within 2 hours of the addition of hormone (Figures 1–4) and marked within 2 to 4 hours (Figures 5 and 6). After 10 hours of treatment with gibberellic acid, there is no synthesis of those proteins characteristic of the minus hormone control tissue (Figures 3 and 4). The major labeled protein band in the salt-soluble proteins after treatment of the layers with gibberellic acid is α-amylase. The identity of the proteins in the other bands is not yet known. The addition of abscisic acid along with gibberellic acid prevents the gibberellic-acid-dependent shift in kinds of proteins being labeled (data not shown).

AMINO ACID SYNTHESIS IN THE EMBRYO

INTRODUCTION

Deuterium oxide in nontoxic concentrations has been used for the *in vivo* labeling of macromolecules such as nucleic acids and proteins. This is possible because of the nonexchangeability of the deuterium of the deuterium–carbon bonds that are formed during the biosynthesis of the monomeric precursors of such macromolecules. Deuterium oxide has the advantage as an *in vivo* label of freely and quickly entering all subcellular compartments. We here report a study of the biosynthesis in intact tissue of a number of amino acids using deuterium oxide as a tracer and the Gas–Liquid Chromatography–Mass Spectrometer (GLC–MS) system to determine the labeling of the individual amino acids. The extent and pattern of deuteration of each amino acid allow conclusions not only about the biosynthesis of the individual amino acids but also about their exposure to aminotransferases, dehydratases, dehydrogenases and other enzymes that might introduce deuterium into specific positions of the amino acids.

We have grown barley seedlings in the dark in deuterium oxide under two conditions (Figure 7): (a) germination and growth with the endo-

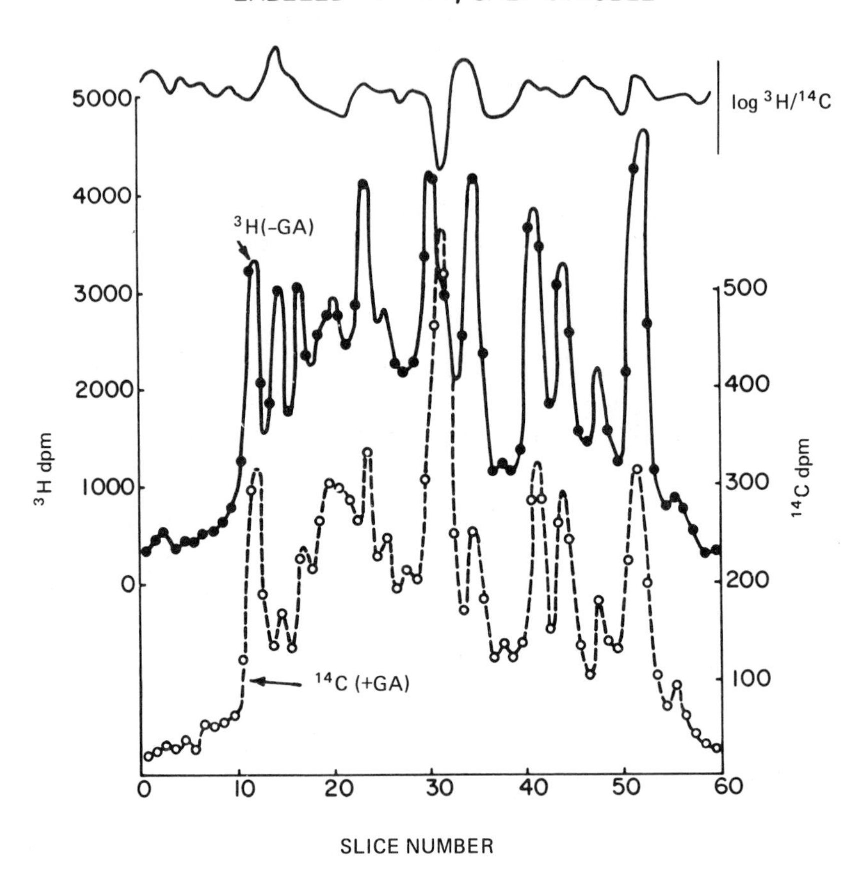

FIGURE 1 Labeling of the salt-soluble proteins of control and GA-treated aleurone layers. The layers were incubated with labeled amino acids for 2 hours beginning 4 hours after the start of the incubation of the layers in buffer or in buffer plus GA. The slice numbers refer to the slices obtained from the SDS gels following electrophoresis. Details were as described in the section on materials and methods.

sperm attached and (b) germination and growth of the excised embryo (no endosperm). Under the first condition we might expect the biosynthesis of at least some of the amino acids to be repressed or inhibited by the amino acids resulting from the hydrolysis of endosperm protein. Under the second condition we might expect all amino acid biosynthetic pathways to be functional.

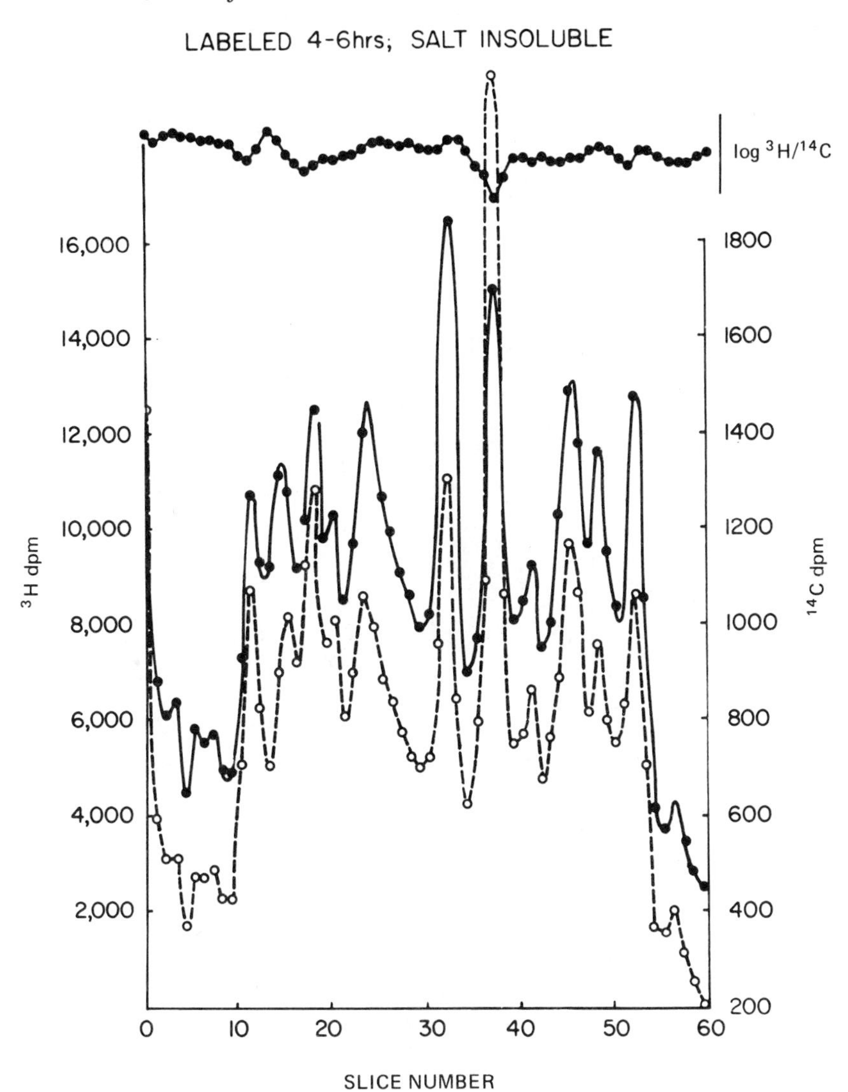

FIGURE 2 Labeling of the salt-insoluble proteins of control and GA-treated aleurone layers. The layers were incubated with labeled amino acids for 2 hours beginning 4 hours after the start of the incubation of the layers in buffer or in buffer plus GA. The slice numbers refer to the slices obtained from the SDS gels following electrophoresis. Details were as described in the section on materials and methods.

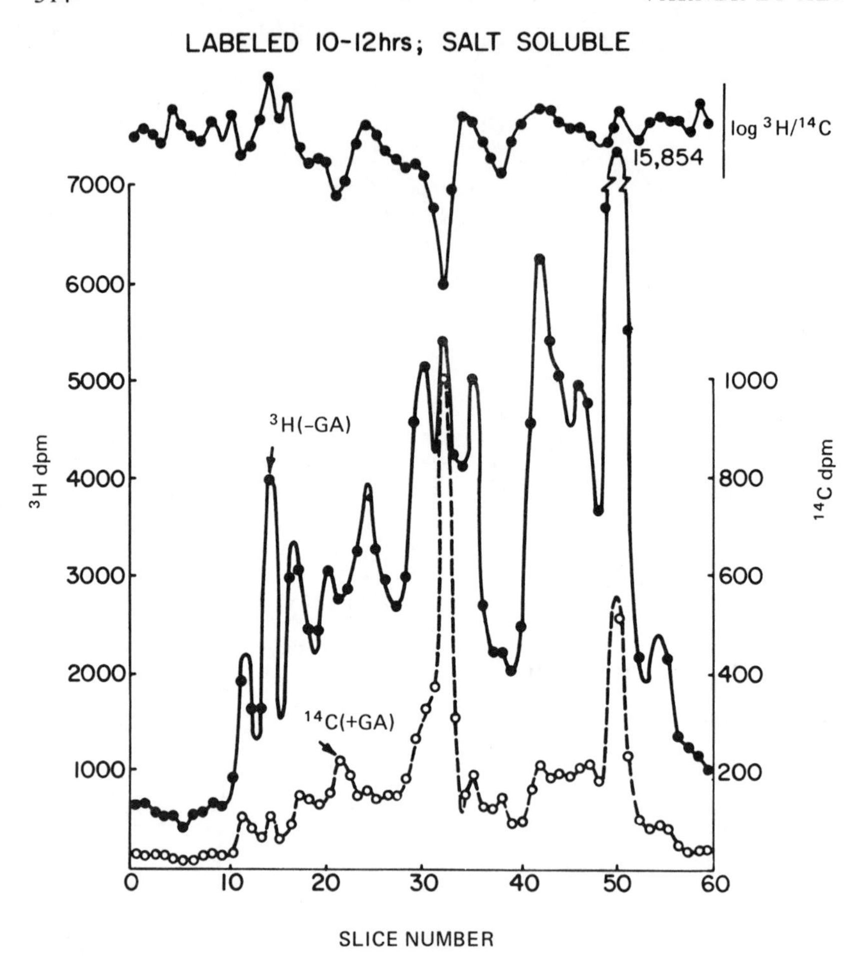

FIGURE 3 Labeling of the salt-soluble proteins of control and GA-treated aleurone layers. The layers were incubated with labeled amino acids for 2 hours beginning 10 hours after the start of the incubation of the layers in buffer or in buffer plus GA. The slice numbers refer to the slices obtained from the SDS gels following electrophoresis. Details were as described in the section on materials and methods.

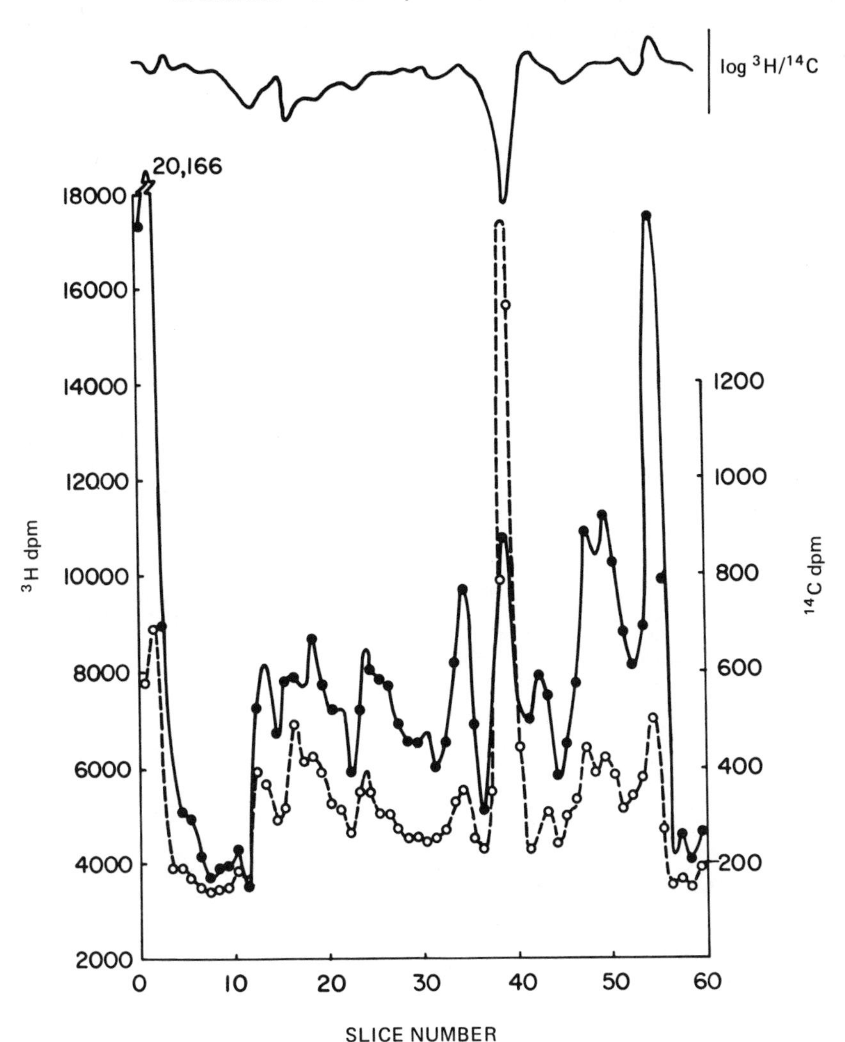

FIGURE 4 Labeling of the salt-insoluble proteins of control and GA-treated aleurone layers. The layers were incubated with labeled amino acids for two hours beginning 10 hours after the start of the incubation of the layers in buffer, or in buffer plus GA. The slice numbers refer to the slices obtained from the SDS gels following electrophoresis. Details were as described in the section on materials and methods.

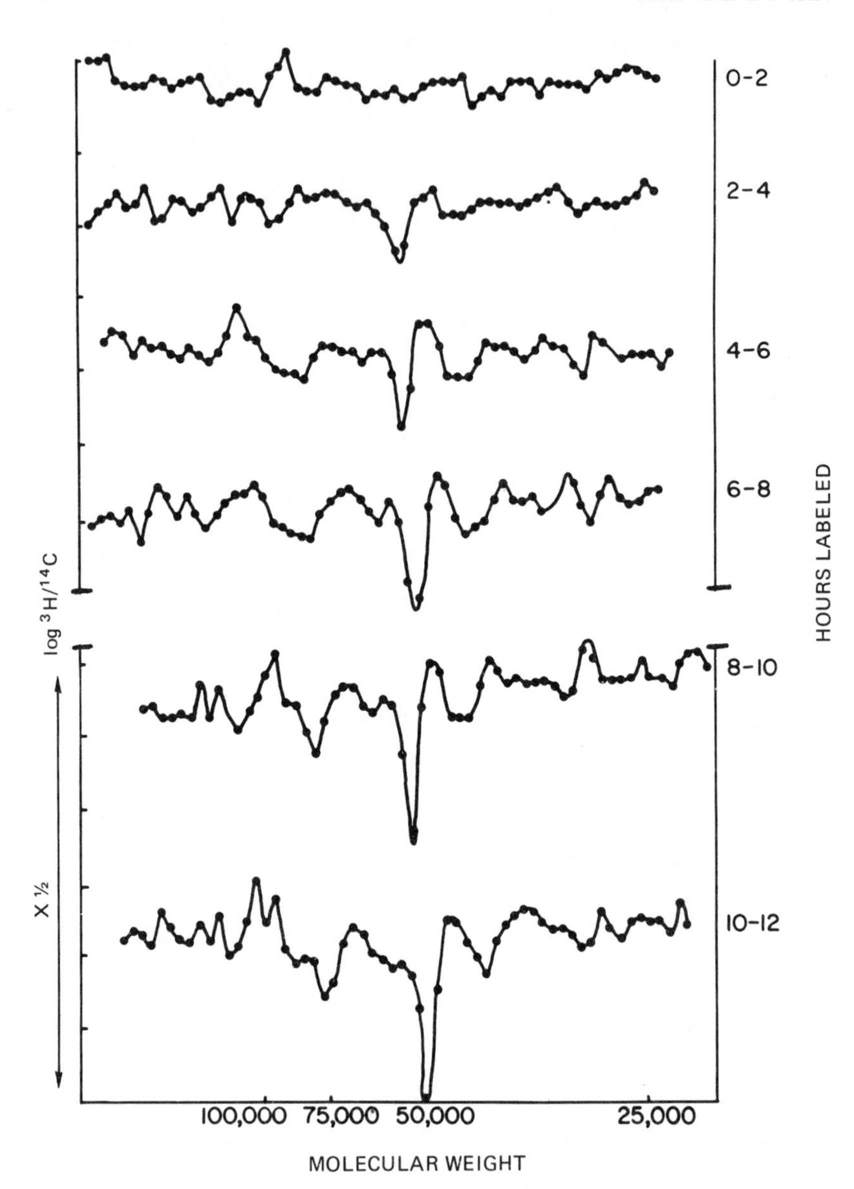

FIGURE 5 Labeling of the salt-soluble proteins of the control and GA-treated aleurone layers expressed as ^3H (control) to ^{14}C (GA-treated) ratios at incubation times up to 12 hours. Details of the experiments were as in Figures 1 and 3.

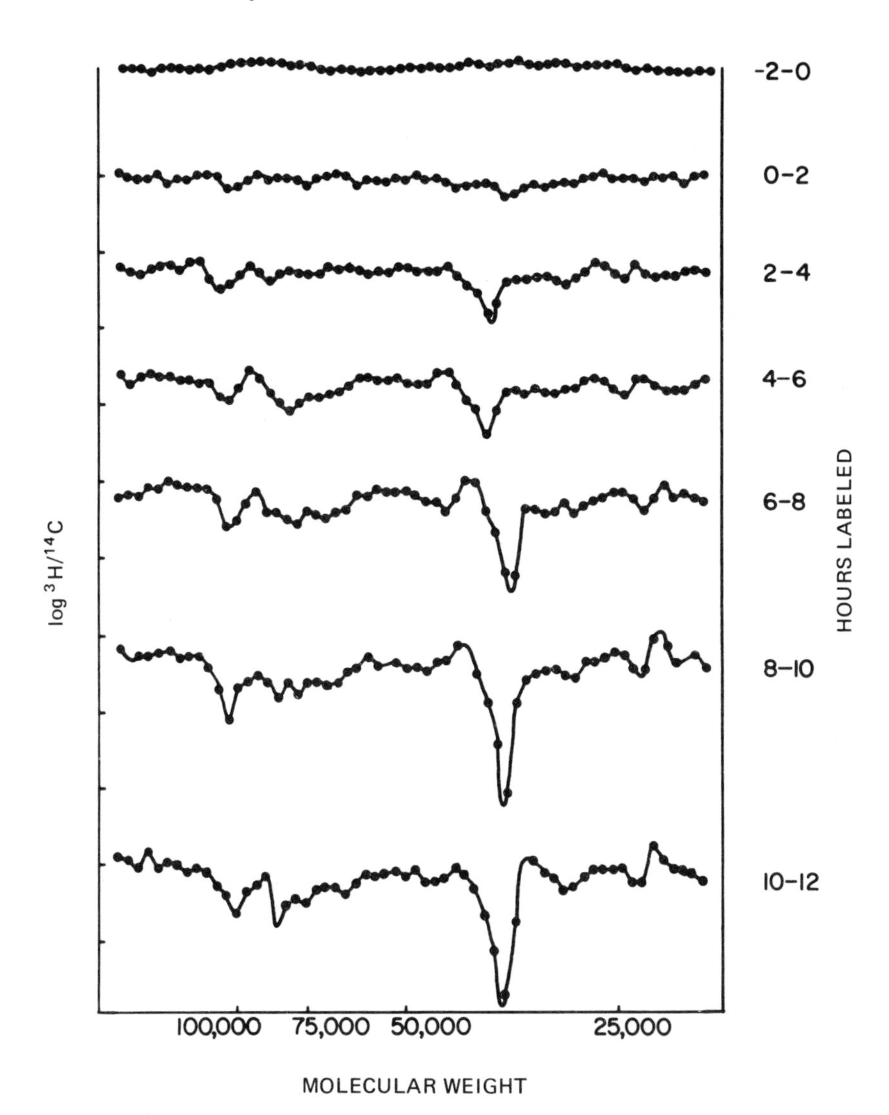

FIGURE 6 Labeling of the salt-insoluble proteins of the control and GA-treated aleurone layers expressed as ^{3}H (control) to ^{14}C (GA-treated) ratios at incubation times up to 12 hours. Details of the experiments were as in Figures 2 and 4.

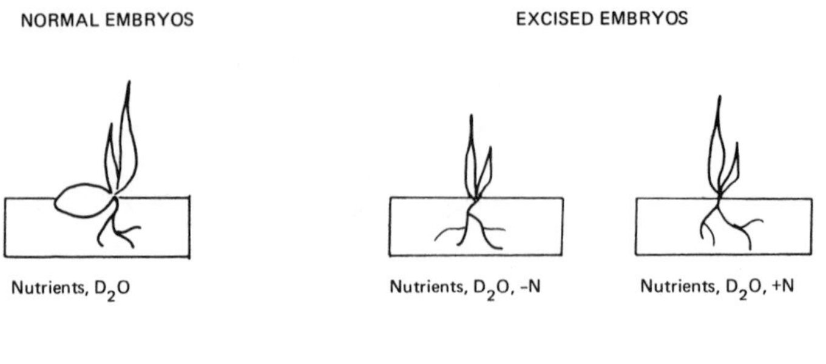

NORMAL { Root, Coleoptile }
EXCISED { -N, +N }
Amino acid extraction, derivatization with TMS → GLC → Mass spectrometer → Stored data → Print out

FIGURE 7 Experimental procedure for determining amino acid synthesis in the embryo. Barley seedlings are grown in the dark in deuterium oxide under the two conditions shown.

METHODS

Germination of Seedlings

Barley seeds (*Hordeum vulgare* L. Cv. Himalaya) were treated with 20 percent Clorox (commercially available bleach) for 20 minutes, rinsed with sterilized water placed in sterilized petri dishes lined with Whatman No. 3 filter paper in 0 percent or 40 percent deuterium oxide and allowed to germinate in the dark. Roots and coleoptiles (containing the first leaf) were dissected after 6 days of growth.

Culture of Excised Embryos

Barley seeds, after Clorox treatment and washing with water as above, were soaked in sterilized water for 12–15 hours. Embryos were dissected away from the endosperm and placed on 1 percent agar medium containing 200 m*M* sucrose (with or without 10 m*M* KNO_3) and 40 percent deuterium oxide (Quail and Varner, 1971). They were allowed to grow in the dark for 6 days.

In 40 percent D_2O the growth of the seedlings for both the normally germinating and excised embryos was about 18 percent inhibited on the basis of fresh and dry weight measurements. Otherwise, the pattern of growth appeared normal.

Extraction of Total Protein

After 6 days of germination the intact seedlings were dissected into roots and shoots for extraction of protein. The 6-day-old embryos were extracted without dissection. [Amino acid hydrolysates were prepared for analysis using the methods of Putter *et al.* (1969), except that the decolorization step was omitted.]

The plant material was homogenized by hand with a mortar and pestle with 10 percent TCA, using a volume 10 times that of the plant material. The homogenate was heated to 70°C for 30 minutes, cooled and centrifuged at 10,000 *g* for 15 minutes at 4°C. The residue was extracted with 10 percent TCA and centrifuged. The TCA-insoluble residue was extracted twice with 95 percent ethanol and twice with acetone. Each time, the residue was recovered by centrifugation. The residue was dried under an infrared lamp.

Hydrolysis and Purification of the Hydrolysate

About 60–70 mg of TCA-insoluble material was hydrolyzed in a sealed tube in 5 ml 6 *N* HCl at 110°C for 24–26 hours.

The hydrolysate was filtered through glass fiber filter paper (Whatman GF/C) under suction and evaporated to dryness in a vacuum desiccator. The last trace of HCl was removed by dissolving the hyrdolysate in water and subsequent evaporation.

Dowex-50W [H^+ form, mesh 50–100 (Sigma)] resin was used to free the mixture of amino acids from sugars and organic acids. The resin was regenerated with 1 *N* HCl and washed with water. A glass column of 1 cm × 30 cm dimensions was used with 5.75-ml bed volume of resin. The hydrolysate was dissolved in 5 ml of 0.1 *N* HCl and passed through the resin at a flow rate of 0.5 ml/min. An additional 10 ml of 0.1 *N* HCl was also passed through the resin. Then three 10-ml portions of water were passed successively at a flow rate of 0.8 ml/min. The amino acids were eluted with 20 ml of 3 *N* NH_4OH at a flow rate of 0.5 ml/min. The NH_3 was evaporated in the presence of concentrated H_2SO_4 and Drierite in a vacuum desiccator. The residue was dissolved in 2 ml of 0.05 *N* HCl.

Derivatization

The trimethylsilylation of the amino acids and their separation in the gas chromatography system followed the methods of Gehrke and Leimer (1971), with the important exception that the reaction time was 0.5 instead of 2.5 hours.

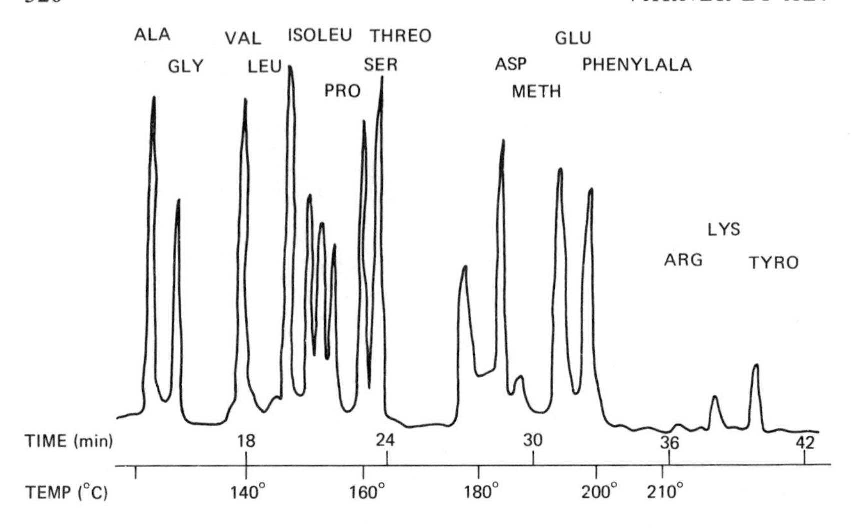

FIGURE 8 Separation of mixture of amino acids by gas–liquid chromatography.

The retention time for each amino acid derivative was determined by running each one separately through the GLC column. The resolution of a known mixture of amino acids is shown in Figure 8.

GLC-MS *Analysis of Amino Acids*

The combined GLC-MS analysis was obtained with the LKB-Model 9000 instrument fitted with a computer for recording the scans and storing the data on tape (Holmes *et al.*, 1973).

Scanning was generally at peak height—ascending and descending sides of the peak were also scanned for determining isotope fractionation. (Under the derivatization conditions used, arginine was resolved poorly in the GLC, and its mass spectrum was different from the expected spectrum. The so-called glycine-3 peak, which appears before that of proline, has no similarity to the mass spectra of glycine.

Interpretation of Mass Spectra

Four major fragmentation pathways were observed for the trimethylsilyl (TMS) derivatives of the amino acids as reported earlier (VandenHeuval and Smith, 1970; VandenHeuval *et al.*, 1970) (Figure 9). They are (a) the loss of a methyl group from the molecular ion M (M-15), (b) the loss of a methyl group and CO with retention of the silyloxy group (M-43),

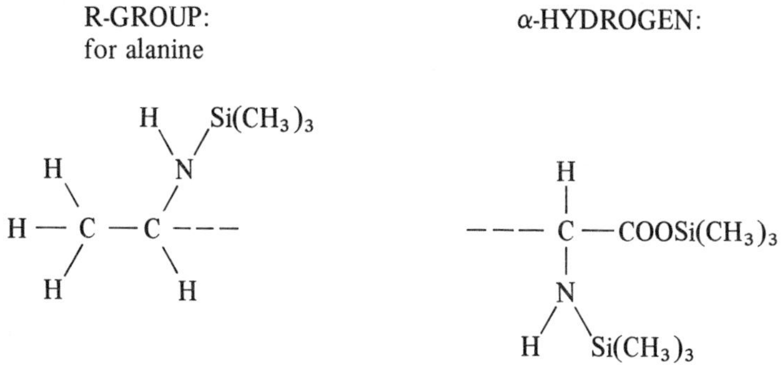

```
R-GROUP:                                α-HYDROGEN:
for alanine

        H   Si(CH3)3
         \ /
  H       N                                  H
   \     /                                   |
H — C — C ---                      --- C — COOSi(CH3)3
   /     \                                   |
  H       H                                  N
                                            / \
                                           H   Si(CH3)3
```

FIGURE 9 Typical TMS fragments.

(c) the loss of carbotrimethyl silyloxy (*M*-117), and (d) loss of the amino acid side chain group *R* to yield a fragment of mass/charge (m/e) 218. Signals occur at m/e 73 and m/e 147 due to the fragments $[Si(CH_3)_3]^+$ and $[(CH_3)^2Si = OSi(CH_3)_3]^+$, respectively. For information regarding the incorporation of deuterium into the amino acids, we considered in particular the m/e (*M*-117) and m/e 218 fragments. The m/e 218 fragment provides information regarding the extent of transamination of the amino acid, since the α-H can be replaced only through this mechanism. The replacement of hydrogen by deuterium will result in an increase of mass 1, resulting in a signal at m/e 219.

Correction of the Raw Data for the Natural Abundance of ^{2}H, ^{29}Si, ^{15}N, ^{18}O *and* ^{13}C

The data printout from the PDP-12 computer attached to the LKB-9000 GLC-MS consists of two columns listing the mass/charge values and their corresponding relative intensities. The abundance of a particular ion depends primarily on the activation energy for fragmentation, and it is therefore necessary to consider relative intensities of the peaks rather than their absolute intensities (Waller, 1972). Interpretation of the data is based on the assumption that the peak height ratios are proportional to the abundance of the molecular species from which they are derived (VandenHeuval and Smith, 1970).

To obtain pertinent information from the data (i.e., amount of deuterium incorporated into the various amino acids), all the components of each mass peak must be considered and corrections made for naturally occurring isotopes. A detailed procedure for doing this is given by Mitra *et al*. (1976).

RESULTS AND DISCUSSION

Deuteration of the following amino acids was examined: alanine, glycine, valine, leucine, isoleucine, proline, serine, aspartate, glutamate, methionine, phenylalanine, and tyrosine. All of these amino acids became deuterated to some extent both in the excised embryos and in the normally germinated seedlings.

Alanine, aspartate, serine, glycine, methionine, and glutamate were the most heavily deuterated (Table 1). Thus, although the shoots and roots of the normally germinating seedlings may have had access to already formed alanine, aspartate, serine, glycine, methionine, and, especially, glutamate from the hydrolysis of endosperm proteins, these preformed amino acids were not incorporated directly into protein but at the least underwent extensive transamination. The introduction of ^{2}H into the $n + 1$, $n + 2$, and $n + 3$ fragments of alanine need not involve any enzyme other than glutamate–pyruvate transaminase (GPT) because GPT catalyzes the exchange with the medium not only of the α-H of alanine but also of the β-Hs (Babu and Johnston, 1974). Therefore, we cannot draw conclusions about the extent of the biosynthesis of the carbon skeleton of alanine under the different conditions of germination of the embryo. If, as seems likely, transaminases equilibrate the β-Hs as well as the α-H of other amino acids, then we can conclude nothing about the synthesis of the carbon skeleton of aspartate and serine.

Because GPT catalyzes exchange with the medium of only one of the Hs of glycine (Babu and Johnston, 1974), the appearance of any signal at the $n + 2$ position is indicative of the involvement of the carbon skeleton of glycine in some metabolic reaction in addition to transamination. This occurs in the excised embryos given nitrate and sucrose.

The signal in the $n + 4$ and $n + 5$ positions for methionine could be due to deuterium on either the methylene carbon or the methyl carbon. Because the MS cannot distinguish between these two possibilities on the fragment examined, we can conclude little about the biosynthesis of the carbon skeleton of methionine from the deuteration pattern. Appearance of a signal at the $n + 4$ and $n + 5$ positions of glutamate indicates involvement of the glutamate carbon skeleton in metabolic reactions in addition to transamination. Such metabolism occurred under all conditions examined.

Valine, leucine, and isoleucine were not extensively deuterated. Transamination probably accounts for most of the deuterium incorporated into these amino acids except for the leucine in the excised embryos growing on nitrate and sucrose where the increased signals at

TABLE 1 Incorporation of Deuterium into Amino Acids[a]

Amino Acid Fragment	Mass Number of the Fragment	Normal Embryo		Excised Embryo	
		Root[b]	Coleoptile	–N	+N
Alanine	n[c]	115	115	115	115
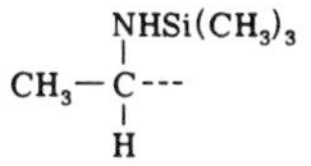 $CH_3-C(H)(NHSi(CH_3)_3)$---	$n+1$	153	120	82	112
	$n+2$	100	58	46	100
	$n+3$	26	11	13	38
	$n+4$	3	1	2	0
Aspartate	n	132	132	132	132
$(CH_3)_3SiO-C(=O)-C(H)(H)-C(H)(NHSi(CH_3)_3)$---	$n+1$	95	120	75	123
	$n+2$	32	39	28	18
	$n+3$	2	2	4	1
Serine					
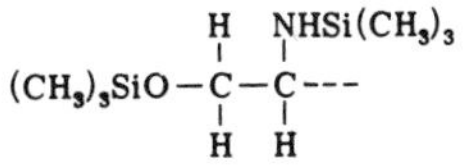 $(CH_3)_3SiO-C(H)(H)-C(H)(NHSi(CH_3)_3)$---	n	132	132	132	132
	$n+1$	53	73	51	52
	$n+2$	21	13	16	13
	$n+3$	4	2	5	0
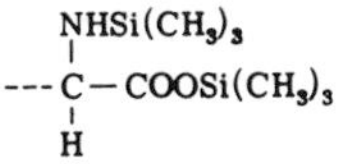 ---$C(H)(NHSi(CH_3)_3)-COOSi(CH_3)_3$	n	133	133	133	133
	$n+1$	30	25	23	28
Glycine					
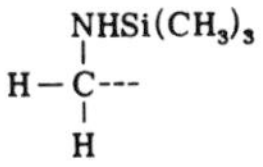 $H-C(H)(NHSi(CH_3)_3)$---	n	115	115	115	115
	$n+1$	44	35	36	78
	$n+2$	5	3	6	11
Methionine	n	131	131	131	131
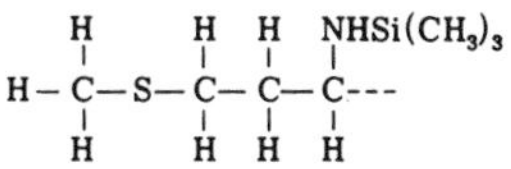 $H-C(H)(H)-S-C(H)(H)-C(H)(H)-C(H)(NHSi(CH_3)_3)$---	$n+1$	107	96	106	110
	$n+2$	80	31	80	109
	$n+3$	23	7	34	68
	$n+4$	1	1	10	24
	$n+5$	0	0	2	5
Glutamate	n	118	118	118	118
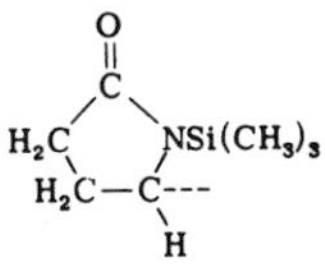 ring: $O=C$, H_2C, $NSi(CH_3)_3$, $H_2C-C(H)$---	$n+1$	118	163	80	78
	$n+2$	96	53	71	55
	$n+3$	60	12	36	20
	$n+4$	4	2	9	7
	$n+5$	1	0	1	1

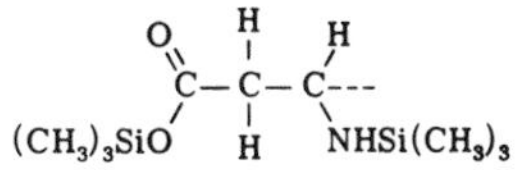

TABLE 1 (continued)

Amino Acid Fragment	Mass Number of the Fragment	Normal Embryo Root[b]	Normal Embryo Coleoptile	Excised Embryo -N	Excised Embryo +N
Valine 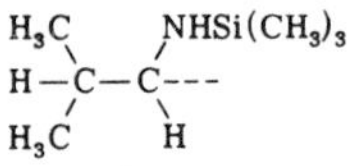	n	118	118	118	118
H_3C $NHSi(CH_3)_3$	$n+1$	22	20	16	23
H—C—C---	$n+2$	6	3	5	11
H_3C H	$n+3$	6	3	8	8
	$n+4$	1	1	1	4
$NHSi(CH_3)_3$					
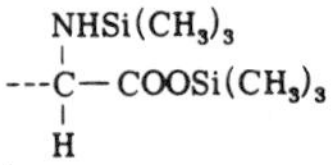 ---C—$COOSi(CH_3)_3$	n	133	133	133	133
H	$n+1$	20	20	20	19
Leucine 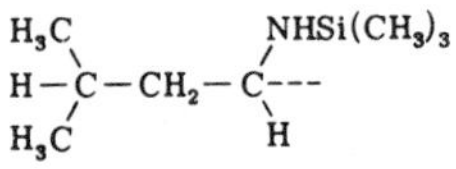	n	122	122	122	122
H_3C $NHSi(CH_3)_3$	$n+1$	54	52	36	40
H—C—CH_2—C---	$n+2$	8	19	15	28
H_3C H	$n+3$	6	11	9	20
	$n+4$	3	5	5	12
$NHSi(CH_3)_3$					
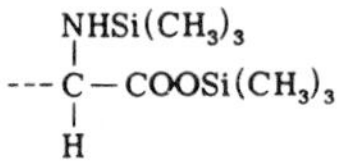 ---C—$COOSi(CH_3)_3$	n	131	131	131	131
H	$n+1$	45	37	34	38
Isoleucine 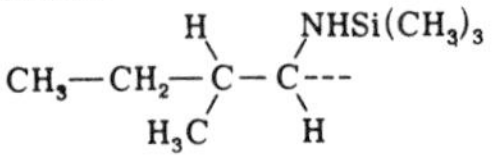	n	119	119	119	119
H $NHSi(CH_3)_3$	$n+1$	36	45	45	39
CH_3—CH_2—C—C---	$n+2$	6	6	20	16
H_3C H	$n+3$	2	2	10	8
	$n+4$	0	0	4	3
$NHSi(CH_3)_3$					
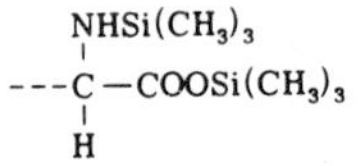 ---C—$COOSi(CH_3)_3$	n	129	129	129	129
H	$n+1$	26	38	31	30
Tyrosine					
H H	n	130		130	130
C=C	$n+1$	48		38	44
---CH_2—C C—$OSi(CH_3)_3$	$n+2$	13		15	14
C—C	$n+3$	1		0	1
H H	$n+4$	0		0	0
Phenylalanine					
H H	n	126	126	126	126
C=C $NHSi(CH_3)_3$	$n+1$	48	34	43	66
H— C C—CH_2—C---	$n+2$	34	18	38	72
C—C H	$n+3$	18	10	18	42
H H	$n+4$	5	4	7	13
	$n+5$	0	2	2	6

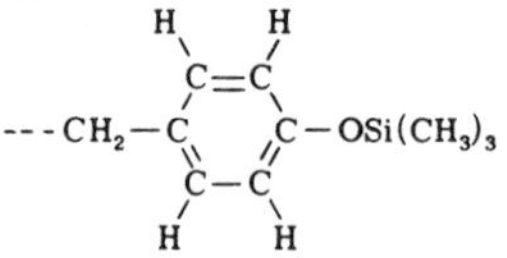

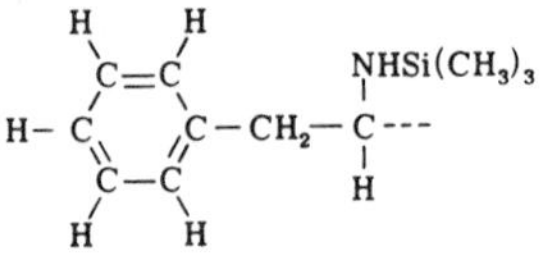

TABLE 1 (continued)

Amino Acid Fragment	Mass Number of the Fragment	Normal Embryo		Excised Embryo	
		Root[b]	Coleoptile	-N	+N
Proline	n	116	116	116	116
	$n+1$	7	5	13	22
	$n+2$	3	3	15	35
H_2C—CH, H_2C, $NSi(CH_3)_3$, CH_2	$n+3$	1	2	12	33
	$n+4$	0	1	5	20
	$n+5$	0	0	4	12

[a]Data are given for R-group fragments and the α-H fragment (if available) obtained from normally germinating and excised embryo tissues grown 6 days in 40 percent deuterium oxide medium.
[b]All numbers are the corrected relative signals at each mass-to-charge (m/e) value.
[c]n is the fragment containing only hydrogen; $n + 1$, $n + 2$, and so on are the fragments containing one, two, and more deuterium atoms.

the $n + 4$ and the $n + 5$ positions and the decreased signal at the $n + 1$ position can be accounted for only by the biosynthesis of the carbon skeleton of leucine. The deuteration labeling patterns in valine and isoleucine are consistent with the possibility that amino transferase with these two amino acids cannot—perhaps because of steric hindrance—equilibrate the β-H with the medium. (Compare the valine and isoleucine deuteration patterns with glycine, in which only one H is exchanged by transaminase, and with tyrosine, in which both Hs are exchanged.)

Tyrosine shows deuteration of the β-carbon but no deuteration of the ring carbons. Therefore, there could not have been biosynthesis of tyrosine from carbohydrate precursors. The deuterium must have been introduced into the β-position by aminotransferase. Thus, in the first 6 days of germination and growth of the seedlings, the tyrosine used for the synthesis of new proteins must have come from existing proteins rather than from the synthesis of new tyrosine.

If we accept as likely the possibility that amino transferase equilibrates the β-Hs of phenylalanine with the medium, then most of the signal observed in the $n + 1$, $n + 2$, and $n + 3$ positions of the phenylalanine fragment from the normally germinating seedlings is due to deuterium introduced by aminotransferase. On the other hand, there is considerable deuteration of the ring carbons of the phenylalanine from those embryos grown on sucrose and nitrate. This presumably is the result of the biosynthesis of phenylalanine from its carbohydrate precursors and is in marked contrast to the lack of biosynthesis of tyrosine in the same seedlings.

The difference in the extent and pattern of deuteration of proline from the excised seedlings, compared to proline from the normally germinating seedlings, makes it clear that the presence of the endosperm almost entirely prevents proline biosynthesis and that removal of the endosperm permits extensive synthesis of proline—even in the absence of added nitrate. Preliminary results indicate that 5 m*M* proline inhibits the biosynthesis of proline by excised seedlings growing on mineral salts, sucrose, and nitrate (Mitra, unpublished observations).

REFERENCES

Ashford, A. E., and J. V. Jacobsen. 1974. Planta 120:81.

Babu, U. M., and R. B. Johnston. 1974. Biochem. Biophys. Res. Commun. 58:460.

Bennett, P. A., and M. J. Chrispeels. 1972. Plant Physiol. 49:445.

Chrispeels, M. J., and A. J. Termer. 1973. Planta 113:35.

Chrispeels, M. J., and J. E. Varner. 1967. Plant Physiol. 42:1008.

Filner, P., and J. E. Varner. 1967. Proc. Nat. Acad. Sci. U.S.A. 58:1520.

Gehrke, C. W., and K. Leimer. 1971. J. Chromatogr. 57:219.

Holmes, W. F., W. H. Holland, B. L. Shore, D. M. Bier, and W. R. Sherman. 1973. Anal. Chem. 45:2063.

Jacobsen, J. V., and R. B. Knox. 1974. 115:193.

Jacobsen, J. V., and J. E. Varner. 1967. Plant 42:1596.

Jones, R. L. 1969a. Planta 88:73.

Jones, R. L. 1969b. Planta 87:119.

Jones, R. L. 1972. Planta 103:95.

Jones, R. L. 1973a. *In* Annual Review of Plant Physiology (W. R. Briggs, P. Green, and R. Jones, ed.). Annual Reviews, Inc., Palo Alto, California, p. 571.

Jones, R. L. 1973b. Plant Physiol. 52:303.

Laemmli, U. K. 1970. Nature 227:680.

Mitra, R., J. Burton, and J. E. Varner. 1976. Anal. Biochem. 70:Jan. 1976.

Putter, I., A. Barreto, J. M. Markley, and O. Jardetsky. 1969. Proc. Nat. Acad. Sci. U.S.A. 64:1396.

Quail, P. H., and J. E. Varner. 1971. Anal. Biochem. 39:344.

VandenHeuval, W. J., and J. L. Smith. 1970. J. Chromatogr. Sci. 8:567.

VandenHeuval, W. J., J. L. Smith, I. Putter, and J. S. Cohen. 1970. J. Chromatogr. 50:405.

Vigil, L., and M. Ruddat. 1973. Plant Physiol. 51:549.

Waller, G. R. (ed.). 1972. Biochemical Applications of Mass Spectrometry. Wiley-Interscience, New York.

Yomo, H., and J. E. Varner. 1971. *In* Current Topics in Development Biology, Vol. 6 (A. Monroy and A. A. Moscona, ed.). Academic Press, Inc., New York, p. 111.

DISCUSSION

DR. MIFLIN: What possibility is there that deuterium is getting into methionine by methyl transfer between homocysteine and methionine back and forth? The only comment I have about tyrosine biosynthesis is that one of the few cases where we have got isoenzymes that seem to be different in regulatory properties in higher plants is in the mutase that leads to phenylalanine and tyrosine. It is conceivable that what these isoenzymes mean is that the two pathways are separate. It is not a branch pathway, but you have isoenzymes, one going to one and one going to the other. So they need not necessarily be working together.

DR. VARNER: Your point about methionine being labeled in the methyl group is a very good one. There are a number of ways you can introduce deuterium into threonine, threonine deaminase, threonine dehydrotase, also in cystine by cystine dehydrotase and into serine by serine dehydrotase, and so on. I do not think that it is impossible to examine biosynthesis of those amino acids by this method, but you have additional ways of introducing label that complicate it.

I mentioned at the start that this work had been done by Dr. Ranjit Mitra, and for those of you who have not met him, he is an International Atomic Energy Agency Fellow in my lab this year. He will be taking this problem with him back to Bombay and pursuing it there. We are just seeing that the technique works. With the inhibition of proline biosynthesis, for instance, the next experiment to be done is to germinate these things in agar with added proline, proline only. At the moment, all we can say is that endosperm inhibits proline biosynthesis. Is it proline that inhibits proline biosynthesis? From Ann Oaks' data we would say that it is, but we want to check it out our way. That will tell us that the method will work to sort out known pathways and it ought to work to sort out unknown ones.

DR. MARCUS: Can you use cycloheximide and get no inhibition?

DR. VARNER: At 30 minutes, if you would apply cycloheximide, you get 90 percent inhibition of amino acid incorporation in the protein and no inhibition of the GA-dependent increase in glyceride transferase.

DR. MARCUS: So you do not really have to use all those analogs to do the experiment?

DR. VARNER: I like the analogs better even at 30 minutes. No, it is a necessary experiment to do. I think if cycloheximide did not inhibit it in 30 minutes, it would have been embarrassing. It would not have destroyed our case, but it would have been embarrassing.

DR. HARDY: It would certainly be interesting to identify the actual site of replacement of the D with the H in some of your products. For example, is NMR a technique that might be used to answer the question, "Is it methyl protons or methylene protons that are being replaced?"

DR. VARNER: This can be done. I think every last deuterium can be located by looking at different derivatives, not only trimethylsilyl here but other derivatives. The ethyl esters, tributyl esters, and trifluorobutyl esters and so on give different fragmentation patterns, and this fact plus the fact that you can put all of this in a high-resolution mass spectrometer where you can determine molecular weights or mass numbers to the third decimal point, I think, would allow us to determine the position of nearly every deuterium.

DR. HARDY: Will the fragmentation be such as to distinguish the methylene deuterium and the methyl deuterium?

DR. VARNER: The difference between methylene and methyl almost always pops up because it is easier to fragment off the methyl group than it is to split a methylene group out of the center of the chain. I think this can be done, but that is an order of magnitude more expensive and more time-consuming. These runs take 40 minutes. Once you have your trimethylsilyl derivatives, it takes 40 minutes to run it through the gas chromatograph. The data come out of the mass spec on the tape, and then at night you stick the tape in the computer, it prints out everything for you the next day. You have it the next day. So, there is not a lot of man time or machine time involved in it at this level. If we go to high-resolution mass spectrometer, the increase in expense is sufficient that one really would not want to do this except for particular amino acids. It would be interesting to look at methionine, as Dr. Miflin mentioned, or at tyrosine and phenylalanine if one suspected a precursor product relationship there.

DR. WILSON: I can see one potential conflict. You have here one experiment with an interesting implication if all of the amino acids are being reworked in the developing seed. Remember Sodek's (L. Sodek and C. M. Wilson. 1970. Arch. Biochem. Biophys. 140:29–38) experiment in which ^{14}C leucine was recovered as ^{14}C leucine? This might mean, though, that normally leucine does not get transported into the seed, but if it happened to be there it was used. However, the bulk of the leucine might have been newly synthesized.

DR. VARNER: I think before drawing any final conclusions one would like to do the other half of this experiment—let leaves photosynthesize in the presence of deuterium oxide and see if any deuterated amino acids arrive and are incorporated in the bean. That can be done.

LEON S. DURE III

Messenger RNA Synthesis and Utilization in Seed Development and Germination

During the past several years we have collected evidence that most of the protein synthesis necessary to begin germination in cotton cotyledons is directed by mRNA that is transcribed toward the end of embryogenesis (Waters and Dure, 1966; Ihle and Dure, 1969) but is prevented from being utilized during embryogenesis by the presence of the plant growth regulator abscisic acid (Ihle and Dure, 1970, 1972a). We utilized two specific germination enzymes in this study, one of which was purified to homogeneity as part of this work (Ihle and Dure, 1972b,c). Furthermore, by precociously germinating very young embryos we have been able to induce prematurely the transcription of the mRNA for germination enzymes and have suggestive evidence for what signals the synthesis *in vivo* in a gross sense (Ihle and Dure, 1972a). In view of these results, a number of questions can be asked of our system, and many of them can apparently be investigated fruitfully with the methodology at hand.

If, indeed, a large body of mRNA is synthesized in late embryogenesis whose translation will inaugurate germination, and if the growth regulator, abscisic acid, acts to prevent its premature functioning in protein synthesis until the mature, dessicated seed is formed, the question arises which, asked in its simplest form, is "How is the functioning of the mRNA prevented?" Subsequent questions that naturally follow are "What is the role of abscisic acid in preventing the functioning of this mRNA?" and "What is the physical form of this body of stored mRNA during the last stages of embryogenesis and in the dry seed?" The pursuit of the answers to these questions must be done against a background of what is known in eukaryotes about the steps between cistron transcription and the arrival of functional mRNA in the cytosol. Although there is no uniform acceptance of some of the assumed steps involved in what is now called mRNA processing, Figure 1 represents a composite though simplified view of some of the current concepts of this process.

In this figure the unit of DNA containing the cistron for a polypeptide chain is seen surrounded by DNA that presumably does not code for polypeptide sequences. Evidence obtained from experiments with HeLa cells suggests that much of this "nonfunctional DNA" (nonfunctional in the sense of coding for protein) is transcribed by the mRNA-producing RNA polymerase (Lindberg and Darnell, 1970). Such

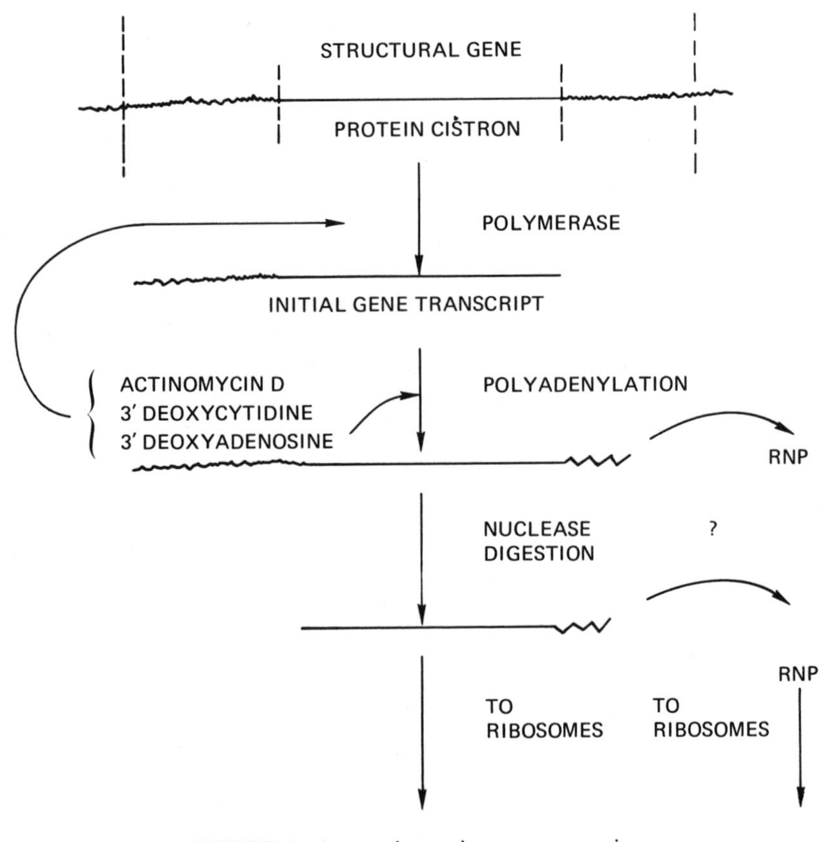

FIGURE 1 Assumed steps in mRNA processing.

an initial gene product would be a considerably larger molecule than that required for the functional mRNA. Short pulses of isotopically labeled RNA precursors have allowed a fraction of large RNA polymers to be observed in HeLa cells that is appropriately called "heavy nuclear RNA" (hnRNA). A similar body of RNA has been demonstrated in the slime mold *Dictyostelium* (Firtel and Lodish, 1973), although the molecular size is somewhat smaller than that of HeLa cells. Figure 1 also shows that the nonfunctional DNA is at least partially comprised of repeated sequence or reiterated DNA. This is strictly conjectural at this point, yet sequences of DNA that occur hundreds of times scattered throughout the genome have been found to exist in several eukaryotic organisms (Davidson *et al.*, 1973; Lee and Thomas, 1973). Although there is some disagreement on how far apart repeated sequences are spaced in a genome, in a general sense they can be visualized as separating functional cistrons and hence may be transcribed as part of the hnRNA of the eukaryote nucleus. A particularly interesting calculation arrived at for the repeated sequence DNA of *Xenopus laevis* is that there are about 2,000 different repeated sequences that occur on the average of 1,600 times each in the genome of this organism (Davidson, 1974). This is a tantalizing finding in attempting to visualize possible recognition mechanisms operating to regulate the pattern of gene expression that brings about ontogeny and development.

Figure 1 illustrates that nonfunctional DNA is present on both the 5′ and 3′ sides of the protein cistron. There is no hard evidence for this or for the alternative possibility that it is present primarily on the 5′ side.

The first step in mRNA processing, as shown in Figure 1, appears to be the enzymatic addition of AMP residues to the 3′ end of the mRNA polynucleotide chain. This addition is not thought to be DNA-directed since it is insensitive to actinomycin D and since no poly-T regions corresponding in chain length to the poly-A region of the mRNA have been found in DNA. Enzymes distinct from RNA polymerase have been isolated that catalyze the addition of AMP residues to many types of RNA (Edmunds and Abrams, 1960; Mans and Walter, 1971; Niessing and Sekeris, 1972).

The second step in the processing of the mRNA contained in the hnRNA [also coming to be known as pre-mRNA in other organisms (Ryskov *et al.*, 1973)] is the enzymatic digestion of most of the RNA transcribed from the nonfunctional DNA. Again, most of the evidence for this comes from HeLa cell experiments in which it has been found that most of hnRNA does not leave the nucleus, has only a transient existence, and that the pulse-labeled RNA that does reach the cytoplasm has an average molecular weight much less than that of hnRNA

(Lindberg and Darnell, 1970). A similar observation has been made with *Dictyostelium* (Firtel and Lodish, 1973), where, in addition, most of the nucleotide loss appears to occur on the 5′ end of the mRNA nucleotide chain.

The mRNA, pruned of most of its extraneous nucleotides and with its additional poly-A sequence, is the mRNA that apparently emerges in the cytosol of eukaryotes, and that can be isolated from polysomes. The list of organisms in which mRNA–poly-A has been isolated lengthens weekly, yet there appears to be no standard poly-A chain length, which varies from an average of 200 A residues in HeLa cells (Mendecki *et al.*, 1972) to an average of 100 A in slime molds (Firtel and Lodish, 1973) and in cotton to 70 A in yeast (McLaughlin *et al.*, 1973). The two mRNAs for the hemoglobin subunits have 40 and 60 poly-A residues, respectively (Darnell, 1974). The function of the poly-A sequence is even more baffling at the moment, since it apparently is *not required* for translation. Some hemoglobin subunit mRNA has no poly-A sequence *in vivo,* and certain plant RNA viruses, whose RNA functions directly as mRNA, do not have a poly-A sequence (Shih and Kaesberg, 1971; Hall, 1974).

Figure 1 also indicates that at some point the pre-mRNA probably combines with protein and exists as a ribonucleoprotein particle (RNP). Such particles allegedly containing mRNA have been isolated from a large number of tissues and organisms and have been called informosomes (Spirin and Nemer, 1965), informofers (Samarina *et al.*, 1968), or simply RNPs (Henshaw, 1968). Yet a precursor–product relationship between the RNP–RNA and polysomal mRNA has proven difficult to demonstrate unequivocally.

Notice in Figure 1 that three inhibitors can be brought to bear on the putative processing sequence. Actinomycin D is probably the best known inhibitor of DNA-directed RNA synthesis. Cordycepin (3′-deoxyadenosine) is a more recently identified inhibitor, and in mammalian tissue it is thought to specifically and exclusively inhibit the addition of the poly-A sequence to mRNA (Henshaw, 1968). Based on its structure, it is thought that cordycepin, after becoming triphosphorylated, acts as a chain terminator by virtue of its lack of a 3′OH. However, we have found that this inhibitor apparently is incorporated into DNA-directed RNA by nucleoplasm and nucleolar polymerases as well in cotton, since all RNA synthesis is stopped by this compound (Harris and Dure, unpublished data). Substrate levels of this inhibitor (10^{-3} M) are required for complete inhibition, indicating that its mode of action is as presumed—i.e., its triphosphorylated derivative must compete with ATP and, once incorporated into the growing chain, terminates the chain. The mode of action of 3′deoxycytidine is somewhat conjectural as yet (although at the moment we have preliminary evidence that it mimics 3′deoxyadenosine in blocking DNA-directed RNA synthesis at $10^{-3} M$). However, in theory it should have no effect on poly-A addition to mRNA. Its potential use will be indicated subsequently.

It is against this background, incomplete, conjectural, and probably highly oversimplified though it may be, that our current experiments are placed.

In order to explain how the mRNA for the germination enzymes is prevented from being translated when it is initially transcribed in embryogenesis, we have tested the hypothesis that the stored mRNA of the dry seed cotyledon has not completed the processing sequence, which in turn must be carried out at the onset of germination.

As a crude test of this possibility, we have recently examined the effect of cordycepin on the appearance of the two germination indicator enzymes, the carboxypeptidase and isocitratase, and on visible germination itself. If this inhibitor stops germination and the appearance of the germination enzymes, it would suggest that the stored mRNA required poly-A addition—provided that cordycepin inhibits exclusively poly-A addition in cotton cotyledons. The results show a complete inhibition of germination and of the appearance of the two enzymes when the inhibitor is given the embryos at substrate levels ($10^{-3} M$) in the initial imbibition fluid and continuously thereafter during the germination period (4 days). Oddly, if the inhibitor is given the embryos at later times in germination, some visible germination takes place and some activity of the two enzymes develops. For example, if cordycepin is given the embryos 32 hours after imbibition is started, normal germination of the cotyledons is observed for 4 days, and the activity of the two enzymes develops with the same kinetics of appearance as it does in controls, reaching a maximum at about 86 hours after imbibition commenced. If the inhibitor is supplied at 15 hours after imbibition, germination is markedly curtailed, and only about 50 percent of the maximum activity of the two enzymes develops. After performing many experiments in which embryos were challenged with the inhibitor at different times after imbibition, we amassed data that suggested that the process inhibited by cordycepin that resulted in the failure of the cotyledons to germinate and to develop the two indicator enzyme activities begins about 6 hours after imbibition and is completed by 32 hours. This in turn suggested that the stored mRNA for the germination enzymes required the poly-A addition step in mRNA processing and that this step was completed by 32 hours of germination. In support of this was our subsequent finding that cordycepin supplied at the beginning of germination inhibited the incorporation of radioactive amino acids into proteins 75 percent during the 24–48 hours of germination when compared to controls. It should be remembered that actinomycin D does not inhibit protein synthesis during the first 3 days of germination at all, although it inhibits RNA synthesis more than 95 percent.

It may be argued that the triphosphorylated derivative of cordycepin inhibits the appearance of the germination enzymes and visible germination by inhibiting protein synthesis *per se*. It is conceivable that triphosphorylated cordycepin could do this in two ways. First, 3′deoxy ATP could interfere with amino acid activation and transfer to tRNA. Second, 3′deoxy ATP could participate in the turnover of the C–C–A terminus of tRNA, leading to the formation of C-C-3′deoxy A tRNA. Transfer RNA species modified in this fashion could not participate in protein synthesis since the 3′hydroxyl is necessary for the formation of the aminoacyl-tRNA ester. However, we do not feel that cordycepin interferes with either of these aspects of protein synthesis, since the period of cordycepin sensitivity and the period of enzyme synthesis during germination are quite distinct. Enzyme synthesis takes place mainly between the thirtieth and seventieth hour of germination, yet cordycepin has little effect on the amount of enzyme activity produced if added to cotyledons after 30 hours of germination. Furthermore, removing cordycepin after 30 hours of germination does not restore the cotyledons' ability to produce enzyme activity subsequently. Thus it would appear that cordycepin inhibits events that take place prior to the actual synthesis of the germination enzymes.

Unfortunately, these preliminary experiments lost some of their excitement when we measured the effect of 10^{-3} M cordycepin on RNA synthesis during the early stages of germination and found that the synthesis of all species of RNA is almost totally stopped by the inhibitor. Apparently, the triphosphorylated cordycepin is incorporated in RNA by both types of nuclear RNA polymerases and terminates the growing chains before they attain much size. In this respect plant systems differ from mammalian systems, where cordycepin is thought to specifically inhibit the poly-A addition enzyme (Penman *et al.*, 1970). A similar inhibition of nuclear RNA polymerases has been found in soybean tissues (Joe Key, personal communication).

Thus the cordycepin inhibition could not be said to substantively demonstrate a requirement for poly-A addition to the stored mRNA for it to become available for translation. There is always the possibility—though we consider it very remote—that the mRNA for the germination enzymes *is* made during early germination but is not accessible to actinomycin D inhibition for unknown reasons but is accessible to inhibition by cordycepin. (In our opinion the removal of abscisic acid inhibition of enzyme synthesis in precocious germination by actinomycin D and the actinomycin D sensitivity of enzyme synthesis in very young embryos strongly indicates that this is not the case.)

Since the cordycepin experiments at least partially suggest a requirement for processing of the stored mRNA and further suggest that this takes place by 32 hours of germination, we have attempted to demonstrate first that poly-A is incorporated into nonribosomal non-tRNA during the early hours of germination and second that some of the incorporated poly-A is incorporated into RNA that is pre-existent in the dry seed, i.e., the stored mRNA, during this period.

Our methodology involves incubating cotton cotyledons from the twelfth to sixteenth hour of germination in $^{32}PO_4$ and ^{3}H adenosine, extracting total nucleic acid, precipitating high-molecular-weight RNA with 2 *M* NaCl to lose DNA, 5S RNA and tRNA, and then passing the high-molecular-weight RNA through a column of poly-U sepharose. This column matrix in theory retains only RNA molecules containing long chains of poly-A that interact in a helical fashion with the poly-U that is covalently attached to the sepharose. The bound material is eluted by elevating the temperature to 50°C. The molecular weight distribution of this bound fraction has been found to be 0.5–3 million daltons as determined by SDS–polyacrylamide gel electrophoresis. This material (allegedly mRNA~poly-A is next digested with a combination of T_1 RNase and RNase-A that attacks essentially all the nucleotide linkages save those of the poly-A. The poly-A nucleotide chains are then collected on a millipore filter (type HA), eluted, and the average chain length is determined by the ratio of ^{3}H adenosine to ^{3}H AMP given by the electrophoresis of alkaline digests. All fractions are routinely monitored by SDS–polyacrylamide chromatography, and the base compositions and $^{32}P/^{3}H$ ratios are determined for all fractions as well. This procedure has worked very well and routinely yields radioactive poly-A chains with an average length of 100 AMP residues.

The next aspect of testing our hypothesis has involved the demonstration that some of the poly-A chains are put on pre-existing RNA. In this we have used actinomycin D to greatly reduce the amount of radioactivity in all RNA fractions in order to accentuate the poly-A radioactivity. The key to these experiments is the demonstration that the radioactivity found in the mRNA portion of the mRNA~poly-A is far too low relative to the radioactivity found in the poly-A. This suggests that not all the mRNA onto which the poly-A has been attached is synthesized during the germination period. This we can do in several ways, as shown by some preliminary data.

First, it must be assumed that the average length of mRNA should be at least 1,000 nucleotides, which in turn could code for protein subunits that average about 30,000 molecular weight. The following data can be experimentally determined:

1. The average chain length of poly-A
2. The percentage A in mRNA~poly-A
3. The percentage A in mRNA alone
4. The percentage $^{32}PO_4$ of mRNA~poly-A that is contained in poly-A
5. The percentage ^{3}H adenosine of mRNA~poly-A that is contained in poly-A

Thus, if the poly-A chain length averages 100, the percentage A in mRNA is 25 percent, and the mRNA has an average chain length of 1,000, then there should be 350 adenosines out of 1,100 total nucleotides. This predicts that percentage adenosine should be about 32 percent when the base composition of the mRNA~poly-A is determined. If this percentage is 35 or 40 percent, it indicates that the mRNA chain is less than 1,000 (very unlikely) or that *not all of the mRNA chains are radioactive*.

Alternatively, if the amount of radioactive PO_4 occurring in the poly-A moiety is known, the amount in the mRNA moiety known, and the average chain length of the poly-A known, these facts can indicate the average chain length of the mRNA. Again, if this predicted length is much smaller than 1,000, some of the mRNA chains that have radioactive poly-A chains attached to them must themselves be nonradioactive—i.e., pre-existent mRNA. Finally, once the average poly-A chain length, the percentage ^{3}H adenosine in mRNA, and the percentage of ^{3}H adenosine in mRNA~poly-A that is in poly-A are known, yet another determination of the mRNA chain length can be made.

Assuming that some new mRNA is synthesized during early germination and, along with the stored mRNA, is processed during early germination, then mRNA chain length determinations arrived at by the processes outlined above should yield numbers less than 1,000. The amount that they are less than 1,000 should suggest the ratio of newly synthesized mRNA to stored mRNA being processed.

If the experiments are performed with cotyledons germinated in the presence of actinomycin D, the radioactivity in mRNA should be drastically reduced, whereas the radioactivity in the poly-A chains should be reduced much less—i.e., reduced only to the degree that actinomycin D inhibits *new* mRNA synthesis, since there would be no new mRNA available for polyadenylation. Thus a second rough measure of the amount of stored mRNA relative to newly synthesized mRNA can be attained by measuring the degree to which actinomycin D decreases the amount of radioactive poly-A relative to controls.

Table 1 presents the data that we have obtained along these lines. These data show that about 20 percent of the total radioactive mRNA~poly-A is poly-A. This mRNA–poly-A fraction is very high in ^{3}H adenosine, and the poly-A chain length is 100 nucleotides. Thus, these data predict the poly-A–mRNA chain length to be about 400 nucleotides—far too small to code for common-sized polypeptide subunits—so many of the mRNA chains must not be radioactive (i.e., synthesized prior to germination). This, of course, is the sanguine interpretation of the data to date.

Further, our data indicate that actinomycin D inhibits mRNA synthesis about 70 percent but decreases poly-A addition only 30 percent. Table 1 shows that ^{3}H adenosine in the mRNA–poly-A here is 45 percent, which indicates that most of mRNA chains are not radioactive. This is further indicated by the other calculations, which predict an mRNA chain length between 100 and 200 nucleotides. Thus the data collected from actinomycin-D-treated cotyledons also indicate the polyadenylation of pre-existing mRNA.

However, two additional types of experiments must be performed to either add credence to our assumptions or show them to be faulty.

The first of these will involve precociously germinating very small

TABLE 1 Experimentally Determined Chain Length of RNA

	Control	Actinomycin D Treated
Data		
Average chain length poly-A	100	100
% A in mRNA~poly-A	39	45
% A in mRNA	22	23
% ^{32}P of mRNA~poly-A in poly-A	20	50
% ^{3}H adenosine of mRNA~poly-A in poly-A	49	68
Calculated Chain Length of mRNA		
Based on poly-A chain length, % A in mRNA~poly-A, % A in mRNA	364	296
Based on poly-A chain length, % ^{32}P of mRNA~poly-A in poly-A	400	100
Based on poly-A chain length, % A in mRNA, % ^{3}H adenosine of~mRNA poly-A in poly-A	473	218

embryos in which none of the putative germination mRNA has yet been synthesized (i.e., has no stored mRNA) and measuring the incorporation of radioactive $^{32}PO_4$ and ^{3}H adenosine into mRNA~poly-A. Here the percentage of radioactive ^{32}P in poly-A should constitute only about 8 percent of that in mRNA~poly-A, and the percentage of adenosine in these molecules should be about 30 percent. Data of this sort would predict a mRNA~poly-A chain length of over 1,000 nucleotides.

The second set of experiments that should strengthen or upset our premise that the stored mRNA must be processed during early germination involve the inhibitor 3′deoxycytidine. This compound should behave analogously to cordycepin in inhibiting RNA synthesis, but should affect poly-A addition in no way whatsoever. That is, it should behave identically to actinomycin D as far as its effect on RNA synthesis, protein synthesis, the appearance of the germination enzymes, and visible germination, while bringing about these effects through an entirely different mode of action (premature RNA chain termination). We have examined the effect of 3′deoxycytidine (1×10^{-3} M) on visible germination (which it does not affect), on protein synthesis (which it does not inhibit), and on the appearance of the two germination enzyme activities (which it also does not inhibit). We have yet to perform the crucial experiment, which is to see if it stops RNA synthesis. The great scarcity of this compound restricts us in these experiments.

If, after considerably more experimentation, we should prove correct in much of the above and unequivocally demonstrate that the stored mRNA of cotton cotyledons is polyadenylated during the first 30 hours of germination, this immediately confronts us with the problem of how this processing is prevented in embryogenesis, and this in turn brings us back to abscisic acid.

It should not be difficult to determine if ABA inhibits poly-A addition to mRNA. Since ABA does not affect the germination of mature cotton seeds, we must test its effect on poly-A addition in precocious germination, which is totally sensitive to ABA at 10^{-6} M. In this case we shall use embryos that have already accumulated much of the stored mRNA (100–110-mg embryos) and attempt to quantitate the amount of polyadenylation that takes place in precociously germinated embryos versus embryos inhibited by ABA through the use of radioactive RNA precursors. (The synthesis of rRNA and tRNA is not affected by ABA in this system.)

If this polyadenylation is much reduced in the cotyledons of embryos incubated in ABA, a number of possible experimental avenues open up. ABA conceivably could prevent polyadenylation by (a) directly interacting with the poly-A polymerizing enzyme in an allosteric fashion, (b) interacting with a receptor protein to form a complex that interacts in turn with the poly-A polymerizing enzyme, (c) affecting a process that precedes the poly-A addition step in time or upon which poly-A addition is dependent. The first two possibilities can be readily tested by developing a protocol for the purification of the poly-A polymerizing enzyme. This enzyme has been demonstrated and partially purified from a number of organisms (Mans and Walter, 1971; Edmunds and Abrams, 1960), and we have detected its activity in crude cotyledon extracts. In cell-free systems, the enzyme does not show a great deal of specificity for RNA acceptors. Transfer RNA is polyadenylated, as is degraded rRNA. This fact makes it uncertain that the enzyme activity assayed is in all cases the enzyme that carries out the mRNA adenylation *in vivo*.

Should ABA be found to influence the rate of polyadenylation of mRNA, we intend to attempt the purification of the poly-A polymerizing activity we have found in cotyledon extracts and to test the effect of ABA on the purified enzyme. Testing the effect of an ABA~receptor protein complex on this activity will be somewhat more difficult since it entails identifying and partially purifying protein(s) that bind ABA. We anticipate using affinity chromatography (ABA covalently bound through an extender molecule to a sepharose matrix) to isolate proteins with a high affinity for ABA.

At the moment there is no obvious approach to the third possible ABA influence on polyadenylation. A year from now there may be one.

REFERENCES

Darnell, J. E. 1974. Personal communication.
Davidson, E. H., B. R. Hough, C. S. Amenson, R. J. Britten. 1973. J. Mol. Biol. 77:1.
Davidson, E. H. 1974. Personal communication.
Edmunds, M., and R. Abrams. 1960. J. Biol. Chem. 235:1142.
Firtel, R. A., and H. F. Lodish. 1973. J. Mol. Biol. 79:295.
Hall, Timothy. 1974. Personal communication.
Henshaw, E. C. 1968. J. Mol. Biol. 36:401.
Ihle, J. N., and L. S. Dure III. 1969. Biochem. Biophys. Res. Com. 36:705.
Ihle, J. N., and L. S. Dure III. 1970. Biochem. Biophys. Res. Com. 38:995.
Ihle, J. N., and L. S. Dure III. 1972a. J. Biol. Chem. 247:5048.
Ihle, J. N., and L. S. Dure III. 1972b. J. Biol. Chem. 247:5034.
Ihle, J. N., and L. S. Dure III. 1972c. J. Biol. Chem. 247:5041.
Lee, C. S., and C. A. Thomas, Jr. 1973. J. Mol. Biol. 77:25.
Lindberg, U., and J. E. Darnell. 1970. Proc. Nat. Acad. Sci. U.S.A. 68:1331.
Mans, R. J., and T. J. Walter. 1971. Biochim. Biophys. Acta. 247:113.
McLaughlin, C. S., J. R. Warner, M. Edmunds, H. Nakazato, and M. H. Vaughn. 1973. J. Biol. Chem. 248:1466.

Mendecki, J., Y. Lee., and G. Brawerman. 1972. Biochemistry 11:792.
Niessing, J., and C. E. Sekeris. 1972. FEBS Lett. 22:83.
Penman, S., M. Rosbash, and M. Penman. 1970. Proc. Nat. Acad. Sci. U.S.A. 67:1878.
Ryskov, A. P., G. F. Saunders, V. R. Farashyan, and G. P. Georgiev. 1973. Biochim. Biophys. Acta 312:152.
Samarina, O. P., E. M. Lukanidin, J. Molnar, and G. P. Georgiev. 1968. J. Mol. Biol. 33:251.
Shih, D. S., and P. Kaesburg. 1973. Proc. Nat. Acad. Sci. U.S.A. 70:1799.
Spirin, A. S., and M. Nemer. 1965. Science 150:214.
Waters, L. C., and L. S. Dure III. 1966. J. Mol. Biol. 19:1.

COLIN H. DOY

Asexual Approaches, Including Transgenosis and Somatic Cell Hybridization, to the Modification of Plant Genotypes and Phenotypes

Reports in the media both in Australia and internationally have implied that in my laboratory we have transferred genetic information from animals to plants, that the process of nitrogen fixation has been introduced to crop plants not previously able to utilize atmospheric nitrogen so that nitrogenous fertilizers are no longer required, and that the quality of plant protein as a food for man can now be dramatically improved. Not one of these statements is true, and they represent journalistic elaboration of responses to the question, "Has your work any practical value?"

The above notwithstanding, the objective of modifying the total plant genome by asexual means, including the transfer of genes from entirely foreign sources, may be nearer. Such experiments have importance as fundamental science but also have practical potentialities for mankind. In fashionable terms, we are ultimately concerned with the genetic engineering of plants.

Classical plant genetics and breeding are both protected and limited by natural barriers against universal hybridization. For some fundamental studies and for the breeding of completely new kinds of plants, one would like to breach these barriers. This paper is concerned with newly developed approaches to that end. Associated with this is the utilization of cell and tissue lines in culture with, by analogy with microorganisms, advantages in using haploid lines that might also become homozygous diploid seed-bearing plants.

I think Dr. Sprague said in discussion that homozygosity was never a problem to achieve. However, it is my understanding that it can take 8 or 10 years to be certain of homozygosity by classical breeding. A haploid cell line or embryo, obtained by another culture or some other means, has the advantage of the potential to (a) immediately express recessive as well as dominant genes, including mutations; and (b) to obtain homozygosity for all genes of the haploid line by chromosome doubling.

For some lines enough is now known to regenerate complete seed-bearing plants from embryos or cells, and one can expect that basic research will eventually lead to similar success with any line. Of course, there will be the necessity for field testing, but one will start from a basis of homozygosity. The haploid to homozygous diploid approach is not selective for genes, except that normally recessive lethals are eliminated; in other examples one gets homozygosity for undesirable as well as desired genes.

Another limitation is that any one clonal line contains only the genes in a particular set of chromosomes, yet desirable genes are certain to be in the alternative chromosome set. One, of course, also has the opportunity to isolate a line derived from this set and, indeed, any of the nonlethal variations existing in the original haploid segregation. Incidentally, unless cells are cloned, hapoid cell lines from anther culture are likely to be mixed lines.

From discussions with plant breeders, I have gained the impression that these difficulties and complications tend to alienate them from the haploid approach. They are concerned about the genetic variation that may be expressed because it is felt that this is likely to be almost all "bad." However, genetic variation is also the pool for spectacular changes in breeding. I have seen homozygous diploid barley growing in a greenhouse of the Waite Institute of Agricultural Research, Adelaide, Australia. These plants looked vigorous, with a fine head of grain, and therefore I am encouraged to remain open-minded about the haploid approach to plant breeding.

The key to the future is the development of selective methods for what one wants at the earliest stage possible. The newer approaches, such as cell hybridization, suggest means to then recombine and study characteristics *in vivo*. Clearly, there is the possibility of developing somatic cell genetics such as complementation and dominance tests *in vitro*. Transgenosis offers the prospect of introducing and testing the expression and inheritance of genes from completely unrelated cell lines or organisms.

Haploidy is not new, but its extension from a few to many genera and species is, and in recent years we (Gresshoff and Doy, 1972a,b; 1974) have demonstrated with the previously intransigent species *Arabidopsis thaliana*, *Lycopersicon esculentum* (tomato), and *Vitis vinifera* (grape) that an approach based on genetic variation combined with a simple, defined medium and controlled environment is successful. In practice we have had no failures. The haploid utopia would be achieved if, for all plants, pollen itself could be induced to form large numbers of haploid embryos from which one could generate cell lines as well as haploid and homozygous diploid plants at will.

I am excited by the potential of haploidy, cell fusion, and transgenosis as tools for the study of plant biology. To me they represent the thinking and approaches of microbiology, biochemical genetics, and molecular biology in application to higher plant biology. As propositions for the practical man they are, in the immediate future, of lesser attraction, but I think he would be foolish to ignore them, and he should support their development and assessment. In terms of the topic of this workshop, one can offer no immediate prospect that the new approaches will improve seed proteins. They might if anyone could think of a means to express the relevant genes in culture and then select for desirable changes. What the new approaches do offer is the prospect of a better understanding of the basic processes of plant biology, including the expression of genes and the regulation of the function of gene products.

My impressions gained at the 1973 Gordon Conference on Plant Cell and Tissue Culture, and at a symposium and discussion held March 1974 at the University of Minnesota, are that plant breeders are beginning to be seriously interested. I have no desire to oversell the new techniques; rather, I think they need to be critically evaluated and the present limitations stated. I find that people expect too much, too quickly from transgenosis.

I shall discuss transgenosis and cell fusion, comment on matters related to their further development, and conclude with a "warning" of the possible biological dangers of manipulations leading to genetic engineering. It should be realized that these dangers have been with us for some time but have not been discussed in relation to plant research.

TRANSGENOSIS

The following is adapted from a review of transgenosis (Doy, 1975), to which you are referred for further information and discussion. In the context of the present meeting I think it is important to communicate why the term transgenosis was introduced and my own view of what it may come to mean.

The term transgenosis was coined in response to reaction to results obtained in conjunction with my then graduate student, Peter Gresshoff, and colleague Barry Rolfe. We had changed the general phenotype of certain plant cells in defined culture from nongrowth to growth by exposing the cells to bacterial viruses (specialized transducing phage) carrying specific bacterial (*Escherichia coli*) genes known to encode functions for converting the nonutilized carbon source to an available one. It seemed to us that the simplest explanation was that the bacterial genes were transferred into the plant cells and then transcribed and translated to a functional gene product using the machinery of the plant cells (Doy *et al.*, 1972a,b). In a later example, where growth became nongrowth due to the use of phage carrying a nonsense suppressor, only transcription was required (Doy *et al.*, 1972b, 1973a,b).

Although we hoped that the new phenotypes might persist indefinitely (and they did) and that the transgenote might be inherited, we did not claim this. Our colleagues tended to be either extremely skeptical or else to claim for us that inheritance occurred, and moreover to ascribe their own image of mechanisms through the use of terms such as *transformation* and *transduction*. We were opposed to this, and consequently the term *transgenosis* was introduced (Doy *et al.*, 1973a,b). Essentially, it is meant to cover transfer and all processes, such as transcription and translation, needed for phenotypic expression, but does not automatically imply a claim for inheritance, although inheritance may subsequently be proved. The term is also meant to emphasize that for many systems one is at present observing phenomena and offering hypotheses without any knowledge of mechanism except by inference against the background of general biological knowledge.

The new approaches to biology and genetic engineering are of interest to scientists of diverse backgrounds and disciplines. The term transgenosis has been introduced also to help avoid confusion in the use of terms such as transformation and transduction. The molecular biologist associates these terms with definite results and mechanisms from phage and bacterial genetics. At the same time, the botanist and zoologist may have an interpretation based on morphological changes, and I have found recently that transduction also has a neurological meaning.

In the course of discussion in Australia and overseas I have come to think of transgenosis as the transfer of genes by an asexual process, resulting in a change of phenotype but with the potential for a "permanent" change of genotype to form a hybrid. The phenotype may appear before a process of inheritance is established or may appear or disappear after genetic incorporation. I have therefore no objection if the term transgenosis is retained to describe examples where an inherited genotypic change is proven. In the long term, transgenosis may prove useful as a generalization for asexual transfer, expression, and inheritance of genetic information between donor and recipient organisms (cells) widely separated by evolution.

I should like to emphasize that the definition of transgenosis does not restrict one to a particular donor, vector, recipient, or host. Thus the donor/vector could be nucleic acid, a virus, a plasmid, a cell, or a protoplast, and the recipient/host could be many kinds of protoplasts, cells, tissues, or organisms, including those of animals as well as of higher plants.

Present examples of transgenosis come under the general heading of DNA-mediated transgenosis and phage-mediated transgenosis.

DNA-MEDIATED TRANSGENOSIS

This is discussed in detail elsewhere but, in summary, apart from the work of Hess (Hess, 1969a,b; 1970; 1973) there seems to be much controversy about the reproducibility of the results.

The experiments of Hess are very similar to transformation in the bacterial geneticists' sense. DNA extracted from one line of *Petunia hybrida* is used as a transgenote to change the phenotype (usually color) and genotype of another line of *Petunia hybrida*. There was no segregation of the transgenote, and I am not sure if the location of the introduced genes in the recipient and subsequent progeny is firmly established. The color of petunia flowers is determined by many genes (Wiering and de Vlaming, 1973) and is influenced by pH. It has been shown recently that mosaic flowers occur by spontaneous somatic mutation at a similar frequency to the transformation results (Bianchi and Walet-Foederer, 1974).

Speaking strictly, the above is only a marginal example of transgenosis, since the donor and recipient are so closely related. However, the normal sexual route is avoided, and therefore it is an example of the new kind of thinking.

Claims for the persistence, expression, and indefinite inheritance of bacterial DNA are typified by the work of Ledoux and his colleagues since 1961 (Doy, 1975). This group claims that the bacterial DNA becomes integrated into the plant chromosomal DNA in a double-stranded end-to-end junction. The hybrid DNA is inherited through the seeds, and the transgenote of bacterial origin is claimed to correct nutritional deficiencies, such as thiamine auxotrophy, in the plants—for example, in *Arabidopsis thaliana*. I have discussed the work in detail elsewhere (Doy, 1975), and I think nothing is to be gained by doing so again. In general terms, workers outside Ledoux's laboratory have been unable to repeat the work. My own view is that the case for DNA-mediated transgenosis would be strengthened enormously if a nonselected genetic marker originating in the transgenote could be demonstrated in plants thought to express and inherit a selected marker and if DNA deleted for the transgenote were ineffective.

If the work of Ledoux and his coworkers becomes accepted, then this group has already used a DNA-mediated transgenote to create a genetically stable hybrid plant.

Assuming this has occurred, my objection to the approach is that much unwanted genetic information has been transferred. In my opinion, the aim should be to transfer only specific genes. It was this philosophy that caused the work with phage-mediated transgenosis to be started in my laboratory.

PHAGE-MEDIATED TRANSGENOSIS

Phage-mediated transgenosis has now been reported by three separate groups, and I understand that other workers are extending the observations. Thus, unlike DNA-mediated transgenosis, similar phenomena are found by workers from independent laboratories.

Details of the published work are discussed elsewhere (Doy, 1975). In this paper I shall reiterate briefly the published results and explain why I have concluded that unmodified phage is unlikely to be the preferred vehicle for long-term transgenosis.

Our reasons for using specialized transducing phage can be summarized as follows. The phage represents a carrier of a few selected and well-characterized bacterial genes (*lac, gal, supF*) as part of a small total genome. Whole phage rather than extracted DNA was used because it was the simplest thing to do and also the coat protein may initially protect the DNA. The phage DNA can circularize, which would help to protect it against enzymic degradation and also give the possibility for maintenance, replication, and inheritance as a plasmid. Phage preparations of very high titer can be obtained representing a purification of the intended transgenote and therefore increasing the chances for success.

In these laboratories we (Doy *et al.*, 1972a,b; 1973a,b,c) have used specialized transducing phage λ and $\phi 80$ as vectors in the transgenosis of various genes derived from *Escherichia coli* K12. Genes of the galactose operon (*gal*) and the lactose operon (*lac*) were used to create growth in situations where certain plant tissue cultures did not normally grow: that is, when galactose or lactose was the sole source of bulk carbon for growth. The recipient tissue cultures were haploid lines of *Arabidopsis thaliana* and *Lycopersicon esculentum*. There is no significance in the haploid state other than that these lines were derived and in use in the laboratory for other reasons. The other transgenote was the $supF^{+}$ gene from a mutant *E. coli* K12. This experiment originated in the results of experiments with *gal*. $SupF^{+}$ specifies a tRNA able to insert tyrosine at UAG, which otherwise is a nonsense codon.

In summary, growth occurred on galactose or lactose only when the phage carried relevant genes. For *gal* a UAG nonsense mutation defeated successful transgenosis of growth on galactose, suggesting that the plant cells could not efficiently correct the bacterial nonsense codon. We speculated that the UAG codon was important to the plant cells—for example, it might be concerned with full stops in the genetic read-out. Assuming the importance of UAG, we predicted that transgenosis of $supF^{+}$ would be harmful to the plant tissue culture. In the event, it killed cultures on optimal glucose-supplemented growth medium, and it was possible to titrate out the dose to zero effect. I shall return to this result when discussing the possible dangers of transgenosis.

Concentrating on the transgenosis of *lac* and specifically gene *Z* for *E. coli* β-galactosidase we were able to support the evidence of growth with enzyme assays and immunological evidence specific for the *E. coli* enzyme. This was of particular importance because the occasional control grew on lactose but never showed the presence of the specific *E. coli* enzyme. Instead they retained the low level of analogous plant function characteristic of negative controls or tissue growing on other carbon sources.

Two other laboratories have obtained similar results. Carlson (1973) found that two gene products of phage T_3 were expressed in protoplasts of *Hordeum vulgare* (barley). The enzymes are *S*-adenosyl methionine cleaving enzyme, and the RNA-polymerase encoded in the early region of the T_3 genome. As in our work, a nonsense mutation in the T_3 RNA-polymerase gene prevented expression of the enzyme. Johnson *et al*. (1973) found that suspensions of *Acer pseudoplatanus* (sycamore) grew on lactose when exposed to λ *plac5* but not on exposure to λ or no phage at all. No enzymological or immunological data were reported.

In my hands the phenotype of growth on lactose in either the dark or light with numerous variations of auxin and kinetin levels has persisted indefinitely for the *L. esculentum* cultures. I had hoped to generate whole plants, including homozygous diploids, and then investigate, by classical means, the possible inheritance of transgenote gene *Z*. In the event, although greening and chloroplasts developed, no whole plants were formed. This also applied to the untreated stock cultures on sucrose or glucose medium of the *L. esculentum* haploid line that originally yielded plantlets (Gresshoff and Doy, 1972b).

After many experiments over the last 2 years, I have been able to conclude that the long-term phenotype of growth on lactose is not evidence of inheritance and expression of gene *Z*. Despite much effort to induce a change, the levels of β-galactoside have remained low in the long-term cultures and the enzyme is not characteristically *E. coli* in the immunological test. This suggests that a spontaneous mutational or epigenetic change has caused the plant enzyme to become effective in making lactose available as a general carbon source. I have recently demonstrated selection of such a line from suspension cultures of nontransgenosed *L. esculentum*. The suspension cultures that grow best on lactose grow at the same rate as suspension cultures derived from transgenosed tissue cultures. The difference between the non-transgenosed and extransgenosed cultures is that on transfer to suspension culture on lactose the extransgenosed material grew without lag, whereas nontransgenosed cultures either never grew or had a long lag before growth was established.

We have shown (Doy *et al.*, 1973c) that *E. coli* gene *Z* β-galactosidase is expressed at high levels in the *L. esculentum* cultures in what may be bursts of synthesis. I have now concluded that whenever a plant tissue culture shows these very high levels of transgenote expression that culture, or rather most of its subcultures, eventually dies. This suggests that in the long term there is counter selection for transgenosed cells and selection for favorable changes in the normal plant genome. I would speculate that this is due to an imbalance caused by nonspecific transgenosis. The evidence for this arises from subcultures originally exposed to neutral phages. For example, *L. esculentum* growing on glucose medium is killed by $\phi 80 supF^+$ but seemingly not affected by $\phi 80$ (Doy *et al.*, 1973a,b). $\phi 80$ is called a neutral phage since it carries no relevant genes. If the culture exposed to $\phi 80$ is then continuously subcultured on glucose medium, the sequential subcultures grow ever more slowly relative to controls never exposed to phage and eventually seem to be dead.

This is exactly what I would expect for successful transgenosis of genes that convey no selective advantage on the recipient. There is no reason transgenosis should be confined to desired *E. coli* genes carried by the phage. Phenotypic expression would divert the resources of the plant cells, and all the more so the more successful the transgenosis, even of desirable genes. These results therefore lead me to conclude that phage in its present form is not the vehicle for long-term transgenosis. At most, transgenosis of gene *Z* was successful for 9 months in culture.

SOMATIC CELL FUSION

Somatic cell fusion was first discovered for animal cells. Similar studies for plant cells are complicated by the fact that they have a cell wall that must first be removed to form a protoplast (or perhaps more strictly a spheroplast in some cases). Cocking (1972) has extensively reviewed this area of research, which he greatly stimulated by devising means to make large numbers of protoplasts by nonmechanical means. The plant cell protoplasts are fused by various methods in which inorganic salts and pH (Power *et al.*, 1970; Keller and Melchers, 1974) are important. Recently, polyethylene glycol (PEG) has been used to enhance cell fusion (Kao and Michayluk, 1974). In all probability it should now be possible to fuse any kind of plant protoplasts, probably even protoplasts of any kind of cell. My own limited experience of cell fusion makes me think that the problem is often to avoid fusion.

Once having made protoplasts and fused them, the problem is to select the desired fusion product and to regenerate cells and whole plants. The complete chain of events from protoplast to a whole seed-bearing plant has been achieved only for a very few species, I think three at present: tobacco, carrot, and petunia. Clearly, much research is needed in this area with the ultimate aim of learning how to regenerate any protoplast or fusion product to the status of a seed-bearing plant. In this discussion I have used the terms *cell fusion* and *fusion product* because, in my opinion, one is justified in claiming hybridization only if characteristics from each cell type are expressed.

In my terminology, cell fusion and hybridization together are cell-mediated transgenosis. Genetically, fusion could result in a heterokaryon or a hybrid single nucleus. I suspect that some of the difficulties associated with success in generating a heterokaryon or single nucleate hybrid may be the result of interactions between different kinds of cytoplasm. If one introduced only a nucleus, this would avoid that problem but might reveal the opposite situation, where factors in both cytoplasms contribute to success.

To my knowledge, only one interspecific hybrid plant has been produced by cell fusion, the amphiploid hybrid between *Nicotiana glauca* and *N. langsdorfii* (Carlson *et al.*, 1972). This is a significant achievement and transgenosis in the full sense of sexual transfer, expression, and inheritance, but this hybrid can be produced sexually. An even more significant and unanswered question is whether hybrid seed-bearing plants not obtainable by sexual means can be generated by fusion. The farther separated by evolution the fused species become, the less likely it seems that a major stable combination of genes will result. This is related to problems of synchronizing cell division and chromosome retention, whether in different nuclei or the same nucleus. Chromosome pairing would readily occur between fused haploids of the same species, and fused diploids would provide for pairing, whatever the origins.

Thus, it seems that at present one has the potential to bring together within the one limiting plasma membrane any genetic information one likes. The question is what can then be done? One can hope for two things: (a) the generation of a new hybrid plant and (b) the development of systems of somatic cell genetics in culture. At the moment, I see most application in (b), especially between haploids of the same or closely related species.

GENERAL REMARKS

At present I regard all forms of transgenosis as a means to gain insight into basic processes at the level of cells or tissues in culture. Much information can be gained, even with transient situations, provided there is phenotypic expression. If one is to breed completely new species, then it seems that there is a need to limit or get rid of surplus genetic information. Phage-mediated transgenosis was intended to be, and is, a beginning in that direction. Intuitively, the chances of getting a desired seed-bearing hybrid seem better if one restricts the transgenote to selected desirable genes and integrates these into an existing genome, either nuclear or cytoplasmic. I shall return to this later.

As already stated in this paper, and in the discussion of this workshop, haploids have many potential advantages. However, a major question remains unanswered, and that is the degree of gene duplication even in a monohaploid. By that I mean that the selection of a normally recessive character such as auxotrophic mutation becomes much easier if there is only one gene copy specifying the function in question. Since many plants are polyploid, even the haploid has more than one chromosome of a kind and therefore more than one gene of a kind. It is only for true diploids that the haploid has one chromosome of a kind. This consideration is why my work with Peter Gresshoff over the past few years was based on haploids of *A. thaliana* and *L. esculentum*. We, and many others, have searched for auxotrophic mutants, and essentially the only success was in the first report of Carlson (1970), and even these were leaky and have not been followed up. Bromodeoxyuridine incorporation into DNA has not been successful as a means of auxotroph selection in my hands, and from discussions during this overseas visit it seems this is the general experience. Mutants for resistance to bromodeoxyuridine turn up in many laboratories. These mutants may prove useful as forcing markers but beg the question of auxotrophic mutants. The important and perhaps crucial question of gene duplication in haploids can, I think, be tested, and I can illustrate my suggestion by reference to the work of Widholm, which also illustrates matters of direct interest to this workshop on improvement of seed protein.

The aromatic pathway in microorganisms, both prokaryote and eukaryote, is well understood with regard to both intermediate metabolism and control. I have played some part in these investigations and see studies of aromatic biosynthesis in plant systems as a potential model for direct and comparative studies with prokaryotic mechanisms of control. Widholm (1972a,b,c) has studied various aspects of the tryptophan specific branch pathway in both *N. tabacum* and *Daucus carota* cells and cell-free extracts. Tryptophan controls the flow of common aromatic intermediates, partly inhibiting the first enzyme of the specific tryptophan pathway, anthranilate synthetase. This inhibition is mimicked by tryptophan analogs—for example, 5-methyltryptophan, which inhibits growth of both prokaryotes and plant cells in culture because it blocks tryptophan synthesis but does not substitute for tryptophan in protein. Thus one can select resistance to the analogs, and one class of mutants can be expected to be defective in feedback inhibition by tryptophan at anthranilate synthetase. Widholm obtained such mutants, and two properties concern us: (a) these cells now overproduce tryptophan, and (b) anthranilate synthetase is now only partially inhibited by tryptophan. This latter is exactly what one would expect if more than one identical gene were specifying polypeptides involved in the recognition of tryptophan by the enzyme. Only if there is one active gene copy would one expect mutants with a total lack of allosteric inhibition at anthranilate synthetase. Thus, if Widholm had used haploid cells, selection for resistance would also be a test for gene duplication or activity. I propose, therefore, that people look for similar resistance mutants in haploid plant cells (but not necessarily only for anthranilate synthetase) and investigate whether it is possible to obtain a class of mutants with an enzyme completely unaffected by the previously regulatory ligand (e.g., tryptophan for anthranilate synthetase). If this class of mutants cannot be recovered, then either the material is not a unihaploid, or else gene duplication occurs in the unihaploid. In the latter event, the chances of getting auxotrophic mutants and similar mutants covering a broad spectrum of biological processes become poor, and indeed this may explain why only certain kinds of mutants have ever been obtained.

As stated, Widholm's mutants overproduce tryptophan. Up to a point, gene redundancy would not prevent selection of resistance mutants, and therefore, depending on ploidy, gene redundancy, reselection of secondary mutants, and so on, one could adjust the degree of overproduction. As may be the rule in most eukaryotes, repression of enzyme synthesis is not a factor in control. In my opinion, control of higher eukaryotes is likely to depend on developmental control of gene expression and molecular control of rates of function by ligands, including products of pathways.

As far as I have been able to discover, although Widholm's mutants accumulate high pools of tryptophan, this is not reflected in an increased level of tryptophan in cell protein. I don't think seeds have been looked at specifically. This makes the point that overproduction of a "quality" amino acid does not ensure an increase in protein quality. Studies like those of Widholm on tryptophan synthesis and of others interested in lysine overproduction (see Miflin, this volume) are invaluable as basic research, but from the point of view of improvement of plant seed protein they represent but one stage of what is required. If a particular amino acid is limiting for the synthesis of a particular protein, only then will its increased production lead to more of the protein. An increased pool of the amino acid is valuable of itself only if the total foodstuff is eaten, as it might be as an animal feed, but is no good at all if cooking habits lead to it being thrown away with the discards or cooking water.

It should also be realized that if one metabolic process is accelerated, there are likely to be demands made on total resources of material and energy that could lead to unfavorable consequences in other directions. Earlier in the workshop we heard from Dr. Hardy how nitrogen fixation can be limited by carbon resources rather than the intrinsic ability to fix nitrogen.

There are many points like these that one could make, and indeed it may not matter if the source of protein, for example seed, is borne on a plant stunted because its resources have been diverted to seed quality. In the end it may turn out that we synthesize the desired products *in vitro*. Our production of pseudofruits may be the first limited example

(Gresshoff and Doy, 1972b; Gresshoff and Doy, 1973). It may be that in some situations and economies, increased production of high-quality amino acids is a nonproblem since they can be synthesized and added and, if necessary, the flavor adjusted to suit.

Finally, in this section I wish to emphasize that in many situations where transgenosis, including cell fusion, is likely to be developed, selection methods are required. The ideal is a forced selection where only what is wanted survives. To that end, forcing markers, such as nutritional requirements and resistance to chemicals (including drugs), are invaluable. The simplest situation is that in which one parent (or donor) contributes one forcing marker and the other parent (or recipient) another. This approach for obtaining a heterokaryon was, for example, useful in our studies of the biochemical genetics of the regulation of aromatic biosynthesis in *Neurospora crassa* (Halsall and Doy, 1969). Resistance markers can be positively selected in cell lines and therefore are relatively easy to obtain. The difficulty in obtaining auxotrophic markers has already been discussed. As far as I know, a selection for increased levels of a particular seed protein is not possible at the level of cell culture, since it is unlikely that the gene(s) concerned will be turned on. If one can learn to regulate expression of the particular genes, the position might change.

THE FUTURE OF TRANSGENOSIS AND THE POSSIBILITY OF BIOLOGICAL DANGERS

I have already said why I think unmodified phage is not suitable as a direct vector for long-term objectives, including inheritance of the transgenote in plant cells and whole plants. That is not to say that phage is not useful in short-term experiments like those from this laboratory. Phages may turn out to be a vital component of stable transgenosis, since they are vehicles for the selection and replication of new DNA recombinants in *Escherichia coli*. In the following paper, Boyer and coworkers will discuss the restructuring of DNA molecules. This is the way transgenosis is likely to go, especially if fundamental and practical objectives require inheritance and incorporation into the existing genome, whatever the organism (see also Doy, 1975). The new techniques, based on enzymes that cut up DNA and others that rejoin DNA, result in the creation of new combinations of genes, including genes from diverse biological origins. In essence, one can rescramble the separations of evolution, and there is now a gigantic gene pool. These DNAs could serve as transgenotes for either DNA- or virus-mediated transgenosis, and the inclusion of DNA homologous to the

DNA of the recipient cell would maximize the possibility that donor genes may become part of the recipient genome.

The above applies whether or not the recipient organism is a plant or an animal cell. Plant material has unique advantages over animal cells: (a) Plant cells or protoplasts may be developed into seed-bearing plants, and (b) plant metabolism has many parallels in bacterial metabolism since both synthesize many end products from minimal nutritional requirements. One can, therefore, on the basis of (a), seek to breed in completely new genes, including different forms of regulation (for example, repression). Assuming reciprocity of transgenosis of eukaryote genes in prokaryotes, one could, on the basis of (b), select for specific plant genes in *Escherichia coli*. Random pieces of plant DNA could be joined to phage or plasmid DNA and introduced into an *E. coli* strain deleted for the gene corresponding to the plant function and gene of interest. The desired hybrid would therefore be selected on the basis of specific transgenosis in *Escherichia coli*. The plant gene could also be purified as the result of transcription in the prokaryote, and the resultant DNA used either as a transgenote for reintroduction into plant genomes or for mapping the location of the particular gene in the plant genome. Hogness has already done this kind of mapping for *Drosophila* (Hogness, 1975). An elaboration of the above suggestion is to mutate plant genes and then select for the gene in *Escherichia coli*. Clearly, the possibilities are enormous and have potentials for important fundamental and practical discoveries. The techniques of the microbiologist and molecular biologist are largely developed, but, for application to plants, the limiting factor is the present lack of basic knowledge and technical skills at the level of plant cells in culture. This is particularly so where the objectives are practical gains for the benefit of mankind, such as more and better-quality food. In this context an improved tobacco, carrot, or petunia is not very relevant. From the point of view of improved seed proteins, an approach might be based on the isolation of messenger RNA molecules when the protein is being synthesized and then the construction of DNA by reverse transcriptase for use as a transgenote.

The possibilities of this field are going to attract a lot of workers because of the science, but also because of the possibility of grants and fame and fortune, and realistically, I think in some quarters, because of the possibility of creating agents of biological warfare. I find among many scientists, and plant biologists in particular, either a lack of understanding or a reluctance to acknowledge the possibilities of accidental disaster or deliberate evil. Scientists working with transgenosis in animal systems and viruses have recently clearly recognized the possible dangers of some forms of genetic engineering and have recommended a moratorium on certain work (see Nature 250:175, 1974; and Berg *et al.*, 1974). I shall not reiterate the arguments, but I would urge that plant biologists participate in the proposed discussions. The potential dangers of the human and animal work appeal to the emotions and therefore have an immediate impact, especially in the media. To my knowledge the analogous possibilities in plant biology have not been discussed. I do not think that ideas can be suppressed, nor do I think that possibilities for understanding and good should be forgone because of possibilities for evil which might then go on in secret.

My own involvement with the possibility of biological danger in transgenosis and genetic engineering came with the death of plant cells resulting from exposure to the *E. coli* nonsense suppressor $supF^+$ in the form of $\phi 80supF^+$ (Doy *et al.*, 1973a,b). This and analogous phages have been about for a long time. We tested to see whether plant cells more normal than our material were susceptible, but fortunately they were not (Doy *et al.*, 1973a,b). I drew attention at the 1973 Gordon Conference on Plant Cell and Tissue Culture to the possible dangers of transgenosis suggested by this work, but without provoking much reaction; indeed, the first positive and supporting reaction to a similar warning was given by Hogness (1975) at the Eukaryote Chromosome meeting in Canberra in May 1974.

A possible danger with plant systems is the involvement of plant viruses as vectors. Most plant viruses are RNA viruses, but these are also potential sources of biological danger. In discussions I have found that plant virologists do not readily recognize the danger. At present I think it is fairly common to recombine the different RNA components of an RNA plant virus with a consequent change in the host range. This in itself might represent a danger but is usually defended on the basis that the viruses need an insect vector for infection. But what about a greenhouse escape? It should also be appreciated that those interested in transgenosis and genetic engineering are likely to use the protoplast route, so avoiding the need for a vector. In fact some very good work has already been done by Takebe and his associates using tobacco mosaic virus in protoplasts (see, for example, Otsuki *et al.*, 1972). Thus a recombinant plant virus could possibly be introduced into whole plants generated from protoplasts, and at that stage exposure to natural vectors might intervene to spread the virus. In this situation there is an element of security since the infected plant could only be made as a deliberate act.

In this laboratory we have considered the possibility of using transgenosis for many kinds of genetic engineering, including the transfer of bacterial nitrogen fixation into plants directly or by new kinds of biological associations. This would seem to be an important practical goal if the recipient were an important crop plant, such as rice. It would also seem to be safe, since the nitrogen cycle would presumably regenerate atmospheric nitrogen.

As part of this research, we have considered it important to investigate what genetic and environmental factors are important in nitrogen fixation, and indeed to challenge the dogmas of the subject. To that end we proposed (Doy *et al.*, 1973b) to transfer *nif* genes from *Klebsiella* to *E. coli* K12, but a similar result, transfer to *E. coli* C, was reported by Dixon and Postgate (1972) before we implemented our scheme. We (Mary Warren Wilson, Rolfe, and Doy) have now achieved the goal of transfer of anaerobic nitrogen fixation to *E. coli* K12 in a stable form from both *Klebsiella pneumoniae* and *Rhizobium trifolii* and, in the process, have found that certain *E. coli* genes are essential for expression of the *nif* transgenote. Further, we have evidence to suggest that an *E. coli* K12 mutant may be able to fix nitrogen without the introduction of foreign genes and that *Rhizobium trifolii* may have the potential to fix nitrogen in the absence of plant cells. Some of these data suggest that nitrogen fixation may be induced to occur aerobically.

If some or all of these preliminary conclusions can be confirmed, then this is a major advance in the understanding of nitrogen fixation and the possibility of transgenosis to plants. Our studies were possible only because of the construction of hybrid forms of DNA and the use of forcing markers, such as drug resistance, and the replication of these DNAS in bacteria. I believe that these and similar experiments and goals are important, and any suggestion that such work should be banned must be fully discussed and justified, and plant biologists must be involved. My laboratory has, for the last 2 years, observed a voluntary moratorium on some developments of transgenosis because I felt that local facilities were not adequate for safe and experienced experimentation.

REFERENCES

Berg, P., D. Baltimore, H. W. Boyer, S. N. Cohen, R. W. Davis, D. S. Hogness, D. Nathans, R. Roblin, J. D. Watson, S. Weissman, and N. D. Zinder. 1974. Science 185:303.

Bianchi, F., and H. G. Walet-Foederer. 1974. Acta Bot. Neerl. 23:1.

Carlson, P. S. 1970. Science 168:487.

Carlson, P. S. 1973. Proc. Nat. Acad. Sci. U.S.A. 70:598.

Carlson, P. S., H. H. Smith, and R. D. Dearing. 1972. Proc. Nat. Acad. Sci. U.S.A. 69:2292.
Cocking, E. C. 1972. Annu. Rev. Plant Physiol. 23:29.
Dixon, R. A., and J. R. Postgate. 1972. Nature 237:102.
Doy, C. H. 1975. *In* The Eukaryote Chromosome (W. J. Peacock and R. D. Brock, ed.). Australian National University Press, Canberra.
Doy, C. H., P. M. Gresshoff, and B. G. Rolfe. 1972a. Proc. Aust. Biochem. Soc. 5:3.
Doy, C. H., P. M. Gresshoff, and B. G. Rolfe. 1972b. Search 3:447.
Doy, C. H., P. M. Gresshoff, and B. G. Rolfe. 1973a. Proc. Nat. Acad. Sci. U.S.A. 70:723.
Doy, C. H., P. M. Gresshoff, and B. G. Rolfe. 1973b. *In* The Biochemistry of Gene Expression in Higher Organisms (J. K. Pollak and J. Wilson Lee, ed.). Australia and New Zealand Book Co., Sydney, p. 21.
Doy, C. H., P. M. Gresshoff, and B. G. Rolfe. 1973c. Nat. New Biol. 244:90.
Gresshoff, P. M., and C. H. Doy. 1972a. Aust. J. Biol. Sci. 25:259.
Gresshoff, P. M., and C. H. Doy. 1972b. Planta 107:161.
Gresshoff, P. M., and C. H. Doy. 1973. Aust. J. Biol. Sci. 26:505.
Gresshoff, P. M., and C. H. Doy. 1974. Z. Pflanzenphysiol. 73:132.
Halsall, D. M., and C. H. Doy. 1969. Biochim. Biophys. Acta 185:432.
Hess, D. 1969a. Z. Pflanzenphysiol. 60:348.
Hess, D. 1969b. Z. Pflanzenphysiol. 61:286.
Hess, D. 1970. Z. Pflanzenphysiol. 63:461.
Hess, D. 1973. Z. Pflanzenphysiol. 68:432.
Hogness, D. S. 1975. *In* The Eukaryote Chromosome (W. J. Peacock and R. D. Brock, ed.). Australian National University Press, Canberra.
Johnson, C. B., D. Grierson, and H. Smith. 1973. Nat. New Biol. 244:105.
Kao, K. N., and M. R. Michayluk. 1974. Planta 115:355.
Keller, W. A., and G. Melchers. 1974. Z. Naturforsch. 28:737.
Otsuki, Y., T. Shimomura, and I. Takebe. 1972. Virology 50:45.
Power, J. B., S. E. Cummbins, and E. C. Cocking. 1970. Nature 225:1016.
Widholm, J. M. 1972a. Biochim. Biophys. Acta 261:44.
Widholm, J. M. 1972b. Biochim. Biophys. Acta 279:48.
Widholm, J. M. 1972c. Biochim. Biophys. Acta 279:48.
Wiering, H., and P. de Vlaming. 1973. In *Petunia hybrida* I. The gene *Gf*. Genen Phaenen 15:35.

DISCUSSION

DR. KEY: Thank you, Colin. I think that Dr. Doy made a point at the end of his discussion that certainly the Organizing Committee had in mind when we set up the program, namely, that one is going to need to manipulate, modify, and restrict the kind of DNA that one is using if one is going to achieve transgenosis in developing clones and lines for whatever plant materials.

DR. GABELMAN: I would like to comment on the carrot. I think you said somewhat incorrectly, and I do not think you intended to, that Widholm worked with polyploids in both cases. The carrot is a simple diploid.

DR. DOY: It is still polyploid in my sense.

DR. GABELMAN: We are interested in the biosynthesis of carotenes. For many years we wanted to study this in cell culture. In 1971 the Japanese demonstrated the role of 2,4-D in normal carotenoid synthesis (N. Sugano, S. Miya, and A. Nishi. 1971. Plant Cell Physiol. 12:525). Now that we have carrot cultures, we are looking at this from the standpoint of both diploid and haploids. We think we can generate haploids very readily in carrots, and we do have callus growing from these. We grow callus and free-cell cultures from the diploid, and there is no reason to assume we cannot in the haploid. I think that in 10 to 15 years we will handle cells of carrots much as we are handling *E. coli*.

DR. DOY: Let me make it clear that I am not criticizing Widholm's work but using it to make a point. I feel that the work of those seeking to understand metabolic control in plants or just to overproduce amino acids such as tryptophan and lysine is important as fundamental science and also provides information necessary for progress toward practical goals. It is clear from this workshop that synthesis of more of a particular end product is not necessarily the limiting factor in improvement of seed protein quality. At best, it is one step in a chain of necessary events, and one must also realize that a change favorable in one aspect may upset metabolic and energy balances in unfavorable ways as well.

HERBERT W. BOYER, ROBERT C. TAIT,
BRIAN J. McCARTHY, *and*
HOWARD M. GOODMAN

Cloning of Eukaryotic DNA as an Approach to the Analysis of Chromosome Structure and Function

The structural genetic complexity of the genomes of higher organisms usually obviates the reduction of these systems to the type of *in vitro* analysis that has been so productive for the genomes of prokaryotic organisms. A great deal of effort, however, has been devoted to the methodology of gene isolation from higher organisms (see Brown and Stern 1974, for review) with the objective of having purified genes for *in vitro* experimentation. The most successful procedures have relied on the relative fraction of the genome that the gene comprises and physical differences between the gene of interest and the rest of the genome. For example, one can isolate the amplified ribosomal and 5S DNA, reiterated histone genes, or reiterated satellite DNA. More recent innovations have used "reverse transcriptase" to synthesize DNA from pure messenger RNA obtained from specialized cells. These elaborate efforts to extend the methodology of gene isolation underscore the significance of having purified genes available in sufficient quantity for *in vitro* analyses.

We will outline a new methodology for the isolation and purification of genes of higher organisms that has developed in the last 2 years (see Cohen *et al.*, 1973; Morrow *et al.*, 1974; Hershfield *et al.*, 1974).

MOLECULAR CLONING OF DNA

We refer to the methodology to be described here as the molecular cloning of DNA (or genes). The enzymatic basis of the procedure involves only two enzymes, a restriction endonuclease (Boyer, 1974) that generates cohesive termini in DNA (Hedgpeth *et al.*, 1972; Mertz and Davis, 1972) and polynucleotide ligase to covalently join the reassociated hydrogen-bonded termini. The site-specific endonucleolytic activity of DNA restriction endonucleases has made them popular tools for molecular biologists. This property and the observation that some of these endonucleases generate staggered phosphodiester bond cleavages at these sites provides the basis of the cloning procedure. Three representative DNA restriction endonucleases and their substrates are presented in Table 1. The availability of three such endonucleases with different specificities provides flexibility to the approach since the substrates for these enzymes are distributed in DNA in a random fashion. Thus if the *Eco* RI endonuclease cleaves a gene of interest one could use another endonuclease to generate the intact gene in a DNA fragment.

The first procedural step involves the *in vitro* recombination of *Eco* RI endonuclease-generated fragments of DNA with a small bacterial DNA plasmid that has been cleaved by the *Eco* RI endonuclease to a linear structure. In principle, the other two endonucleases listed in Table 1 could be used in the same way, but most of the methodology has been worked out with the *Eco* RI endonuclease. Interaction of the *Eco* RI-generated termini is sufficiently stable at temperatures of 5° to 15°C to be ligated with polynucleotide ligase to a covalently recombined structure (Mertz and Davis, 1972; Dugaiczyk *et al.*, 1974). These DNA molecules are used to transform a culture of *E. coli* under conditions where one can obtain about 10^4 to 10^5 transformants per microgram of DNA. The transformed cells are selected from the culture by requiring some phenotype associated with the plasmid genome or the *Eco* RI fragment of DNA that has been recombined into the plasmid for survival.

TABLE 1 Representative DNA Restriction Endonucleases

Restriction Endonuclease	Source	Substrate[a]	Reference
Eco RI	*E. coli*	p G↓A A T T C C T T A A↑G p	Hedgpeth *et al.* (1972)
Eco RII	*E. coli*	↓p C C A G G G G T C C p↑	Boyer *et al.* (1973) Bigger *et al.* (1973)
*Hind*III	*H. influenzae*	p A↓A G C T T T T C G A↑A p	Boyer (1974)

[a]The 5′ nucleotides are prefixed with p, and the arrows designate the sites of phosphodiester bond cleavages.

The transformant bacteria are purified, and plasmid DNA is prepared from each independent transformant. The plasmid DNA is cleaved by the *Eco* RI endonuclease and the products analyzed by agarose gel electrophoresis (Figure 1) (Helling *et al.*, 1974) in order to determine if a recombinant plasmid has been isolated. One needs a bacterial culture of only 10-ml volume to obtain enough plasmid DNA for this type of analysis, and a reasonable number of transformants can be screened in 1 day. Once a particular *Eco* RI fragment has been inserted into a bacterial plasmid, it replicates as part of that plasmid and is segregated to daughter cells at cell division. This culture of bacteria can then be used as a constant source of the inserted DNA fragment by growing large cultures of the bacterium containing the recombinant plasmid, followed by purification of the plasmid. If one needs to or wishes to remove the bacterial segment of the DNA, the purified plasmid can be re-treated with the the *Eco* RI endonuclease and the fragments separated by one of several available procedures, such as gel electrophoresis or equilibrium sedimentation in CsCl. If the *Eco* RI fragment was derived from the genome of an organism such as *Drosophila melanogaster* or *Xenopus laevis*, for example, one would have a perpetual source of a given piece of the organism's genome.

PARAMETERS AFFECTING THE *in vitro* LIGATION OF DNA WITH *Eco* RI TERMINI

If one considers a linear fragment of DNA with a molecular weight of about 6×10^6 daltons (about the size of the bacterial plasmid to be described below) with two *Eco* RI termini, it is obvious that the ends of one molecule could reassociate to recircularize it or could reassociate with a second molecule to form a dimer. The recircularized molecule would be unable to participate in any other associations, but the linear dimer could circularize or extend itself by reassociating with a third molecule. The directions of this reaction can be influenced by varying the concentration of DNA (Wang and Davidson, 1966)—e.g., at concentrations below 2.5 μg/ml only circular molecules would be formed, while at concentrations above 25 μg/ml linear polymers of the starting material would be favored. The considerations to be made for this type of calculation are the contour length of the molecule and its flexibility under the conditions employed in the reassociation reaction. We have empirically determined the concentration of DNA of any given length that favors the polymerization reaction rather than the circularization reaction (Dugaiczyk *et al.*, 1975). Under the conditions used for the ligation of *Eco* RI termini, these concentrations were found to be

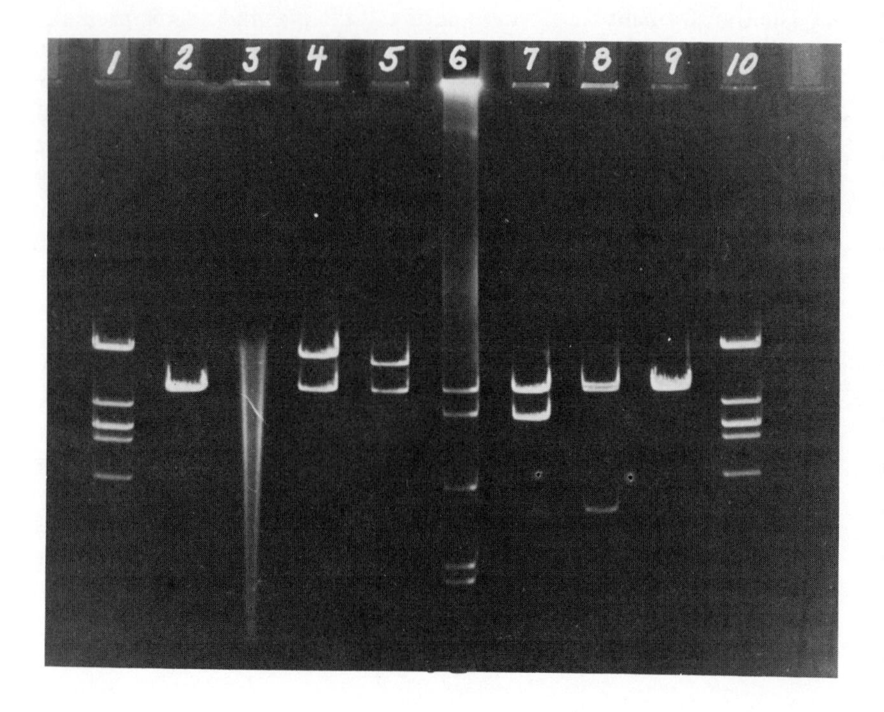

FIGURE 1 Agarose slab gel electrophoresis of DNA digested by the *Eco* RI endonuclease. The digested DNA (about 0.5 μg) was applied to the sample slots in 25 μl volumes containing 10 percent glycerol (w/v). Electrophoresis was carried out at a constant 100 V for 2.75 hours in 0.8 percent agarose (Tris-borate buffer). The slab gel apparatus and buffer system have been described (Helling *et al.*, 1974; Dugaiczyk *et al.*, 1975). The gel was stained in electrophoresis buffer with 0.5 μg/ml of ethidium bromide and photographed on a shortwave UV transilluminator. The fragments of the phage λ DNA generated by *Eco* RI endonuclease were used as standard markers for estimation of molecular weights (Helling *et al.*, 1974). The molecular weight estimates for these fragments (slots 1 and 10) are 13.7, 4.68 (3.7 and 3.56 fragments appear as one band here), 3.03, and 2.09×10^6 daltons. Slot 2: *Eco* RI-cleaved pSC101 DNA, 5.8×10^6 d. Slot 3: *Eco* RI endonuclease digest of *Drosophila melanogaster* DNA. Slots 4 through 9 have *Eco* RI endonuclease-cleaved recombinant plasmid DNA from several representative transformants. The linear pSC101 fragment is common to all of these plasmids that contain, in addition, fragments of molecular weight in millions of daltons of 10.5 (slot 4); 8.3 (slot 5); 4.4, 2.0, 1.0, 0.89, and 0.31 (slot 6); 4.3 (slot 7); 6.0 and 1.5 (slot 8). The cleaved DNA in slot 9 has two *Eco* RI endonuclease-generated fragments of the same molecular weight.

somewhat less than those calculated from theoretical considerations. However, this is to our advantage, since it requires less DNA to obtain the desired intermolecular recombination.

MOLECULAR VEHICLES FOR THE CLONING PROCEDURE

One of the requirements for the cloning procedure is the bacterial plasmid into which the *Eco* RI fragments are recombined. Since this plasmid replicates anything that is covalently attached to it, we refer to it as the "molecular vehicle." The molecular vehicle must have several features in order to be operative in this role. In addition to autonomous replication, it must have a gene that can be employed for selection purposes. Ideally, it should have one site for the *Eco* RI restriction endonuclease, and the insertion of DNA into this site must not disrupt the replication of the vehicle and the gene used for selection of the transformants.

At present two plasmids have been used as molecular vehicles (Morrow *et al.*, 1974; Hershfield *et al.*, 1974). A small (5.8×10^6 d) bacterial plasmid, pSC101, has one *Eco* RI site and a gene whose product inactivates the antibiotic tetracycline. *Eco* RI endonuclease-generated fragments of DNA can be recombined into this molecule and recovered in tetracycline-resistant transformants.

The second plasmid, colE1, brings several new features to the cloning vehicle. This DNA molecule (4.2×10^6 d) is also cleaved once by the *Eco* RI endonuclease. Insertion of *Eco* RI fragments into this molecule at the *Eco* RI site does not alter the ability of the molecule to replicate itself, but such hybrid plasmids no longer produce colicin E1 (Hershfield *et al.*, 1974). Recombinant colE1 plasmids do exhibit immunity to the bactericidal activity of colicin E1, however. Thus, when one uses the colE1 plasmid molecule as a cloning vehicle, one can select for colicin immunity after transformation and score for recombinant hybrid plasmids by the inability of the recombinant plasmid to produce colicin E1. The most attractive feature of the colE1 plasmid as a cloning vehicle is the potential of obtaining large numbers of copies of the plasmid by incubating the colE1-containing cells in chloramphenicol. Ordinarily, there are about 25 copies of the colE1 plasmid per cell, but, when protein synthesis is inhibited, as many as 3,000 copies of the plasmid accumulate per cell (Helinski and Clewell, 1971). Moreover, the amplified colE1 plasmid DNA is free of the protein relaxation complex that interferes with the purification of supercoiled plasmid DNA. It is possible to obtain 2–3 mg of purified colE1 DNA from one liter of cells after 12 hours of amplification in chloramphenicol, and it has been demonstrated that recombinant colE1 plasmids amplify to the same extent as the unadulterated colE1 plasmid (Hershfield *et al.*, 1974).

CLONING OF EUKARYOTIC DNA IN *E. coli*

The amplified ribosomal DNA (rDNA) of *X. laevis*, a South African toad, consists of several hundred tandemly repeated segments containing the rRNA cistrons interspersed with spacer sequences (Dawid *et al.*, 1970). This DNA can be cleaved into two discrete sizes with the *Eco* RI endonuclease and inserted into the *Eco* RI-generated linear form of the plasmid, pSC101 (Morrow *et al.*, 1974). These investigations have shown that by using an appropriate concentration of the *Eco* RI endonuclease-generated plasmid and *X. laevis* DNA in the *in vitro* recombination procedure about one out of four tetracycline-resistant transformants yields plasmids with fragments of the toad rDNA. The hybrid plasmids were shown to contain either one fragment of toad DNA (4.2×10^6 d or 3.0×10^6 d) in addition to the pSC101 plasmid DNA or the two pieces in common with the plasmid molecule. This experiment demonstrated several important facts. First of all, one can recover pieces of DNA covalently inserted into bacterial plasmids without having any functional selection for the nonplasmid piece of DNA. Second, it demonstrated that eukaryotic DNA can replicate in *E. coli* when covalently attached to a bacterial plasmid. In addition to this, it was demonstrated by these investigators that RNA was transcribed from the eukaryotic DNA by *E. coli*.

Of course, in this case the eukaryotic DNA was highly purified and well defined. How does one apply this technique to the cloning of unfractionated eukaryotic DNA? We have found that it is quite straightforward to randomly clone *Eco* RI-generated fragments of eukaryotic DNA. The following experiment illustrates this procedure.

Drosophila melanogaster DNA is cleaved by the *Eco* RI restriction endonuclease to fragments that range in size from 0.5×10^6 d to over 10×10^6 d (see Figure 1). Fragments above and below this range exist, but in low quantities. In a typical cloning experiment, about 16μg of *Eco* RI endonuclease-cleaved *D. melanogaster* DNA was mixed with 1 μg of *Eco* RI-treated pSC101 DNA in a volume of 50 μl and ligated at 12°C for 12 hours. The DNA was applied to a 5–20 percent sucrose gradient and subjected to velocity centrifugation. The gradient was fractionated, and each fraction was used to transform a culture of *E. coli* and tetracycline-resistant transformants selected from the culture. Transformants derived from DNA in the fast-sedimenting region of the gradient yielded about 50 percent of the transformants as hybrid recombinant plasmids. The hybrid recombinant plasmids are recognized by the recovery of plasmid DNA from the transformant, which is larger than the molecular vehicle and contains two or more fragments of DNA after *Eco* RI endonuclease digestion. Some of the plasmids derived from this experiment are degraded by the *Eco* RI endonuclease to yield one, two, three, or more fragments of *D. melanogaster* DNA, although the majority of these yield only one *Eco* RI fragment in addition to the molecular vehicle (see Figure 1). On the basis of the relative mobilities of the *Eco* RI fragments in the agarose gel, one can estimate that the *D. melanogaster* DNA fragments cloned in these plasmids can represent most of the size classes generated by the *Eco* RI restriction endonuclease. From one experiment of this type, it has been possible to generate hundreds of recombinant plasmids.

CHARACTERIZATION OF CLONED *Drosophila* DNA

Each piece of *D. melanogaster* DNA cloned in the pSC101 plasmid can be characterized in several ways. The cloned fragment can be separated from the molecular vehicle by treatment with the *Eco* RI endonuclease, and its molecular weight determined by agarose gel electrophoresis (Helling *et al.*, 1974). The DNA can be enzymatically repaired with radioactive deoxyribonucleotide triphosphates, and standard nucleic acid hybridization techniques can be employed to determine redundancy and expression of that DNA in the organism. Only one fragment has been found to date that has a reiteration frequency of about 15–20; all of the others have been assessed as being unique on the basis of nucleic acid hybridization analyses. In all cases examined, the *E. coli* cells carrying the recombinant plasmids transcribe RNA, which hybridizes to *D. melanogaster* DNA. One can use this (Morrow *et al.*, 1974) RNA or transcribe the DNA *in vitro* to localize the fragment in the *Drosophila* chromosome by *in situ* hybridization techniques (Pardue *et al.*, 1970). With this rather straightforward set of experiments, each cloned DNA molecule can be characterized in terms of its genetic location, redundancy, and expression.

APPROACHES TO THE ANALYSIS OF EUKARYOTIC DNA FRAGMENTS

After characterization of the various recombinant plasmids, one would hope to find several fragments of *D. melanogaster* DNA that are actively transcribed in the organism or in tissue culture cell. These can be used in conjunction with nontranscribed fragments to assay the fidelity of *in vitro* transcription from isolated chromatin in several ways (Axel *et al.*, 1970; Paul and Gilmour, 1968). One would also anticipate that cloned *D. melanogaster* DNA could be reconstituted at the appropriate stoichiometric relations with nuclear proteins to establish site-specific transcription (Paul *et al.*, 1973) and to serve as a physical basis for the purification of regulatory proteins (Albert and Herrich, 1971). The intent of the experimental design, albeit ambitious, would be analogous to that of Squires *et al.* (1973) and Rose *et al.* (1973), who were able to define the role of tryptophan repressor of *E. coli* in an *in vitro* system reconstituted from the tryptophan operon, tryptophanyl tRNA, tRNA synthetase, tryptophan, and RNA polymerase in the synthesis of mRNA from the tryptophan operon. With the availability of milligram quantities of eukaryotic genes, one could examine the transcription of that fragment as a function of numerous parameters, e.g., histone, DNA binding protein, homologous and/or heterologous polymerase, and nonhistone protein from homologous or heterologous sources.

The availability of the cloned DNA will be helpful in elucidating the arrangement of nucleotide sequences in eukaryotic DNA. Current models (Britten and Davidson, 1969; Bonner and Wu, 1973; Davidson *et al.,* 1973) are based on apparent interspersion of redundant sequences and unique sequences. The methodology of nucleic acid hybridization can be applied to the various cloned fragments, and, in conjunction with *in situ* hybridization to localize the arrangement of the cloned DNA, one would anticipate some clarification of eukaryotic genetic structure.

An alternative and promising approach to the random cloning is the application of current technology to the preselection of genes (see Birnsteil *et al.*, 1974) of interest prior to the cloning procedure. For example, it should be possible to obtain a DNA preparation relatively enriched for histone genes. Since sensitive assays are available to detect these genes, one could screen fragments of DNA cloned from this preparation to find those containing histone genes.

SUMMARY

The availability of the cloning procedure will provide the methodology for obtaining eukaryotic DNA containing a few genes in quantities sufficient for *in vitro* experimentation. Genes can be cloned at random, or specific genes can be cloned by the application of various enrichment techniques.

ACKNOWLEDGMENTS

Investigations reported here were supported by U.S. Public Health Service Grants GM14378 and CA14026.

REFERENCES

Albert, B., and G. Herrich. 1971. Methods Enzymol. 21:198.
Axel, R., M. Cedar, and G. Felsenfeld. 1973. Proc. Nat. Acad. Sci. U.S.A. 70:2029.
Bigger, A., N. Murray, and K. Murray. 1973. Nat. New Biol. 244:7.
Birnsteil, M., J. Telford, E. Weinberg, and D. Stafford. 1974. Proc. Nat. Acad. Sci. U.S.A. 71:2900.
Bonner, J., and J. Wu. 1973. Proc. Nat. Acad. Sci. U.S.A. 70:535.
Boyer, H. 1974. Fed. Proc. 33:1125.
Boyer, H., L. Chow, A. Dugaiczyk, J. Hedgpeth, and H. Goodman. 1973. Nat. New Biol. 244:40.
Britten, R. J., and E. Davidson. 1969. Science 165:349.
Brown, D., and R. Stern. Annu. Rev. Biochem. 1974. 43:667.
Cohen, S., A. Chang, H. Boyer, and R. Helling. 1973. Proc. Nat. Acad. Sci. U.S.A. 70:3240.
Davidson, E., B. Hough, C. Amenson, and R. Britten. 1973. J. Mol. Biol. 77:1.
Dawid, I., D. Brown, and R. Reeder. 1970. J. Mol. Biol. 51:341.
Dugaiczyk, A., H. Boyer, and H. Goodman. 1974. Biochemistry 13:503.
Dugaiczyk, A., H. Boyer, and H. Goodman. 1975. J. Mol. Biol. 96:171.
Hedgpeth, J., H. Goodman, and H. Boyer. 1972. Proc. Nat. Acad. Sci. U.S.A. 69:3448.
Helinski, D., and D. Clewell. 1971. Annu. Rev. Biochem. 40:899.
Helling, R., H. Goodman, and H. Boyer. 1974. J. Virol. 14:1235.
Hershfield, V., H. Boyer, M. Lovett, C. Yanofsky, and D. Helinski. 1974. Proc. Nat. Acad. Sci. U.S.A. 71:345.
Mertz, J., and R. Davis. 1972. Proc. Nat. Acad. Sci. U.S.A. 69:3370.
Modrich, P., Y. Anraku, and I. Lehman. 1973. J. Biol. Chem. 248:7495.
Morrow, J., S. Cohen, A. Chang, H. Boyer, H. Goodman, and R. Helling. 1974. Proc. Nat. Acad. Sci. U.S.A. 71:1743.
Pardue, M., S. Gerbi, R. Eckhardt, and Gall. 1970. J. Chromosoma 29:268.
Paul, J., and R. Gilmour. 1968. J. Mol. Biol. 36:305.
Paul, J., R. Gilmour, G. Affara, C. Birnie, P. Harrison, A. Hell, S. Humphries, J. Windass, and B. Young. 1973. Cold Spring Harbor Symp. Quant. Biol. 38:885.
Rose, J., C. Squires, C. Yanofsky, H. Yang, and G. Euboy. 1973. Nat. New Biol. 245:131.
Squires, C., J. Rose, C. Yanofsky, H. Yang, and G. Zubay. 1973. Nat. New Biol. 245:131.
Wang, J., and N. Davidson. 1966. J. Mol. Biol. 15:111.

DISCUSSION

DR. DOY: To me this is the way transgenosis must go. I want to make two points from the plant person's point of view. Metabolic pathways are similar both in

E. coli and in plants. Thus, one might chop up plant DNA, put it into lambda or a plasmid and then put it back into *E. coli,* deleted for that gene, and detect that you have a specific plant gene. It might then be possible to put this gene back into the same or a different plant.

Similarly, you could hybridize plant to plant DNA, if one of the plant DNAS were from the plant for which you wish to introduce new information into the chromosome. The hybrid DNA might then pair with a chromosome of the plant that you are going to put it back into and so help to integrate the foreign DNA you want to get in. This is my concept of how we might be able to go with plants.

I think we are moving toward a situation that Dr. McClintock mentioned to me yesterday, where the test tube may replace the plant in the long run, but the finish line is not yet in sight.

IV

REVIEW AND SUMMARY

HAMISH N. MUNRO

Interpretive Summary and Review

Dr. Altschul stated the target at the outset: "The point of this meeting on genetic improvement of seed proteins is to explore the possibility of reducing both the land and energy constraints by genetic intervention in the photosynthetic process, particularly in the conversion of inorganic nitrogen to protein." To this objective can be added that of changing the amino acid pattern of the mixed seed proteins to approximate as closely as possible the optimum pattern for human utilization.

At the end of the workshop, at which nearly 20 papers ranging from plant biochemistry to molecular biology and genetics were presented, it is possible to assess the emerging picture in the form of partial answers to three questions. These questions are

What are the nutritionally desirable objectives?

What are the rate-limiting factors in synthesis of amino acids by plants?

What are the potential control points regulating the types of proteins accumulating in seed grains?

NUTRITIONAL OBJECTIVES OF SEED PROTEIN IMPROVEMENT

While progress continues to be made in improving grain for human consumption, advances are also occurring in our knowledge of man's requirements for nutrients. In the context of the present workshop, it is important to note that our current understanding of man's need for protein has been summarized in a recent FAO/WHO report (1973). Three features of human protein requirements are especially relevant to the workshop: (a) how much protein is needed? (b) What are the essential amino acid requirements of man? (c) What are the best available measures by which to judge the quality of the seed protein for human use?

PROTEIN REQUIREMENTS

Estimates of protein requirements are largely based on nitrogen balance studies. In the case of adult human subjects, this represents the least amount of protein required to bring the subject into nitrogen equilibrium. In the case of the average young adult, this is achieved on about 0.44 g high-quality protein per kg body weight per day. However, allowance has to be made for individual variation, and for this reason the estimate has been raised 30 percent to cover individuals with needs up to two standard deviations above the mean. This raises the safe level to 0.57 g per kg body weight for men and 0.52 g per kg for women. A correction is then applied to allow for quality of dietary protein less than 100 percent. For example, if the quality of the mixture of proteins in the diet is 80 percent of that of the standard (e.g., whole egg protein), then the safe level of protein intake on such a diet would be raised to 0.7 g per kg ($0.57 \times 100/80$), or 49 g of protein daily for a 70 kg man.

It will be noted that 49 g of protein provides 196 kcal. If the subject's daily requirements for energy are 2,500 kcal, his protein needs will be almost 8 percent of this requirement. Since voluntary food intake is determined primarily by energy needs, this means that a diet providing less than 8 percent of its energy content in the form of protein is unlikely to meet the upper limit of protein requirement of all members of a population. This is relevant to the proportion of protein to starch stored in seed grains used as food.

For the growing child, the criterion of requirement is growth. Such requirements are well established for infants. Up to 3 months of age, 2.4 g protein/kg/day is needed for maximal growth, and this falls to 1.4 g/kg by 9–11 months. During later childhood, estimates of requirements are less securely based and are largely obtained from a consideration of the amounts of protein deposited daily during different stages of growth. For example, from ages 10 through 12, an acceptable requirement would be about 0.8 g protein per kg.

ESSENTIAL AMINO ACID REQUIREMENTS

By means of nitrogen balance studies, and more recently on the basis of measurements of changes in blood amino acid levels at different levels of intake, the requirements of adults and of children for essential amino acids have been assessed. Despite the wide variations from subject to subject, the basic information is reasonably well established (see FAO/WHO, 1973). At 6 months of age, about 39 percent of the 1.85 g of protein per kg body weight needed daily by the human infant has to be in the form of the nine essential amino acids; this falls to 15 percent of the daily requirement of 0.55 g protein per kg needed by adults (Table 1). Data for children between infancy and adulthood are less satisfactory. It can, however, be concluded that the essential amino acid pattern of a dietary protein source is more significant for the infant and rapidly growing child than for the adult.

EVALUATION OF PROTEIN QUALITY

From what has just been said it follows that our knowledge of human amino acid needs can be used to work out the concentrations of

TABLE 1 Daily Protein and Amino Acid Requirements of Human Subjects[a]

Requirement	Age of Subject			Scoring Pattern (% of Ideal Protein)
	3–6 mo.	10–12 yr.	Adult	
Protein (g/kg)	1.85	0.80	0.55	100%
Amino acids (mg/kg)				
Isoleucine	70	30	10	4.0
Leucine	161	45	14	7.0
Lysine	103	60	12	5.5
Met + Cystine	58	27	13	3.5
Phe + Tyrosine	125	27	14	6.0
Threonine	87	35	7	4.0
Tryptophan	17	4	3	1.0
Valine	93	33	10	5.0
Total	714	261	84	36.0
Percentage of protein need	39%	33%	15%	36%

[a]Data taken from FAO/WHO Report (1973). Note that the estimates of requirements are calculated for 98 percent of the population at each age and thus express the upper limit of the range of requirements found in a healthy population.

essential amino acids in an ideal protein that will be fully utilized. The FAO/WHO report (1973) provides such a scoring pattern, and it is obvious from inspecting it (Table 1) that this pattern does approximate the needs of the preschool child. This is appropriate since this is the major group at risk, whereas the adult has a low requirement for essential amino acids. Using such a pattern, one can compare the amino acid composition of dietary proteins and assign them a score based on the least abundant essential amino acid relative to requirements.

Tests of protein quality based on animal growth and nitrogen retention are an important additional means of evaluating proteins. A detailed discussion of such tests is given in a revised version of the 1974 NRC report, *Improvement of Protein Nutriture*. The protein efficiency ratio (PER) is no longer regarded as adequate, and a dose-response slope assay is recommended as the best routine test. Such procedures require considerable amounts of each source of protein for testing. Methods suitable for small amounts of protein and for rapid testing are desirable alternatives at some stages in testing new varieties of seed grains. In this connection, the new colorimetric procedures devised by Dr. Rabson and the dye-binding method for basic amino acid content (notably lysine) used by Dr. Nelson and reported in this volume should prove useful adjuncts in this area of rapid testing of small amounts of material.

RATE-LIMITING FACTORS IN AMINO ACID BIOSYNTHESIS BY PLANTS

It seems of great importance to develop means of selecting grain plants on the basis of their ability to utilize the environment as efficiently as possible in the production of seed protein. Thus the biochemical features of amino acid synthesis by plants need to be thoroughly understood, notably the factors regulating the flow of nitrogen through the system to amino acids and their efficient utilization to make seed proteins. Some sites of such regulation are shown in Figure 1. It would be desirable to have a knowledge of the kinetic factors operating in this sequence of metabolic events, giving data as detailed as that for glycolysis (e.g., Newsholm, 1973). This demands information about (a) the specific activities in μmoles/g tissue for each enzyme in a pathway; (b) K_m values for each enzyme and the operational levels of reactants in plant tissues for comparison with these K_m values; (c) the concentrations of reactants and products for each enzyme *in vivo* and *in vitro*, in order to determine which enzymes are not operating in equilibrium *in vivo* and are thus potential sites of regulation; and (d) the response of

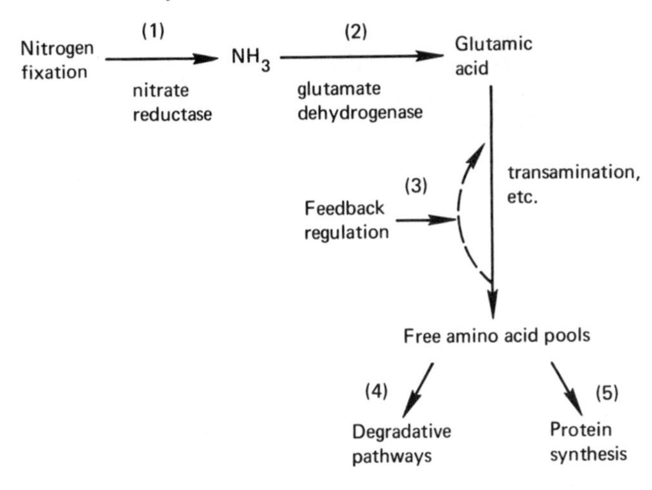

FIGURE 1 Sites by regulation of amino acid synthesis and utilization by plants.

reactants and products of each enzymic process to changes in substrate flow through the pathway, in order to identify sites of regulation. At present, such information is fragmentary or unavailable.

SITE OF REGULATION

In Figure 1, four major potential sites of regulation have been identified. Each of these could be rate-limiting in the accumulation of protein by the plant. First, there is the fixation of nitrogen involving nitrate reductase. Increased nitrate availability in the soil results in increased amino acid accumulation and more seed protein as Dr. Hageman has shown. He has also demonstrated that nitrogen reductase levels in the developing grain plant can be correlated with nitrate availability and that this enzyme could in consequence be rate-limiting in the utilization of soil nitrogen right up to the level of protein deposition in the plant. The question we must ask is whether the levels of this enzyme in different cereal strains can be correlated with protein yield in the seedlings. The genetic factors controlling the levels of nitrate reductase have been investigated in fungi (*Neurospora* and *Aspergillus*) and are complex. Dr. Hageman's group has identified genotypic differences in levels of nitrate reductase in different corn seedlings, but it has proved a little more difficult to demonstrate that selection of strains on the basis of nitrate reductase activity is predictive of high protein yields in the mature grain.

The transfer of the nitrate fixed as ammonia to amino acids is dependent on glutamic dehydrogenase, but the overall rate-limiting role of this enzyme seems not to have been established. However, Dr. Miflin has shed some light on factors regulating the biosynthesis of individual amino acids, including feedback control by allosteric effectors similar to those observed in bacterial pathways of amino acid biosynthesis [labeled (3) in Figure 1]. Two points emerge from a consideration of such information. First, the biosynthesis by plants of amino acids essential to man is under considerable regulatory control and could thus be a limiting factor in the synthesis of seed proteins of high biological value. Consequently, mutants lacking such controls might be useful in promoting the synthesis of proteins containing large amounts of key essential amino acids. Second, the synthesis of several amino acids nonessential to man is responsive to nitrogen supply without apparent regulatory control. Moreover, accumulation of these amino acids favors the synthesis of storage proteins that are low in lysine and other essential amino acids. Thus, as Bishop (1929) observed nearly 50 years ago, seeds with high protein yields due to abundant fertilizer treatment accumulate proteins low in lysine and other essential amino acids, such as the prolamine storage proteins.

Even less well defined is degradative removal of amino acids and the regulation exerted over this process [labeled (4) in Figure 1]. Finally, the removal of amino acids as protein is influenced by additional factors inherent in the regulatory machinery for protein synthesis [labeled (5) in Figure 1]. These are discussed in the next section.

From this survey of rate-limiting factors in amino acid biosynthesis, we see that a combination of increased nitrogen from fertilizers together with mutants showing relaxation of essential amino acid synthesis might be most beneficial in ensuring high yields of seed proteins of good quality. Other more general factors involved in optimal retention of nitrogen as seed protein were discussed by Dr. Canvin and by Dr. Hardy and his colleagues, who demonstrated the importance of available CO_2 for optimal nitrogen retention.

REGULATION OF PROTEIN BIOSYNTHESIS IN PLANTS

As pointed out by Dr. Dure, the amount of protein in plant tissues is a balance between synthesis and breakdown. In the rapidly growing plant and in the seed, it is assumed that synthesis is the main factor in accumulation of proteins. Synthesis can thus determine the amount of protein laid down and also the types of protein.

The mechanism of protein synthesis depends on transcription of the genetic message encoded in the DNA of the cell nucleus and its subsequent translation in the cytoplasm. In this workshop, basic aspects of the mechanism of transcription and its control have been dealt with for bacteria by Dr. Zubay and for animal cells by Dr. Means, while the mechanism of cytoplasmic translation in plants was discussed by Dr. Boulter, and posttranscriptional regulation of protein synthesis was dealt with speculatively by Dr. Clark. These last two contributors have identified in extensive detail the potential sites of regulation, and I shall not recapitulate their presentations but rather shall try to select the mechanisms most likely to be important in control of quantity and type of seed protein. Figure 2 is an attempt to classify such major control sites into those affecting the supply of amino acids for peptide chain elongation and those affecting the machinery of protein synthesis, notably messenger RNA availability.

FACTORS AFFECTING SUBSTRATE AVAILABILITY

The supply of amino acids at the site of protein synthesis [labeled (1) in Figure 2] is known from studies on animal cells not only to be a factor in rate of protein synthesis, but also to affect synthesis of individual proteins selectively. There is evidence (a) that proteins made on free ribosomes in the liver and retained within the cell are more responsive to amino acid supply than are proteins made by membrane-attached

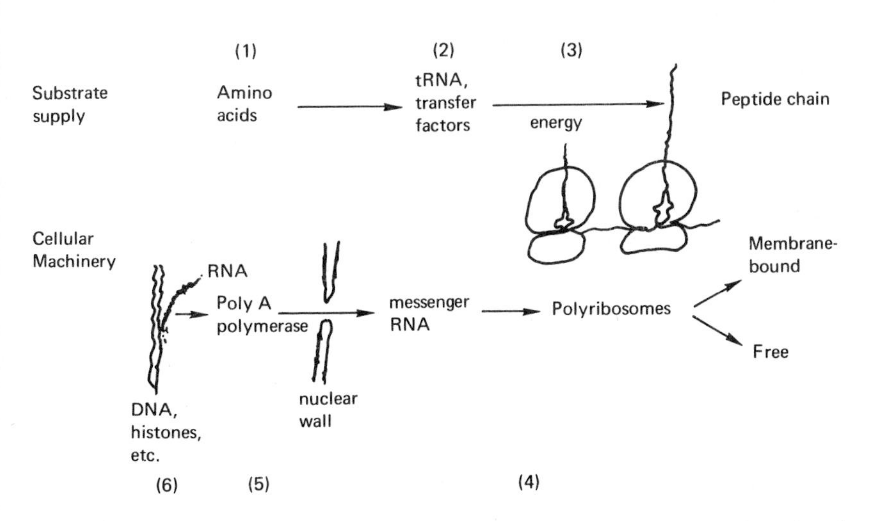

FIGURE 2 Potential sites of regulation of protein synthesis in the production of seed proteins.

ribosomes and secreted by the liver (Peters and Peters, 1972); (b) that administration of methionine or cystine to sheep selectively stimulates synthesis of wool proteins with a high cysteine content, thus giving the wool a different texture (Broad *et al.*, 1970); and (c) that the rates of formation of α- and β-chains of globin within the reticulocyte can vary independently according to the most limiting amino acid in the plasma in which cells are suspended (Hunt *et al.*, 1969). This may also obtain in the case of the storage proteins of seeds. As pointed out earlier, abundance of nitrogen in the soil can result in excess production of amino acids nonessential to man and also in synthesis of storage proteins rich in these nonessential amino acids but low in essential amino acids. This apparent causal relationship between the spectrum of free amino acids in the pool and the composition of the seed protein laid down requires more direct verification, but the evidence seems suggestive.

The different isoaccepting species of tRNA [labeled (2) in Figure 2] have often been invoked in regulating the capacity of individual messenger RNAs to be translated. For example, there is good evidence that the spectrum of isoaccepting tRNAs can alter in response to the demand for synthesis of new proteins by a cell (e.g., synthesis of egg-yolk proteins by chick liver following estrogen administration). However, it is usual for the spectrum of tRNA isoaccepting species to change quantitatively rather than qualitatively, and it seems probable that such alterations are part of a coordinated program of responses to hormones or other effectors. Indeed, the soluble factors involved in protein synthesis (e.g., synthetases, elongation factors) appear to undergo coordinated changes brought about by self-regulating mechanisms. For example, we have noted the mutual stabilization of aminoacyl-tRNA and the elongation factor transferring the amino acid from the tRNA to the ribosome (Hubert *et al.*, 1974). Finally, wide changes in supply of high-energy phosphate bonds [labeled (3) in Figure 2] seem to be tolerated without appreciable alteration in protein synthesis capacity (Oler *et al.*, 1968). From this survey, we may conclude that amino acid concentration is the most likely component among the soluble factors in the cell to limit rate of protein synthesis.

FACTORS AFFECTING THE MACHINERY OF PROTEIN SYNTHESIS

Turning to regulation of the machinery of protein synthesis, there is still much to be learned about the control of protein formation by free ribosomes versus their formation by membrane-attached ribosomes, which include ribosomes making the proteins stored in the aleurone grains of seeds and discussed by Dr. Dieckert. It is still not possible to say how the messenger RNA for a protein retained in the cytoplasm is translated only by free ribosomes whereas the messenger RNA of a protein destined for secretion or deposition in storage vacuoles is translated by membrane-attached ribosomes. However, such a division of function implies some form of regulation that might be the site of selective translation [labeled (4) in Figure 2]. This complex situation is reviewed elsewhere (Munro and Steinert, 1974). The transfer of messenger RNA from the nucleus to the cytoplasm depends in most cases on attachment of polyadenylic acid to the RNA [labeled (5) in Figure 2], but this seems to be nondiscriminatory and thus not likely to alter the type of messenger passing into the cytoplasm, though the total amount of messenger RNA transferred to the cytoplasm could be determined by the availability of polyadenylic acid. In this connection, we have observed that an enzyme making polyadenylic acid in mitochondria is sensitive to amino acid supply (Jacob and Munro, 1974). This would suggest coordination of availability of polyadenylic acid in mitochondria with an increased influx of substrate for protein synthesis—namely, amino acids. Many other factors involved in the mechanism of protein synthesis undergo coordinated changes in response to variations in amino acid supply. These include the RNA polymerases of nuclei and mitochondria, ribosomal RNA synthesis and maturation and release of ribosomes into the cytoplasm, and also rate of ribosomal RNA breakdown.

Finally, there is regulation of protein synthesis at the chromosomal level [labeled (6) in Figure 2]. The synthesis and amount of a specific protein in grain seed could be determined at the chromosome level (a) by the presence or absence of the appropriate genetic information in the genome, (b) by the number of copies for transcription, and (c) by the availability of the DNA of the gene to RNA polymerase. Regarding changes in the proportions of the storage proteins in mutants of maize, barley, and sorghum reported by Dr. Mertz, it may be noted that these nutritionally beneficial changes represent reduction but not complete deletion of the grain content of storage proteins low in lysine. This suggests that the amounts of messenger RNAs coded for these low-lysine proteins are probably diminished in the mutants through reduction in number of gene copies or inhibition of their transcription by RNA polymerase. In the case of maize, any one of nine mutations on six different chromosomes can cause a reduction in the amount of lysine-poor zein in the seed, a finding that is in keeping with alterations in regulation of expression of chromosomal transcription. The molecular basis of such control mechanisms has been persuasively developed at this workshop by Dr. Zubay on the basis of bacterial operon regulation. He showed how the superimposition of viral infection on the bacterial genome can add a sophisticated sequential control system that may be taken as a model of similar control systems in eukaryotes, such as the effects of progesterone and estrogen on protein production by the hen oviduct described by Dr. Means. One aspect of genetic control that might be elucidated is whether the developmental programming of high-lysine seed mutants is altered by the mutation. In other words, are the low levels of zein and other lysine-poor storage proteins due to a delay in their time of expression in the mutant grain? Or, as Dr. Dure has suggested, is the messenger RNA present but untranslated? These questions can be answered using messenger isolation and other available methods.

CONCLUSIONS

Our knowledge of factors controlling the amount and types of storage proteins in seeds is presently based on empirical observations. However, biochemical studies on nitrogen fixation, amino acid biosynthesis, and protein synthesis in plant tissues are beginning to fill in the gaps. Much of this work relates to germinating seeds, whereas the factors influencing deposition of protein in the developing grain are of more importance to this workshop. Nevertheless, there is reason to hope that a molecular basis for genetic improvement of seed proteins is emerging. The first requirement is to identify the biochemically desirable features to be selected, such as increased nitrogen fixation, more rapid synthesis of key amino acids, greater protein deposition in the seed, and the essential amino acid pattern of these storage proteins. Selection for such features can be achieved by recognized breeding procedures, but the methods of restructuring DNA molecules may one day provide a means of direct intervention. Such an approach will depend on a large increase in our understanding of the structural features of DNA relevant to control of expression of genetic information.

REFERENCES

Bishop, L. R. 1929. J. Inst. Brewing 35:316.
Broad, A., J. M. Gillespie, and P. J. Reis. 1970. Aust. J. Biol. Sci. 23:140.
FAO/WHO. 1973. WHO Technical Report Series No. 522, WHO, Geneva.
Hubert, C., B. S. Baliga, C. A. Villee, and H. N. Munro. 1974. Unpublished results.
Hunt, R. T., A. R. Hunter, and A. J. Munro. 1969. Proc. Nutr. Soc. 28:248.

Jacob, S. T., and H. N. Munro. 1974. Unpublished results.

Munro, H. N., C. Hubert, and B. S. Baliga. 1974. *In* Alcohol, Nutrition and Protein Synthesis (M. A. Rothschild, ed.). Pergamon Press, New York.

Munro, H. N., and P. Steinert. 1974. *In* International Review of Science. Biochemistry Series, Vol. 7 (ed. H. R. V. Arnstein). Medical Technical Publishing Co., Oxford.

NRC. 1974. Improvement of Protein Nutriture. National Academy of Sciences, Washington, D.C.

Newsholm, E. A. 1973. Regulation in Metabolism. Wiley Interscience, New York.

Oler, A., E. Farber, and K. H. Shull. 1969. Biochim. Biophys. Acta 190:161.

Peters, T., Jr., and J. C. Peters. 1972. J. Biol. Chem. 247:3858.

DISCUSSION

DR. EVERSON: How are we going to get at increasing the utilization of protein in plants? If, in our diet, we eat 49 grams of protein, that is not all available to the human. Much of it goes through the monogastric gut complexed. I thought that sometime in this conference somebody would address the problem of the phytates, the proteolytic inhibitors that probably are complexing proteins and making them unavailable.

DR. MUNRO: These estimated requirements take account of incomplete digestibility, which in general is not high. You are quite right in drawing attention to unavailable amino acids. It would be desirable to include this in evaluation tests, especially on breakfast cereals and other products that are subjected to heat or to prolonged storage in heat and damp, and in particular the reaction lysine undergoes with sugars that makes it nonavailable.

In general, digestibility has not proved a major problem. I would say that not more than 10 percent of the protein intake is not absorbed in the average diet, and it is only under technological conditions (for example, heat-drying milk) that you run the risk of creating complexes that prevent the amino acid from being digested, absorbed, and used.

DR. DIECKERT: I believe that as we tamper with the genetics of plants, we will have to pay attention to protein and nonprotein components that interfere with the metabolism of the animal either directly or indirectly by tying up some essential amino acid in the proteins. A case in point is cottonseed, with gossypol. We have just recently found out that one way it poisons an animal is by uncoupling oxidative phosphorylation. Another way it has an effect is that it ties up lysine.

DR. MUNRO: That is right. This is one of the ways that you get nonavailable lysine.

DR. DIECKERT: So the more you tamper with these plants genetically, the greater the possibility of generating some of these factors that are not going to be to your advantage. So they need to be looked at from that point of view.

DR. MUNRO: There is a conference in Mexico next month that will address itself to, among other things, how you test grain for quality. How do you evaluate the quality of the protein?

DR. ELLIOTT: I just wanted to add a comment to that, Dr. Munro. We have been having conferences like the one you are talking about on an average of about once a year for the last 5 years. We have not settled these issues, nor do we expect to break through these barriers very quickly.

DR. MUNRO: I think the quality of decision making in this field has not been the best. The truth is that there are not enough new data coming in. There are just a few people who are trying to improve these tests.

For example, what can you do with *Tetrahymena*? This is a unicellular organism that requires essential amino acids. It looks ideal for testing protein quality, but it has problems.

OLIVER E. NELSON

Interpretive Summary and Review

The invitation to discuss and summarize the genetic aspects of this workshop I found an interesting challenge. I hope to pull together some of the threads of the discourse, which have been somewhat at variance, and persuade some of my colleagues in plant breeding that now is the time to be thinking about what the studies of plant protein synthesis could mean for improvement of plants with respect to protein quality and quantity.

The small group of thoughtful men who planned this workshop had an intuitive grasp of what I think has emerged clearly from these discussions. With some small amount of progress in protein improvement behind us, our advances are circumscribed by lack of basic information about the systems that we would attempt to modify. We know very little about plant protein-synthesizing systems.

In the cereals, maize, barley, and sorghum, simply inherited variants have been detected that have desirable effects in raising the level of the limiting essential amino acids. We know, as Dr. Mertz pointed out, that they do so by depressing the synthesis of the prolamin fraction and secondarily raising the proportion of the synthesized proteins that are albumins and globulins. They enhance the content of both lysine and tryptophan. These mutants are, I think, regulatory mutants on steps in the chain of protein synthesis. Phillips *et al.* (1971) have shown in maize, for example, that there are strains with a double nucleolus organizer and that such strains have a polysome profile that is much enhanced over that of the strains with a single nucleolar organizer.

Clearly, if number of ribosomes were a limitation on protein synthesis, here is a method by which it could be alleviated.

Let me return to my inquiry as to where we stand with respect to protein improvement in the cereals and legumes. With the legumes, investigators are only nicely launched on programs exploring the existant variability. It will be several years before we know what the situation really is. Dr. Boulter suggested some changes that could take place. I am sure he would agree with me that, with respect to the legumes, increases in yield would be as important as changes in their amino acid composition. With wheat, I have already alluded to the increases that have been found in protein content. At the same time, there have been no amino acid variants of the magnitude of those found in maize, barley, and sorghum. This is probably due to the genetic structure of wheat, which is a hexaploid. It may, indeed, be necessary to go back to the diploid progenitors of wheat, isolate such variants, and then, by a complicated job of genetic engineering, reconstitute hexaploid wheat carrying the genes responsible for variations in protein content.

Rice has quite good nutritional value, but the protein content is very low. There have been no amino acid variants of which I am aware. Current efforts are concentrating on attempts to breed rice with higher protein content.

With the species in which useful variation has already been found, it is logical to ask where one proceeds from our present position. Table 1 summarizes the net effects of gene substitution on protein production in those cereals in which desirable mutations have been found. In the

TABLE 1 Protein Fractions[a] of Normal and Mutant Maize, Barley, and Sorghum

Solubility Fraction	Maize Normal	Maize Mutant (opaque-2)	Barley[b] Normal	Barley[b] Mutant (R1508)	Sorghum[c] Normal	Sorghum[c] Mutant (hl)
Albumin[d] + globulin	6	18	27	46	15	22
Prolamin	59	26	29	9	53	33
Glutelin	30	49	39	39	27	38
Unextracted	4	5	5	6	5	6

[a] As a percent of total protein.
[b] Data from Ingversen *et al.* (1973).
[c] Data from Mertz, this volume.
[d] Includes nonprotein nitrogen.

left row is the situation with regard to maize, both the normal and the mutant form; then normal and mutant barley, and in the right row normal and mutant sorghum. It is clear in each species that if one looks at the prolamin fraction, there has been a substantial decrease in the amount in the mutant. There has been a drop from 59 to 26 percent in maize, 53 to 33 in sorghum, and 29 to 9 in barley. Note that this level of prolamin (9 percent of the protein) is close to that found in present rice varieties. At the same time, there is in all these mutants an increase in the amount of salt-soluble and water-soluble proteins. Table 2 shows the changes in overall amino acid composition arising as a result of these altered proportions of storage proteins in the mutant strains as compared to normal.

I want to draw your attention to an interesting but possibly fortuitous correspondence here. The mutations in maize and sorghum have reduced the content of the alcohol-soluble proteins to about the content existent in normal barley. The mutation in barley has reduced the content of prolamin to about the content of prolamin in normal varieties of rice or oats. At the same time, the increases in albumin and globulin in mutant maize and in mutant sorghum have increased their content of these proteins to approximately that of normal barley. The mutation in barley has increased the content still farther.

I suggest that rice has already incorporated mutations homologous to those we have been detecting in maize, sorghum, and barley. Indeed,

TABLE 2 Marked Changes in Concentration[a] of Amino Acids in Mutant and Normal Endosperms

Amino Acid	Maize Normal	Maize Mutant (opaque-2)	Barley[b] Normal	Barley[b] Mutant (1508)	Sorghum[c] Normal	Sorghum[c] Mutant (hl)
Lys	1.6	3.7	3.7	5.3	1.3	2.6
Arg	3.4	5.2	5.3	5.6	2.7	4.4
Asp	7.0	10.8	6.8	9.7	6.1	7.2
Gly	3.0	4.7	4.3	6.0	2.4	3.4
Trypt	0.3	0.7	—	—	0.9	1.7
Glu	26.0	19.8	27.5	18.1	28.0	20.8
Prol	8.6	8.6	12.4	6.6	8.4	6.4
Leu	18.8	11.6	8.1	7.5	15.9	12.6
Ala	10.1	7.2	4.6	5.9	10.5	8.7

[a] g/100 g crude protein.
[b] Data of Ingversen *et al.* (1973).
[c] Data of Singh and Axtell (1973).

barley constitutes an intermediate situation. Normal barley is between normal maize and rice in prolamin content. The mutation (1508) in barley induced at Risø has reduced the prolamin content to that characteristic of rice or oats.

In the case of opaque-2, which we have known longest and which has been most studied, it is clear that there are some specific enhancements of certain water-soluble proteins. Ribonuclease [as Wilson and Alexander (1967) and Dalby and Davies (1967) showed], alkaline phosphatase, and trypsin inhibitor (Halim *et al.*, 1973) all are known to increase disproportionately in endosperms that are homozygous for the mutant gene.

At the same time, it is possible to say from work both at Purdue and in Yugoslavia by Denic (1970) that the differences in the synthetic properties of normal and opaque-2 maize are not due to the transfer RNAs they contain.

It is not clear whether or not these latter observations apply to the barley mutant R1508 and the sorghum mutant, but it is true that their net effect is produced in the same manner—through a depression of the prolamin fraction and an enhancement of other fractions that have more desirable amino acid compositions from the standpoint of nutrition of monogastric animals.

Further, we know that in wheat, although no significant variation in amino acid pattern can be observed, it is possible to breed cultivars that have enhanced protein content (several percent units higher), without substantial change in amino acid composition.

Once the mutants are identified, the plant breeders can use them in their breeding programs. Given analytical assistance, they can proceed with the job of breeding them into hybrids or cultivars that will be economically desirable and in which no or minimal yield reduction will be found. They can proceed to do this with no knowledge of the biochemical mechanism by which the mutations are producing their effect. After all, there is a large hybrid seed corn industry, based on the phenomenon of hybrid vigor, although the genetic basis of hybrid vigor is not clearly established even at the present time.

But I am concerned, for the moment, only with what these mutants might show us about plant protein-synthesizing systems. Where constraints arise, we should take stock on an occasion such as this one. We should ask about those species in which desirable variants have been found how far improvement can be carried. What does one do about economically and nutritionally important species in which there have been analyses of hundreds of thousands of accessions without the identification of exploitable genetic variability? Are there other mechanisms by which amino acid composition can be changed? Is change in amino acid composition always due to a suppression in cereals, for example, of the nutritionally poor prolamin fraction and enhancement of the other fractions, apparently as a consequence?

When we ask questions like these, we wish that we had considerably more basic information at our disposal. Let me illustrate the point with a paradigm: There has been considerable discussion during this meeting about the possibility of plant improvement via somatic cell culture techniques. Clearly, one possibility, as has already been shown, is to select by these techniques variants that have lost a control regulating the level of a particular amino acid. This has been done by challenging with amino acid analogs cells or organisms that have been treated with mutagens.

The mutants surviving the treatment with the analogs, which are toxic to normal cells, have been found to have lost the constraints that regulate the level of amino acids within the system. Presumably, this sort of a change could be made for any species where one has the ability to grow haploid tissue in culture and, as Dr. Doy was emphasizing, where one can, after mutagenesis, apply a selective screen to identify the desired mutant that can be cloned to produce callus and ultimately plants.

For any species where this is possible, I assume that it would be possible to find such variants in amino acid regulation. Let us assume for a given cereal that one wished to raise the lysine content, or for a legume, the methionine content. Take lysine as an example. Would the greater supply of this particular amino acid tend to enforce or encourage the synthesis of proteins that have had a greater content of lysine? My conjecture, and it is at variance with that of Dr. Munro, is that it would not. The basis for my conclusion is that the primary sequence of the seed storage proteins is encoded in the genetic information, just as is the primary sequence for all other proteins.

In the system with which we are concerned—that is, the developing endosperm in cereals or the cotyledon in legumes—we are dealing with developmental programs that ensure the synthesis of certain types of proteins. An experiment that has been cited several times is that reported several years ago by Sodek and Wilson (1970) in which they injected ^{14}C-labeled lysine into the base of self-pollinated ears borne on plants heterozyous for the mutant opaque-2. At various times after the injection of the labeled lysine, they isolated the proteins from normal and mutant kernels and assayed the compounds that were labeled. For the mutant kernels, almost all was in the form of lysine, as it had been applied. For the normal kernels, the longer the interval after the injection of the radioactive lysine, the less the recovery in the form of lysine and the greater the radioactivity in glutamate and proline.

One of the possible interpretations of this experiment is that the opaque-2 tissue lacked the ability to degrade the lysine and that the normal seeds possessed the requisite enzyme systems. An alternative explanation, and one that springs readily to mind for geneticists, is that the opaque-2 endosperms were synthesizing proteins with a high content of lysine and were sequestering the lysine as peptide-bound lysine. In the normal system, where proteins with a lower content of lysine were predominantly being synthesized, the lysine was not incorporated and hence was degraded. There is no definitive evidence to allow a choice between these alternatives. This emphasizes the fact that it is crucial to know before embarking on extensive attempts to obtain mutants derepressed for synthesis of a particular amino acid just what the fate of an amino acid would be under such circumstances. Even if we are able to release the constraints on synthesis of a particular amino acid in the plant, is this change going to be reflected in the proteins that are laid down in the endosperm?

On the subject of possible genetic engineering, it is worth commenting on the definite if oblique sort of confrontation that we have seen between those espousing conventional plant-breeding techniques, on one hand, and those who have been advocating the application of genetic engineering via somatic cell culture systems. Dr. Sprague ably presented the case for emphasizing conventional plant breeding systems. I agree with much he said, particularly with regard to somatic hybridization, and share his pessimism regarding its extensive usefulness. I do foresee, however, enormous experimental leverage for an investigator from the ability to have millions of haploid cells in culture where they can be mutagenized, where selective screens can be applied, and where one can then reconstitute callus tissue and eventually plants displaying the desired variation.

Once the techniques of completing this cycle, from sporophytic tissue (or in some cases from gametophytic tissue) to plant cells in culture and then to haploid plants and ultimately diploid plants, are available for most economically important plants, we will have systems in which advances will be limited only by the ingenuity of the investigator. At the same time, there are situations, particularly with regard to plant protein improvement, where I am not optimistic. I think it will be difficult, if not impossible, to select for or against a gene in a system where it is not being expressed. As far as we know, the genes governing the synthesis of storage proteins are not expressed in sporophytic cells growing in culture. What we know about the action is confined to the endosperm and is not expressed in other tissue. Having said this, I will add that I will be glad to have it shown that it is my imagination that is faulty, and not that the system is inadequate.

With regard to the quantity of protein present in the seed, this appears to be governed by the genotype of the plant on which these seeds are being borne. I think that there is an opportunity to select for greater activity of components of protein synthesis that might be reflected in greater amounts of protein laid down in the seed. For example, nitrate reductase activity might well be selected for in such a system.

On the other hand, one attribute of high-protein plants is that a greater proportion of the protein present in the leaves is, on senescence of a leaf, translocated as amino acids to the developing seeds. I find it difficult to visualize the type of selection that would identify cells capable of differentiating into plants that, on senescence of a leaf, would be able to transport a greater proportion of amino acids into the seeds.

So with all the promise which the system holds, I still feel considerable uncertainty about the possibilities of selecting for characters expressed on the organismal level by applying selective pressures to cells growing in a tissue culture. Critical tests, though, are not far off. Green and Phillips (1974) at Minnesota have recently reported the ability to regenerate reliably from corn callus plantlets and eventually plants. Here is another example in an economically important plant, where we now have the ability to go from the sporophyte to haploid cells growing in culture, and eventually back to diploid plants. So I am looking forward to the efforts of plant geneticists to check these points that have been raised.

I agree with Dr. Sprague on another important point. He asserted that plant breeders do not have the time and possibly not the inclination to carry out these tests of promising new techniques. And as I understood him, he invited molecular biologists to do these tests. They are so involved with their own experimental systems that they predictably are not going to be testing the efficacy of these systems, which are so interesting. I think that such tests should be the domain of geneticists who are committed to research in this area and who should be funded expressly for this particular sort of research. This is an example of one area, among others, that has been left without important support to date.

It is not an overstatement to describe the advances of the last 25 years in biology as constituting a revolution. The cutting edge of scientific advance is moving forward very rapidly. There is an ever-widening gap between the front of research and the point at which it is being applied for human benefit. One most desirable effect of a conference such as this would be a decision to fund efforts to work in the area between what we know or reasonably hypothesize we are capable of in the manipulation of genetic systems and the point at which it becomes applicable.

For what it may be worth—and I hope it reassures plant breeders—I do not see these new techniques as being alternatives to plant breeding programs. They are clearly, at most, sources of new material that would enter the breeding programs. If the fondest dreams of the most ardent exponent of genetic engineering via somatic cell culture were realized, there would not be new varieties arising fully formed from the flask. We are still going to need the same solicitous care from plant breeders to implement the use of hypothetical desirable variants arising via cell culture techniques that has always been required.

I would be remiss if I did not point out that along with advances in our knowledge of protein-synthesizing systems, there could still be a great deal done by orthodox genetic techniques—for example, by cytogenetic manipulation. Chromosome engineering in wheat as practiced by Riley and Kimber (1966) and Sears (1969) and their colleagues has reached the state of a fine art.

There are two corollaries to this hypothesis. The first is that, in both maize and sorghum, a second mutational step could reduce the prolamin content still further and enhance the nutritional quality simultaneously. Second, for rice, oats, and this particular barley mutant (1508) a mutation that reduces further the prolamin content cannot have a substantial effect.

In closing, let me return to one of the points with which I opened. It is my conviction that progress in the field of protein quality and protein quantity improvement is soon to be limited by the lack of basic information on which the geneticists and plant breeders can operate. We have searched for and, in some cases, detected or induced useful variation. Where it is found, such variation has stimulated concerted efforts to breed it into cultivars that will be useful in our agricultural systems. Where variation has not yet been found, we face difficult choices. We can fold our tents and silently steal away. Or we can attempt to do some of the difficult things that one hypothesizes to be possible. We can attempt, for example, in wheat, to find or induce variation of the type we desire in the 2N species. We can attempt, in collaboration with the plant biochemists, to identify control mechanisms in protein synthesis that would enable genetic manipulation of the system. My hope, and I suspect that of the organizers of the symposium is that it would stimulate more intensive research on plant protein synthesis. Certainly, we have mutants that would serve as excellent experimental probes for the biochemists and the plant physiologists.

There has been much said during the course of this symposium concerning research support. I hope that at least some of this week's deliberations will come to the attention of the agencies that support research and that these agencies will be receptive to the view that the subjects discussed here are indeed important ones. They are not only intrinsically interesting from a scientific viewpoint but are essential hedges against the expanding population of our globe.

REFERENCES

Dalby, A., and I. I. Davies. 1967. Science 155:1573.
Denic, M. 1970. *In* Improving Plant Protein by Nuclear Techniques.
Green, C. E., R. L. Phillips, and R. A. Kleese. 1974. AEA, Vienna, p. 381. Crop Sci. 14:54.
Halim, A. H., C. E. Wassom, and H. L. Mitchell. 1973. Crop Sci. 13:405.
Ingversen, J., B. Koie, and H. Doll. 1973. Experientia 29:1151.
Phillips, R. L., R. A. Kleese, and S. S. Wang. 1971. Chromosoma 36:79 and personal communication.
Riley, R., and G. Kimber. 1966. Rep. Plant Breeding Inst. (Cambridge) 1964–65:6.
Sears, E. R. 1969. Annu. Rev. Genet. 3:451.
Singh, B., and J. D. Axtell. 1973. Crop Sci. 13:535.
Sodek, L., and C. M. Wilson. 1970. Arch. Biochem. Biophys. 140:29.
Wilson, C. M., and D. E. Alexander. 1967. Science 155:1575.

DISCUSSION

DR. DOY: Since you announced that Green has grown a complete plant from *Zea mays* in culture, I should say I saw these plants last week. Bill Sheridan from Columbia, Missouri (personal communication), has developed a haploid time culture of maize.

DR. GABELMAN: No one has really talked about the potential that might accrue from hybridizing cytoplasms, either as a genetic tool or as a breeding tool. Do you foresee any?

DR. NELSON: There is so little information that has any predictive value in this particular field that I do not know. Certainly, this is going to be one of the inevitable consequences of hybridization of somatic cells, so I am sure we will be learning a great deal.

DR. MCCLINTOCK: In that connection, DNA recombination occurs between mitochondria from different sources, and in that way you might get something quite new.

DR. PETERSON: Nothing has been said about oats at this conference, and I would just like to mention that we have done protein fractionations of oat groats. The fractional composition of these proteins is almost identical to that of the mutant 1508 of barley that you showed. We do not have any mutant. This is just normal oats.

List of Contributors

AARON M. ALTSCHUL, Georgetown University, Washington, D.C.
CASSIUS BORDELON, Baylor College of Medicine, Houston, Texas
DONALD BOULTER, University of Durham, Durham, England
HERBERT W. BOYER, University of California, San Francisco
JOHN S. BOYER, University of Illinois at Urbana–Champaign
DAVID T. CANVIN, Queen's University, Kingston, Ontario
JOHN M. CLARK, JR., University of Illinois at Urbana–Champaign
W. RONNIE COFFMAN, International Rice Research Institute, Manila
JOHN P. COMSTOCK, Baylor College of Medicine, Houston, Texas
ARTHUR DALBY, Purdue University, West Lafayette, Indiana
M. DALLING, University of Illinois at Urbana–Champaign
JULIUS W. DIECKERT, Texas A&M University, College Station
MARILYNE C. DIECKERT, Texas A&M University, College Station
COLIN H. DOY, Australian National University, Canberra
LEON S. DURE III, University of Georgia, Athens
T. C. ELLIOTT, Michigan State University, East Lansing
EVERETT H. EVERSON, Michigan State University, East Lansing
D. FLINT, Michigan State University, East Lansing
WARREN H. GABELMAN, University of Wisconsin, Madison, Wisconsin
HOWARD M. GOODMAN, University of California, San Francisco, California
RICHARD H. HAGEMAN, University of Illinois at Urbana–Champaign
RALPH W. F. HARDY, E. I. du Pont de Nemours & Co., Inc., Wilmington, Delaware
U. D. HAVELKA, E. I. du Pont de Nemours & Co., Inc., Wilmington, Delaware
STERLING B. HENDRICKS, Silver Spring, Maryland

E. G. Heyne, Kansas State University, Manhattan
J. Inglett, Agricultural Research Service, Northern Regional Research Laboratory, Peoria, Illinois
J. Ingversen, University of Wisconsin, Madison
Neal F. Jensen, Cornell University, Ithaca, New York
Virgil A. Johnson, Agricultural Research Service, USDA, University of Nebraska, Lincoln
T. H. Johnston, Agricultural Research Service, Stuttgart, Arkansas
Joe L. Key, University of Georgia, Athens
E. E. King, Agricultural Research Service, USDA, Stoneville, Mississippi
Lowell Klepper, University of Nebraska, Lincoln
R. J. Lambert, University of Illinois at Urbana–Champaign
Dale Loussaert, University of Illinois at Urbana–Champaign
R. Macke, University of Illinois at Urbana–Champaign
Abraham Marcus, Institute for Cancer Research, Philadelphia
Brian J. McCarthy, University of California, San Francisco
Barbara McClintock, Carnegie Institution of Washington, Cold Spring Harbor, New York
Anthony R. Means, Baylor College of Medicine, Houston, Texas
William C. Merrick, National Institutes of Health, Bethesda, Maryland
Edwin T. Mertz, Purdue University, West Lafayette, Indiana
Benjamin J. Miflin, Rothamsted Experimental Station, Harpenden, England
Ranjit Mitra, Michigan State University, East Lansing
Lars Munck, Carlsberg Research Laboratory, Hellerup, Denmark
Hamish N. Munro, Massachusetts Institute of Technology, Cambridge
Daniel Nathans, The Johns Hopkins University, Baltimore
Oliver E. Nelson, University of Wisconsin, Madison
Ann B. Oaks, McMasters University, Hamilton, Ontario
Bert W. O'Malley, Baylor School of Medicine, Houston, Texas
David M. Peterson, University of Wisconsin, Madison
Bruno Quebedeaux, jr., E. I. du Pont de Nemours & Co., Inc., Wilmington, Delaware
Robert Rabson, Joint FAO/IAEA Division of Atomic Energy in Food and Agriculture, Vienna
L. E. Schrader, University of Wisconsin, Madison
George F. Sprague, University of Illinois at Urbana–Champaign
Robert C. Tait, University of California, San Francisco
N. E. Tolbert, Michigan State University, East Lansing
J. E. Varner, Washington University, St. Louis
D. V. Wettstein, University of Copenhagen, Copenhagen
Curtis M. Wilson, Agricultural Research Service, University of Illinois at Urbana–Champaign
Savio Woo, Baylor College of Medicine, Houston, Texas
Geoffrey Zubay, Columbia University, New York